Praise for *Two Sides of the Moon*

"Sometimes, they were first. Often, they were best. Always, they were colorful. They came back from their missions in space having seen the spirit of themselves, as even more of the human beings they were before leaving our world of air, land, and water to see the Earth below or an earthrise above the Moon. They have been so ever since: Leonov, the artist, and Scott, the engineer/dreamer. The two of them—the Cheaters of Death."
—Tom Hanks, from the introduction

"The lay reader, assuredly, will be left in awe."
—*Chicago Tribune*

"This retelling of the space race through their eyes should become the latest 'must-read' book for anyone interested in the history of the U.S. and Soviet space programs before the advent of the space shuttle."
—*Ad Astra*

"Their book serves as a testimony of the great work that can be accomplished, even between adversaries, given the chance."
—*Tulsa World*

"What was most significant about the lunar voyage was not that men set foot on the Moon, but that they set eye on the Earth."
—Norman Cousins

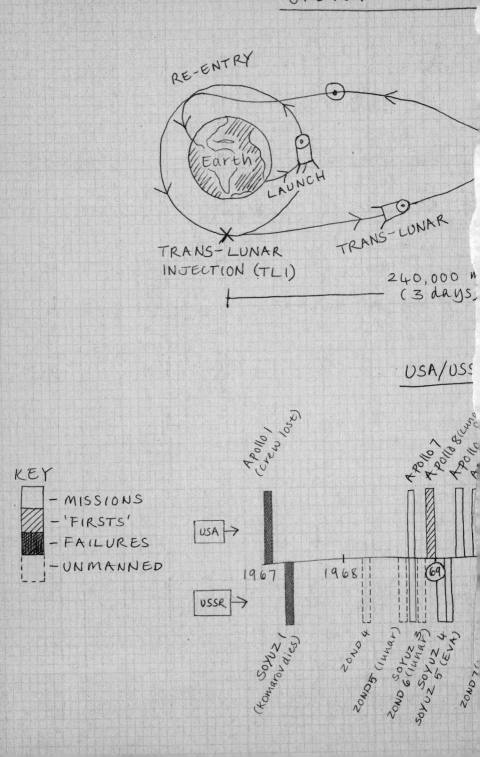

RE-ENTRY

Earth

LAUNCH

TRANS-LUNAR

TRANS-LUNAR
INJECTION (TLI)

240,000 n
(3 days,

USA/USS

KEY

- MISSIONS
- 'FIRSTS'
- FAILURES
- UNMANNED

Apollo 1
(crew lost)

Apollo 7

Apollo 8 (luna

Apollo

USA →

1967   1968   69

USSR →

SOYUZ 1
(Komarov dies)

ZOND 4

ZOND 5 (lunar)

ZOND 6 (lunar)
SOYUZ 3

SOYUZ 4
SOYUZ 5 (EVA)

ZOND 7

LUNAR MISSIONS

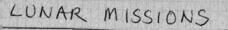

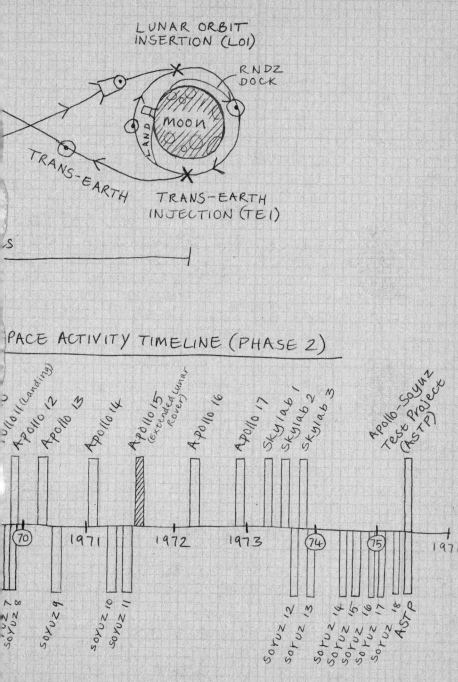

LUNAR ORBIT
INSERTION (LOI)

RNDZ
DOCK

MOON

LAND

TRANS-EARTH

TRANS-EARTH
INJECTION (TEI)

S

PACE ACTIVITY TIMELINE (PHASE 2)

Apollo 11 (Landing)
Apollo 12
Apollo 13
Apollo 14
Apollo 15 (Extended Lunar Rover)
Apollo 16
Apollo 17
Skylab 1
Skylab 2
Skylab 3
Apollo-Soyuz Test Project (ASTP)

70   1971   1972   1973   74   75   1976

Soyuz 7
Soyuz 8
Soyuz 9
Soyuz 10
Soyuz 11
Soyuz 12
Soyuz 13
Soyuz 14
Soyuz 15
Soyuz 16
Soyuz 17
Soyuz 18
ASTP

TWO SIDES OF
THE MOON

Alexei Leonov and David Scott in Soyuz simulator, Star City, June 1973

TWO SIDES OF THE MOON

Our Story of the Cold War Space Race

DAVID SCOTT
and
ALEXEI LEONOV

with Christine Toomey

Thomas Dunne Books
St. Martin's Griffin ⚏ New York

To Charlie Bassett
astronaut extraordinaire, taken before his time

And Jim Irwin
my partner on the Moon, the very best.

D.S.

To Sergei Pavlovich Korolev
outstanding scholar of the twentieth century, great citizen
of Russia, founder of practical cosmonautics

and Yuri Gagarin
my friend with whom we jointly started our way to
space, on the same day, 4 October 1959.

A.L.

THOMAS DUNNE BOOKS.
An imprint of St. Martin's Press.

www.stmartins.com

All photographs courtesy of the authors and NASA

Library of Congress Cataloging-in-Publication Data

Scott, David Randolph, 1932–
    Two sides of the moon / David Scott and Alexei Leonov ; with Christine Toomey
        p. cm.
    ISBN 0-312-30865-5 (hc)
    ISBN 0-312-30866-3 (pbk)
    EAN 978-0-312-30866-7
    1. Space race—United States—History. 2. Space race—Soviet Union—History.
3. Space flight to the moon—History. 4. Outer space—Exploration—United States—
History. 5. Outer space—Exploration—Soviet Union—History. I. Title.

TL789.8.U5 L46 2004
629.4'09—dc22   2004059381

First published in Great Britain by Simon & Schuster,
A Viacom Company

First St. Martin's Griffin Edition: March 2006

10 9 8 7 6 5 4 3 2 1

CONTENTS

FOREWORD

In *Two Sides of the Moon*, David Scott and Alexei Leonov recount the preparation for and execution of the most elaborate non-military competition in history. Each man would play a vital and high-profile role in that contest. In all great competitions success is dependent on a combination of skill, preparation and an element of luck. Certainly all these factors played critical roles for both sides. The reader will gather detailed insights into the strategies, successes, concerns and tragedies of each team. It is unlikely that the space race was the diversion which prevented war. Nevertheless, it was a diversion and provided an outlet that replaced the "brinksmanship" of the early 1950s that might well have led to armed conflict. The competition was intense and allowed both sides to take the high road with the objectives of science and learning. Eventually it provided a mechanism for engendering cooperation between adversaries and it provided an enduring legacy—an enormous improvement of the understanding of our cosmic neighborhood and indeed of our Earth itself.

It is commonly believed that the space age began because of the Cold War between the West, led by the United States, and the Communist Bloc, led by the Soviet Union. That belief is not quite right. The space age actually began because of a scientific event known as the "International Geophysical Year." Sixty-six countries joined together to analyze the planet Earth and its environs: oceanography, meteorology, solar activity, the Earth's magnetic fields, the upper atmosphere, cosmic rays and meteors. The International Geophysical Year was planned for the period 1 July 1957 to 31 December 1958. It was, in fact, the International Geophysical Year and a Half. That particular time was selected because it was the time of maximum sunspot activity. All sorts of unexplained electrical, magnetic and weather phenomena seemed to be somehow related to sunspots.

Soviet and American scientists recognized that if it would be possible to place a manmade object into orbit around the Earth, it would be the perfect platform for sensors and recording instruments to measure many of the characteristics of the natural world for which the International Geophysical Year was created.

They did not realize it at the time but they had started a new competition which would become known as "the space race." The Soviets took an early lead with the successful launch of the first Earth-orbiting satellite, Sputnik. They increased their lead with a series of firsts: first to put a living creature—a dog—into orbit; first to put a man into space; first to have a person exit a spacecraft; first to put a woman in orbit; first to launch a multi-crew spaceship and first to fly unmanned probes to the Moon, Venus and Mars.

The American program was substantially behind but not moribund. The Americans were embarrassed but eager to be a part of the race and determined to succeed. United States President John F. Kennedy asked the National Aeronautics and Space Administration to report on our ability to successfully compete with the Soviets. The NASA officials told the President that the United States could not provide the first space laboratory and had only a slim chance of flying around the Moon first but would have a chance, with a concerted national effort, to land a man on the Moon first.

Although this was an arena in which he knew little, Kennedy concluded that the United States must be in the race and must perform well. He received the support of the Congress and of the American people. The race was underway.

This book is the story of that race as told by two of the principal competing participants. Each side respected the other. They were, in fact, able to understand, perhaps better than anyone, the problems, challenges and risks before them. On rare occasions the competitors met at international events and technical conferences for short periods and found that they liked one another, although, coming from opposing sides, they didn't know whether they could trust one another.

Two Sides of the Moon interweaves the tactics of the Soviet program with those of the American effort. The reader will see the plans being developed in the training and operations centers in Moscow and Houston and will live with the cosmonauts and astronauts inside their spacecraft as they learn to exist and survive in dangerous and inhospitable space.

Dave Scott and Alexei Leonov have each borne the enormous responsibility of commanding spacecraft and of representing their respective countries in the most fascinating and most expensive race in human history. This is their transcendent recounting of that competition.

Neil Armstrong, 2004

INTRODUCTION

With the turn of one or two historical events, Soviet cosmonaut Alexei Leonov might well have spent the 1960s and the 1970s trying to kill American astronaut David Scott and vice versa. Leonov in his MiG and Scott in his Super Saber could have taken flight for combat, each trying to blow the other man out of the sky in a contest between the two strongest nations on Earth for political, military and sociological dominance.

Leonov and Scott were highly skilled and expertly trained pilots familiar with the state-of-the-art air weaponry of their respective arsenals. Each man believed that the country of his birth possessed the fairest standard of living and was the best society possible on Earth. Each man was a driven, competitive careerist who took pride in being first, best or at least most colorful in any endeavor. Luckily for them, and for the rest of the world, events never conspired to have them take to the skies to eliminate the other.

Instead, the Russian and the American took flight in separate, far superior ways. An artist at heart, Leonov became a true cosmonaut; the first human being to float in open space—the only human being at the time to view his beloved Earth and Mother Russia through only the thin faceplate of his pressure suit. Scott, the engineer/dreamer, commanded the first mission to the Moon where exploration entailed more than landing, walking and picking up rocks. He drove a car on the lunar surface, day after day after day, in search of a geologic treasure as valuable to science as the lost ark would be to theologians.

These two men cheated death on hundreds of occasions throughout their lives. From his childhood during the bleakest period in Soviet history, through his early flying career, to his rigorous and extreme cosmonaut training, Leonov could have easily perished from reasons big and small. A poorly machined airfoil, a faulty valve or a miscalculation in the air pressure of his space suit could have killed him as surely as a momentary hesitation of his brain or body. Dave Scott, having already spent thousands of hours flying aircraft both proven and unproven, narrowly escaped death when the Gemini 8 spacecraft went awry soon after lift-off. A precision landing on the Moon could be a recipe for disaster—in fact, it is—but Scott survived his landing colorfully. Both men could have perished in huge balls of fire, the result of their rockets blowing up on the launch pad in either the Baikonur Cosmodrome in the Soviet Union or Cape Kennedy in the United States, without either of them truly earning the title of cosmo- or astronaut.

But once again, neither of these scenarios came to pass and both men did what pilots have been doing since Montgolfier went up in a balloon. They cheated death. Theirs was a gamble taken voluntarily and eagerly with a single-minded pursuit of earning the assignment and then getting the job done. Sometimes, they were first. Often, they were best. Always, they were colorful.

A common generalization about people who go into space, particularly those men who were the participants in the great space race between the USA and the CCCP, is that each is an automaton—a throttle jockey with a crew cut, a mind like a slide rule and a vocabulary of "gee whiz" superlatives and "A-OK" affirmatives. Like all generalizations, this is far from the truth. It is true that to fly airplanes is to have certain shared interests with other pilots, but to earn the right and responsibility to walk in outer space and on the surface of the Moon requires a spectacularly individual drive. You have not only to want to do so for inexplicable reasons, you have to do those things with inexplicable skill.

Leonov and Scott have gone to extra lengths to explain the inexplicable in *Two Sides of the Moon*. And thank goodness they have. The secrets of the Soviet space program are finally put in human terms by Leonov, and guess what? The communist spacemen we were supposed to fear so much took *crayons* into space—all the better to capture the colors of an orbital sunrise. And the Americans who hopped and skipped across the lunar surface with such ease? Getting their suits on took two full hours and they nearly failed to get a drill bit out of the ground.

Nothing routine is accomplished in space and nothing happened on the missions of Alexei Leonov and David Scott that could be called routine. Every second of each of the five missions flown between them was unique, beautiful, fulfilling and dangerously volatile. And yet, each time they returned, neither man claimed to have come back as a changed man who had gone into space and seen . . . *the spirit of the universe.*

No. Even better, they came back from their missions in space having seen the spirit of themselves as even more of the human beings they were before leaving our world of air, land and water to see the Earth below or an earthrise above the Moon. They have been so ever since: Leonov, the artist and Scott, the engineer/dreamer. The two of them—the Cheaters of Death.

Tom Hanks, 2004

"Earth is the cradle of the mind, but one cannot live in the cradle forever."

Konstantin Tsiolkovsky, 1857–1935,
the father of Russian space travel

"Man must rise above the Earth—to the top of the atmosphere and beyond—for only then will he fully understand the world in which he lives."

Socrates, 469–399 BC

PROLOGUE

Major Alexei Arkhipovich Leonov, Soviet Air Force

Pilot, Voskhod 2 Spacecraft

LAUNCHED: BAIKONUR COSMODROME,
SOVIET REPUBLIC OF KAZAKHSTAN, 18 MARCH 1965;
A BITTERLY COLD MORNING WITH LIGHT SNOWFALL

The silence of open space was broken only by the sounds of my beating heart and breathing muffled by headphones inside my helmet. Suspended 500 kilometers above the surface of the Earth, I was attached by just an umbilical cord of cables to our spacecraft, Voskhod 2, as it orbited the Earth at 30,000 kph. Yet it felt as if I were almost motionless, floating above a vast blue sphere draped with a colorful map. Lifting my head I could see the curvature of the Earth's horizon.

"So the world really is round," I said softly to myself, as if the words came from somewhere deep in my soul. For a few moments I felt totally alone in this pristine new environment, taking in the beauty of the panorama below me with an artist's eye.

Then a voice filled the void: "Attention. Attention." It was my commander, Pavel Belyayev, addressing me and, it seemed, the rest of mankind. "Your attention, please," Pasha said, sounding very serious. "A human being has made the first ever walk in open space. He is at this very moment flying free in space." It took me a moment to realize he was talking about me.

Then another voice cut in. "How do you feel, Lyosha?" That voice sounded familiar, too. But I couldn't immediately place it until it continued and I realized that the speaker was Leonid Brezhnev. I was both amazed and exhilarated that he was talking directly to me. No one had warned me I would be receiving such a transmission from the Kremlin, though I knew this was an important moment politically. My mission—man's first steps in open space—marked another triumph for the Soviet Union after my close friend Yuri Gagarin had become the first man in space four years earlier.

Special monitors had been installed in the Kremlin, I found out later, so that the Communist Party's most senior leaders could follow our space flight via a direct link with Mission Control centers in Moscow and the Crimea. Brezhnev's address to me was broadcast almost simultaneously on state radio and television across the Soviet Union. So was my reply, though at the time I hardly knew how to respond.

"Thank you. I'm feeling perfect. I will do my best to fulfil my mission. I'll see you back on the ground," was all I could think to say. It was necessarily a brief exchange because I had work to do and it was proving a great deal tougher than I had thought. Throughout the time I had been floating free of the spacecraft, I had been facing directly toward the Sun. The strength of light was intense and the heat incredible. I felt sweat accumulating in drops on my face and running down under the collar of my shirt.

My pressurized suit was extremely stiff and I had to exert a tremendous pull against the inflated rubber to bend my arms and legs. There was no gravity to give me leverage in the vacuum. Even the slightest movement required extreme effort. It was not planned that I should spend much time outside the spacecraft. So great was concern about the huge psychological barrier it was feared man might face in the void of open space, that I had a web of sensors attached to my body. Information on my pulse, blood pressure and even the alpha-rhythm waves in my brain was being constantly monitored, both by Pasha aboard the spacecraft and by Mission Control.

Before relinquishing my position as the first person to step out into this new domain, however, I wanted to be more adventurous. I gave myself a hefty push away from the spacecraft and immediately started tumbling uncontrollably, rolling head over heels. Only my umbilical cord of cables saved me from drifting off into space. After being yanked to a sharp halt, I had to struggle back to the craft, hauling myself fist over fist along the umbilical.

When I pulled level with Voskhod 2, my body temperature had risen even higher and I was extremely tired. I knew it was time to reenter the spacecraft. I had been in open space for over ten minutes. I still had forty minutes of oxygen left in my autonomous life-support backpack and had a burning desire to stay outside for much longer. But I knew I couldn't take that risk. Not only would it mean disobeying orders, but in another five minutes our orbit would take us away from the Sun and into darkness. I still had to negotiate my way back into the spacecraft through the airlock, a narrow, collapsible set of interconnected rubberized canvas cylinders, which would be difficult enough without the handicap of working in the dark.

But as I edged closer to the airlock's entrance I realized I had a very serious problem. My spacesuit had ballooned in the vacuum to such a degree that my feet had pulled away from my boots and my fingers no longer reached the gloves attached to my sleeves. No engineer had been able to foresee this; the suit had been tested in a pressure chamber simulating a much lesser altitude and no such deformation had occurred. Now the suit was so misshapen that it would be impossible for me to enter the airlock feet first as I had in training. I simply couldn't do it. I had to find another way of getting back inside the spacecraft, and quickly.

The only way it seemed possible was by squeezing head first into the airlock. Even to manage that I would have to reduce the size of my spacesuit by reducing its high-pressure oxygen, via a valve incorporated in its lining, which would expose me to the risk of oxygen starvation. If I consulted Mission Control, I knew it would alarm them and, anyway, there was no time for discussion. I was the only one who could solve the problem, and I knew I had to do it fast.

Reentering the spacecraft this way also meant I would have to perform a somersault once inside the airlock in order to close the outer hatch. It would all take far longer than was scheduled and I was not sure my life-support system would hold out. The exhilaration I had felt just minutes before as I looked down at the Earth evaporated.

Bathed in sweat in the ballooning suit with my heart racing, I knew
I could not afford to panic. But time was running out . . .

Major David Randolph Scott, United States Air Force

Pilot, Gemini 8 Spacecraft

LAUNCHED: CAPE KENNEDY, FLORIDA, 16 MARCH 1966;
CLEAR BLUE SKIES

Day was turning to night. Night would last for just forty-five
minutes as we orbited the Earth at more than twenty times the speed
of sound. This was our fifth sunset since the start of our mission
aboard Gemini 8, and during the period of darkness that followed
we would have little sense of motion in the calm sea of space.

At the break of the previous brief day I had had a clear sense of
movement as I looked through one of our spacecraft's two small
windows and witnessed the spectacular effect of the "air-glow"
caused by sunlight shining through the moisture of the Earth's
atmosphere. As the Sun began to rise beyond the Earth's horizon,
the broad curvature of our planet had picked up a fringe of dark
blue. This quickly gave way to a brilliant rainbow of colors—first
light blue, then purple, red, and orange—before the spectacular
hues were bleached white as the Sun emerged in a brilliant ball.

But now, as our flight trajectory took us away from the east coast
of Africa and out over the Indian Ocean, the contours of the Earth
below disappeared with the fading light. It was then that the first
sign of trouble was transmitted to us from Mission Control via the
Tananarieve tracking station on the island of Madagascar.

"If you run into trouble and the Agena goes wild, just send in
command 400 to turn it off and take control of the spacecraft. Do
you copy?"

Neil Armstrong, my commander, and I exchanged a cautious look
but said nothing. We had spent so much time together in the seven
months since we had been chosen for this mission that we could
communicate without talking a great deal. We could pretty much
read each other's minds, like professional basketball players when

they make a "no-look pass," one tossing the ball without looking because he knows exactly where the other guy is going to be.

We both knew it might signal there was something wrong with the Agena rocket, with which we had just performed the first-ever rendezvous and docking in space. The Agena, now locked to the front of our spaceship, had a troublesome history: several had failed during tests and others had experienced major problems shortly after take-off. Since it was my job to send commands to the Agena, I glanced at the rocket's control panel and acknowledged Mission Control's concern.

"Rog, we understand," I radioed back via Tananarieve.

We then passed out of range of the tracking station and I switched the cabin lights to full on and made a note of the time. It was 7:00 GET, or "Ground Elapsed Time," the amount of time that had passed since lift-off—our reference point for all phases of the mission. Glancing away from the clock I noticed Neil's "8-ball," the instrument indicating our orientation, showed our spaceship was listing by 30 degrees.

"Hey, Neil, we're in a bank," I said and he swiveled his eyes to check my identical 8-ball: both agreed.

"You're right," he confirmed. "We're rolling." The motion was so slow we felt nothing. Below a certain threshold the delicate balance mechanism of the inner ear is unable to detect any rolling movement and we couldn't verify the bank by looking out of the window since we were by then in total darkness. But we knew we should have been steering a smooth course without any rotation or fluctuation. We knew straight away that something was wrong.

Following Mission Control's message we both suspected our rolling was being caused by a problem with the Agena. So Neil told me to turn off its attitude thrusters and he took command of the two joined vehicles with Gemini's control system. It seemed at first we were right, that the Agena was living up to its bad reputation. As the two vehicles stabilized I sent another command to disable all programs on the Agena. But then we started to roll again.

"I thought you disabled the Agena?" Neil queried, and I confirmed that I had.

"Well, turn it back on," he ordered, and I immediately sent the counter-command code 401. Still we continued to roll. The situation was clearly becoming much more serious.

"Turn it off again," Neil told me once more and again I sent the 400 command allowing him to take control of both vehicles with the Gemini. Again the roll stopped and we began closely checking our systems to evaluate what had gone wrong. We could not consult Mission Control for another ten minutes until we came within reach of the next tracking station, the ship *Coastal Sentry Quebec* off the coast of China in the western Pacific. We were still running through our checks when the rolling began once more. Again Neil ordered me to switch off the Agena.

"It's off," I assured him. "But I'll send 400 again."

This time nothing happened. Looking at each other with growing alarm, we realized the speed of the roll was starting to build. We had passed back into daylight by this time and now had visual confirmation of what our instruments were telling us. What we saw was alarming. Round the rim of the Agena locked ahead of us we could see the line of the Earth's horizon protruding like the arms of a giant clock—dark sky to one side of them, the deep blue of our planet to the other. The arms were moving slowly at first, but then started rotating faster, as if the time on the clock were being rapidly adjusted.

Neil tried to fight the motion with the hand controller. But we were rotating in all three axes by this time: the roll was now combined with an up-and-down "pitch" and side-to-side "yaw" movement. The giant timepiece effect blurred as our spacecraft pitched downward, still rolling, and the blue Earth first loomed up to fill the view from our windows then suddenly gave way to dark sky—again and again. It was as if we were peering out from inside an enormous and intricate silver baton which had been hurled through space by a cosmic majorette.

The violent motion was putting severe stress on the docking collar holding Gemini and the Agena together. We knew nothing like this had ever happened, even in a simulation. We were in an uncontrollable tumble in space. And it was about to get much worse . . .

CHAPTER 1

High on Flight

1932–56

Alexei Leonov

Temperatures drop to below –50°C in the small village of Listvyanka, Central Siberia, USSR, where I was born on 30 May 1934. As a small child I used to listen at night for the sound of branches in the wood next to our log cabin, heavy with ice, cracking and splitting. The water from the mountain stream feeding our deep well steamed as we lifted its wooden cover to break off icicles, which my brothers and I sucked for fresh water.

The stars were always so clear in that remote place that I could easily believe the story my mother told me about a thousand torches being lit on the surface of the Moon at the end of each day to illuminate the night.

But I was just three years old when we were forced to leave our home on the outskirts of our small village, Listvyanka, near Mariinsk. It was January 1938, deep winter and unimaginably cold, when neighbors arrived at our home to strip it bare. They took our food, our furniture, even the clothes off our backs. One neighbor ordered me to remove my trousers and left me standing there in only a long shirt.

I remember, vividly, running through the larch forest, frozen and frightened, to meet an older sister, who was coming home to visit that day.

"There's nothing to eat," I wailed. "Mother's crying." When my sister handed me two large loaves of bread I tucked them under my arms and tried to run home with them. But I was small and the loaves were heavy. I kept dropping them, picking them up and running on.

Our family had become destitute after my father, Arkhip Alexeyevich, was accused of being an "enemy of the people." He was sent to prison on the strength of the false testimony of a corrupt co-worker. He was not alone: many were being arrested. It was part of a conscientious drive by the authorities to eradicate anyone who showed too much independence or strength of character. These were the years of Stalin's purges. Many disappeared into remote gulags and were never seen again. We did not know then the extent to which it was happening.

In those days my father supported Stalin and his policy of collectivization. He believed in Bolshevism and the ideals of the revolution. But my family's support of revolutionary ideals went back even further to the first Russian Revolution of 1905, when my mother's father was exiled for revolutionary activities from Lugansk, in the Ukraine, where he worked as a mechanic in a flour mill and organized illegal workers' strikes. From Lugansk my grandfather moved first to Rostov and then to Siberia, where he found work as a miner.

As a young man my father also worked as a coal miner in Shakty, close to Rostov, until he was conscripted into the army at the start of the First World War. When the war was over he also moved to Siberia, then an area of greater economic freedom. There he married my mother and became a peasant farmer on the outskirts of Listvyanka. My parents had twelve children—seven girls and five boys—but two of my brothers and one sister died, leaving nine of us.

My father was eventually elected chairman of the village council of Listvyanka and donated everything he had to the local collective, including his pride and joy, a horse he had bred to be swift and strong enough to thrive in the Siberian winter. Soon afterward the head of the collective, a Tatar as far as I remember, slaughtered my father's horse for meat. When my father swore vengeance the Tatar

HIGH ON FLIGHT 9

concocted a story about my father allowing seeds for the next year's harvest to dry out. My father was thrown into jail without trial.

As we were then regarded as the family of an "enemy of the people," we were branded subversives. Our neighbors were encouraged to come and take from us whatever they wanted. It became a free-for-all. My elder brother and sisters were thrown out of school. We were about to lose our house. We had no choice but to leave our village.

We had nowhere to go but to live with another of my elder sisters, who had recently married and was working on the construction site of a power station in Kemerovo, several hundred kilometers away. She and her husband, who also worked at the plant, had been given one room in a workers' dormitory near the factory. My mother, then pregnant, five sisters, older brother and I, the youngest, were taken by horse and cart to the railway station to join them. I cried bitterly that I did not want to go. We had only a few blankets to keep us warm.

I remember how tenderly my brother-in-law treated me when he met us at the train station. He collected us in a large horse-drawn sledge, laid us children like sardines on the floor and covered us with his overcoat. As we rode toward the factory dormitory he kept asking me "Are you cold, Lonya?" (that was my nickname as a child). "Yes I'm cold," I said and he would stop the sleigh and tuck his sheepskin tighter round me before carrying on.

We lived, eleven of us, in a sixteen-square-meter room for the next two years. I slept, under a bed, on the floor.

My father was eventually absolved of any wrongdoing. His former commanding officer in the Division of Latvian Riflemen, during the First World War, demanded a proper trial be held for my father, whom he had greatly respected as a soldier. Not only was my father paid compensation for wrongful imprisonment but he was offered the position of head of the collective, to replace the brutal and dishonest Tatar. My father chose instead to join us in Kemerovo, and he found a job in the power plant where my sister and brother-in-law worked. My father had a very strong character; he was an individual. He never joined the Communist Party, but he soon became a person of some standing in the local community. The work went well. He was issued with one of the few loudspeakers connected to the community radio station—used to disseminate news and information—as a mark of the high regard in which he was held.

Our family was allocated two extra rooms in the workers' dormitory. With his compensation money my father bought some basic furniture, kitchen implements and an overcoat for each child; he still could not afford shoes for all of us. From then on we were regarded as one of the wealthiest families in Kemerovo. We were the only family in the dormitory who owned a sausage-making machine. We became known as "The Leonovs with the meat mincer." We were not so short of food after that. But life was still hard.

I used to earn extra bread for the family by drawing pictures on the whitewashed stoves of our neighbors' rooms. I loved to draw, and my parents encouraged me, buying me paints and pencils. Paper was in short supply, so I used wrapping paper. Later I earned extra money by painting canvases for our neighbors to cover their bare plaster walls. My painting slowly became a family enterprise; my father helped me stretch bedsheets over a simple wooden frame. I then turned them into rough canvases by preparing them with flour mixed with animal glue, and decorated them with oil paintings of mountains and woodland scenes. From a very early age I had a strong ambition to become an artist.

Then, when I was about six, I was overwhelmed by a competing ambition: to become a pilot. It happened the first time I set eyes on a Soviet pilot, who had come to stay with one of our neighbors. I remember how dashing he looked in his dark-blue uniform with a snow-white shirt, navy tie and crossed leather belts spanning his broad chest. I was so impressed I used to follow him around, admiring him from a distance.

One day he noticed me creeping along behind him. "Why are you following me?" he asked and I told him straight. "I want to be like you, to look like you, one day."

"Why not?" he replied. "There is nothing stopping you if that is what you really want. But in order to become a pilot you must be physically strong. You must study hard and each morning you must wash your face and hands with soap."

Like most little boys I was not too keen on soap and water. But then he asked, "Do you promise to follow this advice?"

I could hardly reply quickly enough. "Yes, I do," I promised and I ran home, took a bar of soap and washed my face vigorously. Then, every time I saw the pilot, I ran to him to show him my clean face and hands. He would smile and nod his approval.

After that I relished watching Soviet films about pilots whenever they were shown in the local House of Culture cinema club. I remember one called *Courage* and another, *Interceptors*, about a fighter pilot and a small boy whose life he saved. I loved those films. I must have seen them more than a dozen times. I started to make models of all sorts of planes in the Soviet Air Force. Later there was a book I came to love too, called *A Novel of a Real Man*, about a fighter pilot whose legs were amputated after a crash landing, who learned not only to walk again but to dance and eventually fly, too. I kept that book by my bed for many years. It taught me that you should never give up under any circumstances.

David Scott

The thirty de Havilland Jenny biplanes flying in tight formation in the cloudless Texan sky spelled out the letters "USA." Bending down toward me, I remember my mother pointing to the tip of the letter "S" and shouting to make herself heard above the roar of the engines.

"There's your dad, Davy!" she said, taking my hand. As I sat on my tricycle in the front yard of our house at Randolph Field, San Antonio, transfixed by the sight of these beautiful open-cockpit airplanes swooping effortlessly above, my mother pulled out a box camera and captured the spectacle on film.

Her photograph of that day took pride of place in my father's den. Though I can't have been more than three when it was taken, it always reminded me of the moment in my childhood when I became hooked on the idea of becoming a pilot, just like my dad.

After that I sometimes slipped on my father's brown leather Army Air Corps flying jacket with its big fur collar and his leather helmet and goggles and imagined I, too, was soaring over our house at Randolph airfield (after which I had been given my middle name). His dress uniform of formal jacket, breeches, Sam Brown belt and ceremonial saber was far too big for me to try on. But that flying jacket seemed a perfect fit to a small boy—even if it did hang down to my ankles.

It was not until I was twelve that my father was able to take me up in a plane with him for the first time; Air Corps regulations were very strict. But sometimes when I was a young boy, he sat me in the cockpit of a plane on the airstrip and let me get the feel of the controls. Other times he would drop little parachutes, each weighted with a small stone, out of the cockpit of his plane as he flew over our back yard. Wrapped round the stone were simple messages to me like "To David, love Dad." Those little notes fluttering down in the breeze further fueled my determination to take to the skies one day for my own airborne adventures.

Yet my father, Tom William Scott, had taken to flying only by chance, following a bet at a Friday night party. He had put himself through college by working in the oilfields of southern California and spent the summers drilling holes and pumping oil to save enough money both for his tuition and to support his mother after she and my grandfather divorced. My grandmother had moved to Fresno, California, and then Los Angeles from Wichita, Kansas, following the breakdown of her marriage and, though he never liked to talk about it, that must have been a tough time for my father.

After graduating from university he'd taken an administrative job in a Hollywood film studio. But one Friday night at a party he and a friend were set the challenge of passing the physical for the Army Air Corps. The following Saturday they took the test, passed and got signed up. Two days later my father quit his job at the studio and went to fly. He was trained as what was known in those days as a "pursuit pilot"; the term "fighter pilot" was not widely used before the Second World War. Yet this was only ten years after the end of the First World War, the time of the Great Depression. The

Army Air Corps didn't have much money and so he flew all the old biplanes, like the de Havilland Jennies. But he loved to fly those old planes. It was his whole life. By the time I was born he had become an instructor and so was home a lot, except when he had to go on tour to promote the Army Air Corps.

My father was of Scots descent, very frugal. We didn't have a lot of money when I was growing up, but we had a good life; the military looked after you. After my parents married and my father was posted to Randolph Field, where I was born on 6 June 1932, we lived in a big, stucco-fronted, two-story house. From the stifling heat of San Antonio, in the days before air-conditioning, we were then moved briefly to an air base in Indiana before my father was posted to the Philippines.

Just before Christmas 1936 we boarded an old army transport ship bound for Manila, in the Philippines, where we were to stay in some comfort for the next three years. With a small army of servants to take care of the house chores, my mother was free to enjoy bridge and mah-jong parties and to travel widely. Both my parents took trips to China, though separately because of the risks involved. Following Japan's invasion of China in 1937, the region was hurtling headlong toward war, as, halfway round the world, Europe was too. But, being a young boy, I was oblivious of all of this. After a mule cart brought me home from school each day I was more concerned with learning how to climb coconut trees.

There was always strict discipline at home, however. My father was very stern, but loving. He had a strong, crisp voice. I always addressed him as "sir" and my mother as "ma'am," and we sat down to a formal dinner every evening. My father set very high standards. He was well organized and neat, and expected me to be the same. He made the rules clear and I followed the rules: you did back then. Any time I stepped out of line I was made to sit in a corner facing the wall. I was quite quiet as a child, not timid but reserved. This sometimes frustrated my father, who'd tell me to "Go out there and mix it up."

I remember, when I was six, attending my first "swim meet" in the Philippines. It was a race, but I didn't know the rules. When the starter gun went off I looked around and waited for all the other boys to dive in the water. Once they were in, I thought, "OK, now it's my turn," and jumped in. I thought that's what I was supposed to do—be polite. Of course I came in last. My father was pretty unhappy about that. He explained in no uncertain terms that we were all supposed to get in the water at the same time.

"Don't wait for the other boys. Go at the start," he told me. My father was a tough guy. He pushed me pretty hard.

Around the same time I remember going out with my father in a small motor launch to meet the ocean liner my mother was returning on from a visit to China. The wind was blowing hard and we were out there bobbing about in a rough sea.

"There's your mom up there at the railing," my father boomed. "Wave to your mom." I really wasn't in any mood to appear cheerful: I still remember how uncomfortable I felt out there in that small boat. But there was nothing else for it but to say "Yes, sir" and wave.

When my father received orders in December 1939 to return to the United States I was seven. I was well aware that the transport ship carrying us home was the last setting sail from the Philippines with military men on board. From then on only women and children were allowed to leave. A lot of my father's friends were left behind and many died in the subsequent fighting. Two years after we arrived home the storm that had been gathering while we were in the Far East unleashed itself on the United States and my small world was turned upside down.

The radio news flash on 7 December 1941 that the Japanese had attacked Pearl Harbor galvanized my father. He put me straight in the car and we drove down to the drugstore to buy a newspaper. It seemed no time at all before he received orders to go "overseas." No one knew where "overseas" meant. I had no idea then that I would not see my father for the next three years.

We were back living in San Antonio, Texas, at that time and, after my father left, my mother decided to sell our big house and move to a smaller place. I had a baby brother, Tom, seven and a half years younger, by then. My mother had a pretty active social life, so I did a lot of babysitting. With my father gone and rationing introduced, my mother bought me a small motor scooter so that I could get to school and home on my own. Tom used to love me taking him for rides on that bike. My mother had a gentler personality than my father did, and it was decided that, while my father was away in the war, I needed more discipline.

So I was enrolled in a small military school run by Episcopalians. It was very strict. We wore uniforms. We marched. We were issued with ranks. If we smarted off to any of our teachers we might pick up a slap across the backside and be held in detention after school.

But there were always the summers to look forward to and they were magnificent. Realizing I needed a male role model while my father was away, my mother packed me off to spend summer vacations with my father's closest college friend, David Shattuck, after whom I had been named. He lived in Hermosa Beach, California. This also got me away from the threat posed by the polio epidemics, which at that time were far more prevalent in Texas than California.

For a ten-year-old boy Hermosa Beach was just about the closest thing to heaven. It took three days on the train to get there from San Antonio. I traveled there on my own with porters keeping an eye on me along the way. Once there I did not wear shoes for three months. Uncle Dave had a place right on the beach, a big sand acre where we could play volleyball and swim. With a group of boys from next door we'd take a packed lunch and go hiking in Pales Verdes or go on down to Venice Beach to visit the Fun House and ride the roller coaster.

Back home in San Antonio I spent every spare hour building model airplanes. I knew every American, British and German airplane in detail. The ceiling of my bedroom was festooned with Messerschmidts, Spitfires, Hurricanes, Mustangs and Lightnings. I don't remember ever having any doubt that I wanted to be a pilot like my dad. I was a big fan of flying films too: *Hell's Angels*, *Dawn Patrol*. I don't know how many times I watched Errol Flynn throw his glass into the fireplace and go into combat. I used to love that stuff. They, of course, were flying First World War planes and, for me, that was the ultimate. That was really flying.

While my father was stationed in Europe during the Second

World War we used to practice blackouts, even in Texas. Although that seemed like fun to a kid, we didn't know whether America would be invaded or not. There were rumors about the Japanese getting into California. We followed the build-up to the Normandy landings. We listened to Churchill's speeches and tuned in to Roosevelt's famous "fireside chats" on the radio. But we heard very little from my dad.

The letters we did receive were all mailed via a post office box in New York. We had no idea where he was. Except for the first Christmas after he left when I received a book about model airplanes with a card inside that read: "Dear David, Merry Christmas, Somewhere in England, Love, Dad."

It wasn't until my father came home at the end of the war that he was able to tell us that he had been stationed as commander of Burtonwood air base near Blackpool, in the north of England. He had been responsible for the maintenance and repair of B-17 and B-24 bombers. He was a lieutenant when the war started and returned a full colonel. By the time he retired in 1957 he was a general.

Alexei Leonov

When the radio in our home crackled into life on the morning of 22 June 1941 we knew straight away that something out of the ordinary was to be announced by Vyacheslav Molotov, then foreign minister. Many neighbors crowded into our room as he started to speak.

"Today, at four o'clock in the morning, German forces crossed the western border of the Soviet Union and began a wide-scale advance from the north to the south of the country. The Red Army is fighting to repel these forces," I remember him saying. The women in the room started to cry. That is how world war interrupted my childhood. It was the beginning of a very dark period for our country.

The front line was several thousand kilometers away. But within a very short time trainloads of wounded soldiers started arriving for treatment at makeshift hospitals in Kemerovo, as they did elsewhere in the country. Other trains also began to arrive with factory workers and disassembled factories, which were to be reassembled far from the front line. The population of Kemerovo doubled in a very short time. Chemical plants sprang up virtually overnight. Our region quickly became the center of the Soviet Union's chemical industry.

Over the next few years, two of the plants were blown up by German sympathizers. Thousands died. Many more were injured, including one of my sisters. There were fears that the Germans would also use air bases in Iran to launch attacks against chemical plants in Siberia. Each family was instructed to dig a trench in their vegetable garden to be used as a makeshift air shelter in case of air strikes. The very somber mood of the country only started to lift after the Red Army defeated Hitler's forces at Stalingrad in 1943.

It was in the autumn of that year that I started school. I had just turned eight. I remember my first day at school with both embarrassment and pride. It was 1 September and the wooden boards that lined the road on my way to school were still covered with frost. My mother, Yevdokiya, dragged me along by the hand. I remember her stopping to greet neighbors along the way and boasting proudly, "Another of my children starts school today." As I stared down at my bare feet—I had no shoes—I could see the ice melting in puddles where I stood.

When we arrived in the yard of Primary School Number 35 I saw a number of others were barefoot, too. To mark the start of a new school year we were lined up and, before classes started, had to chant, "Thank you Comrade Stalin for our happy childhood!" Several days later the school committee delivered to our home a pair of shoes for me. They were dark chocolate in color and had a wonderful smell of fresh leather. But when I looked at them carefully I saw they had not laces but cross-straps: they were girl's shoes. I treasured them anyway.

On that first day of school I remember carrying all of my most precious possessions with me in a gas-mask bag which flapped down about my knees, making it difficult to walk. There were several crayons, fragments of watercolor paints in individual matchboxes, a magazine, reproductions of my favorite paintings and some of my own drawings. Throughout the first lesson I kept wondering whether I should show my new classmates what I had in my bag. The decision was made for me. During break a boy we later nicknamed "Cockroach" snatched my bag, scattering its contents around the classroom.

When the teacher came back she saw my illustrations lying on the floor. "Well, Alexei," she exclaimed. "Look, you are an artist." That was the first time I realized I could do something not everyone else could. It is a happy memory. But those years were full of great sorrow, too.

The terrible loss of life among Soviet soldiers and civilians during the war touched our family, as it did every family in the country. Twenty million Soviet citizens were killed in a few years. My four uncles and six male cousins, who had stayed living in the western part of the country, fought for their country and they were all killed. Many of our neighbors returned from the front terribly maimed.

The day the war ended, in 1945, I remember as vividly as the day it began. I was out tending cows in the field when people started running past me, skipping and shouting. They told me to leave the cows and follow them. When I arrived home people were again gathered around our radio loudspeaker, crying and laughing. Slowly people took out small stashes of food they had saved from their rations. Some had a little salted fish, some cabbage or small portions of bread. There was no meat or fruit during the war, except for one tangerine for a child for New Year celebrations sent from China—our ally at that time. Others had a little alcohol. We had a feast, a modest one, but a feast all the same.

David Scott

D-Day—6 June 1944—was my twelfth birthday. I'd never had a birthday like it; all the parties thrown to celebrate the Normandy landings seemed like a way of celebrating my birthday, too. It was another eleven months before Germany surrendered and a little while after that before my dad was able to return home. While stationed in Europe, he was assigned a US Mitchell B-25 light bomber and he took several weeks to fly it back to the United States via Africa and South America.

By the time he got back I knew I was old enough for him to take me up for my first ride in an airplane. I couldn't wait. That flight was magnificent—if a little alarming at first. Almost as soon as we took off my father began to put the plane through a lot of elaborate acrobatic maneuvers. As we went into the first loop I remember thinking the front of the plane would never stop climbing. It was pretty uncomfortable, and I hadn't expected it. But as we soared high above the clouds and then swooped down toward the ground, watching fields, trees, houses and cars flash below like toys, it was the most exciting thing I had ever experienced. The feeling of freedom was exhilarating. As we came in to land I began to understand why my father felt such a passion for flying. It was to be many years before I experienced the real thrill that comes from controlling an airplane. But I was always sure the day would come.

After the war ended we were on the move again. This time to March Air Force Base near Riverside, California, where I was enrolled in the Riverside Polytechnic High School. It was huge, over 3,000 kids, a far cry from the small military school I'd attended in San Antonio. For a start there were girls at Riverside. They took typing classes and, as a way of getting to know them, a lot of us boys signed up for typing classes, too. Those were good days, full of high-school parties, playing pool and dating. I was not assertive or aggressive at school. I usually found myself a seat at the back of the class and listened. I didn't get into too many arguments or discussions. But, again, my father told me to "mix it up."

"Don't sit in the back of the room," he'd say. "Get in there and argue." When I played basketball he was there on the sidelines. "Get out there and push them around a little bit," he said.

My father had been a keen Boy Scout, so I joined the Boy Scouts, too. He was a very keen golfer and I used to caddy for him. He used to take me to ball games, and also taught me to drive.

When I joined the swimming team at Riverside I started breaking all the records. I hadn't thought I was a particularly strong swimmer, but all that ocean swimming at Hermosa Beach must have been pretty good training. I learned a great deal from my swimming career, as it turned out. It taught me competitiveness, teamwork, endurance, how to perform at my physical best and how to prepare for a major event both physically and mentally. I also benefited from the strong influence of three great coaches, at high school and later at Michigan University and West Point. I was eventually honored by being selected to the US All-American Swimming Team and the Eastern Collegiate All Star Water Polo Team.

So when my father received orders that he was being moved again, this time to Washington, DC, there was a lot of debate over whether I should stay in California to finish high school and continue with my swimming. In the end I moved with my parents and finished my schooling in Georgetown. But I was certain even then that where I really wanted to go was the US Military Academy at West Point. I just wasn't certain I'd get in. I had no connections with any congressmen or senators who could sponsor me. I took the government civil service exam for competitive appointments. But, in the meantime, I won a swimming scholarship to study mechanical engineering at the University of Michigan. In the autumn of 1949 I left home for what turned out to be one short, long year there.

I sometimes wonder why I gave up everything I had going for me in Michigan; I was on one of the best swimming teams in America. I had a roommate who had a car, I had a girlfriend. I was having a great time. But the following spring I was called to take a medical exam for West Point. Shortly afterward I was offered a place. On the Fourth of July, 1950, I boarded a train in Oklahoma, where my parents had moved, bound for West Point in upstate New York.

Although I desperately wanted to become a pilot, I harbored no great ambition to be in the Army. I did appreciate the great sense of camaraderie that came from being in the armed forces. After the Second World War America's military forces had expanded to such a degree that the sense of a close community my father had enjoyed when I was a young boy had gone. Yet my father undoubtedly had something to do with my decision to leave Michigan and head for West Point. He knew I wanted to fly and, because he had enjoyed his own career so immensely, he wanted me to have a career in the

Air Force, too. The best way to get a full military career in those days was to graduate either from the US Naval Academy at Annapolis or from West Point. Though he was not an easy man to impress, my father always gave me good advice. He was very, very happy when I was offered a place at West Point. It pleased him quite a bit.

The best piece of advice I think I was ever given, however, came from a young lieutenant whom I met by chance at the swimming pool when I went to visit my parents that summer. When I told him I was going to West Point he said the most important thing to remember was "Keep your sense of humor." I never forgot that. He was absolutely right. I realized it as soon as I arrived.

There was a lot of pressure at West Point, mental pressure. No one ever beat you up. In fact, there were very strict rules about physical contact; an upperclassman could not touch an underclassman without permission.

"May I touch you, mister?" he'd say if he noticed a piece of lint on your uniform before he was allowed to brush it off. But they orient you very quickly toward respect, discipline and obeying orders. The groundwork is laid in the first two months. Those months are spent in Beast Barracks.

Beast Barracks, oh boy. You were a beast and you were told what to do morning, noon and night. Of course everybody was intimidated and that was the point. Everyone who went to West Point was a high achiever and they all had to be brought down off their pedestals and taught they were just another bottom-of-the-line kid who was going to have to be taught a lot. At the end of the first day we were marched to Trophy Point and sworn in. That seemed pretty good. Then came the hard work. Reveille went at 5:30 a.m., somebody started hollering to get in formation and from that point on you were going full tilt all day; shoes had to be polished, your rifle had to be polished, your uniform had to be kept in order. There was never enough time to do everything. Once we got into the dining hall we had half an hour to eat, but all the time we were eating, someone was chewing at us, asking questions. We had to learn, memorize things like Schofield's definition of discipline from a little handbook about West Point called *Bugle Notes*.

We were taught everything from academics and military history to military hygiene, camping and bayonet drill. Bayonet drill, boy, do I remember that—the toughest physical exercise I ever had. You put on your combat boots and T-shirt, got your rifle, which weighed about 12 lbs with its bayonet sheathed, and then ran "high-post" double-time out to a field about three-quarters of a mile away. Then you went through these bayonet drills—smash, slash, thrust and recover—then double-time back. We were in great shape. But it was absolutely exhausting. It taught us to keep going—that we could drive ourselves to the limit physically and still perform. It was only five years since the end of the Second World War. We were still dominated by that mentality, the mentality of how you went to war. I didn't like it at the time. It wasn't much fun. But I look back on it as probably the most valuable and formative four years of my life.

I learned about discipline and obedience, respect, orderliness, planning ahead and the qualities of leadership. I learned that leadership is all about pulling and not pushing—setting an example. I learned a lot about other people, and about myself. I learned how to make trade-offs and how to respect other people's sensitivities. There have been a lot of accusations; people prepared to knock West Point, over the years. But it has produced a lot of leaders; Dwight Eisenhower, Douglas MacArthur, George Patton. We got some pretty intensive experience of our own in how to train and lead soldiers during our final summer at the academy when we were sent to Fort Jackson, South Carolina. As officers we were responsible for the basic training of young soldiers, who, after sixteen weeks at boot camp, would be going to front-line combat in Korea.

This was serious stuff: it really made you think. Some of them were young kids, straight off the farms, who couldn't read or write. But they were being drafted and sent to war. I learned a lot about the way people think, their motivation, the way they respond, their confusion, the way, often, they do not understand why they are doing what they are doing, but they do it anyway. It helped me, later, to appreciate a lot of the stuff that went on in Vietnam, to understand the resistance to the war. But that lay in the future.

Korea was the overriding conflict of that time. The Korean war broke out the month that I started at West Point. We followed closely, in real time, the speech made a year later by MacArthur in Congress when Harry Truman relieved him of his command in Korea after he tried to expand the war against China. It was required listening. His speech "Old soldiers never die" was marvelous. For an old West Pointer there was great sadness that he was being dismissed. He was a brilliant man. But we were being taught that our democracy comes under civilian, not military, control. The message was clear: just because you are a soldier and a brilliant general doesn't mean that you can run the country. The president runs the country. It was an important lesson about how our system works.

But I had no great interest in politics. I hardly ever read a newspaper. There wasn't much time for that. We mostly studied history. I was aware of the Soviet blockade of Berlin in 1948 and the allied airlift of food into the city. I was aware of Mao Zedong's victory in China the following year. But the Korean war overshadowed all of that for me, especially since my father had been posted as base commander in Okinawa, Japan, early on in the war and I knew he was flying missions over Korea. We all assumed we would be going to the front line as soon as we left West Point. But the war ended a year before I graduated. Our lecturers in military strategy were in no doubt about where the next war would break out: Russia, or the Soviet Union, to be more exact.

Most of our instructors were professional military men on a three-year tour. Almost all of them were Second World War vets. I remember we had one guy who was part of General Patton's 3rd Army. He used to lecture in military tactics. He was a firebrand. He was exciting. He'd get his pointer out and ask us the all-important question, "Where's the next war gonna be?" Then, without waiting for a reply, he'd thwack the stick down on the table and answer himself: "Russia!" Another thwack, "Russia!" Thwack, again, "Russia!"

We read the diaries of German generals on the eastern front. They were really roughly typed, in their original German. We learned what they had to say about the Russians. "Watch out for the partisans," one warned. The Germans feared the Russians. We learned that communism was, by definition, bad—that the communist concept was not one that anyone would want to live under. It was a terrible system. Yet the aspiration of the Soviets and communism was to take over the world. Despite Stalin's death in 1953, there were no illusions that this goal would change. The communists were on the move. They were making inroads everywhere. It was a big threat. The Soviets were the enemy. Even at home there were fears about "enemies within." This reached its height with Senator Joe McCarthy's campaign to "root out the Reds." I was pretty conservative in those days, and believed we had to get rid of communists wherever we found them.

Alexei Leonov

I was convinced when I was growing up that our country was the best in the world. I firmly believed that the Soviet Union produced the best of everything; that Soviet tanks were the best, Soviet planes and trains were the fastest, that no other country could possibly be as good. The propaganda machine ran very smoothly. As one of the best students in my class I was invited to attend the Palace of Young Pioneers, where I received extra art tuition.

Despite what had happened to my father he encouraged us to believe in the ideals of the revolution. "You should not malign the system because of the wrongdoings of some individuals," he told us.

Yet no one was able to speak openly about what was really happening in the country. It was not until our family moved to the Baltic port of Kaliningrad, where my father was transferred to a new job after the war, that I met a boy in my new school, Middle School Number 6, who made me question some of my assumptions. He asked me if I knew how much a worker in America earned and if I knew how good American cars were. It was the first time I started to wonder if there could be another country which dared to be better than we were. I was fourteen by then. I was bright. But I had never come across any books which gave a true description of life in America.

I had read about racial discrimination and the existence of black slaves in *The Adventures of Tom Sawyer* by Mark Twain. But then I became more curious. I began to search our local library for books from which I could learn more about the American way of life. I started to read more widely, including books by Jack London, Theodore Dreiser, James Fenimore Cooper. I did not change my political outlook immediately. I still believed the Soviet Union was the best country in the world. But I did slowly realize that there were countries where people had a higher standard of living.

The death of Josef Stalin during my last year of school was a day of genuine mourning. We all wore black armbands and took it in turns to stand guard by a huge portrait of Stalin in our school hall. I treasured that black armband for quite some time. I took no notice when one of my friends warned me about hanging on to it.

"Mark my words," he said. "The day will come when you will burn that band yourself." It was not until one or two years later that signals started coming from above about the wrongdoings committed during Stalin's rule. Slowly I began to realize what had been going on. I came to the conclusion that Stalin had been a bloodthirsty tyrant, who had done a great deal of damage to my country.

When Nikita Khrushchev emerged as his successor and a statement was issued denouncing Stalin's "cult of personality," I took the black armband I had once treasured and threw it in the fire. I also had a portrait I had painted of Stalin. I was torn, at first, about what to do with it. Should I paint over it or burn it, too? I burned it.

It was years before the country started to change. A whole generation, which had grown up in an atmosphere of fear, could not change their attitude overnight. For at least ten years after the death of Stalin, everyone was very cautious in what they said. "Be careful. Look around," they warned each other. "Pay attention to who is listening before you speak out."

It was as if the whole country continued for a long time to wait for the knock at the door that might signal their arrest. Khrushchev did begin the process of de-Stalinization. He repealed some of the more repressive laws and started to dismantle some of the machinery of the state police. But it was some time before things really started to change.

David Scott

As my time at West Point neared its end I had more pressing matters on my mind than rooting out the Reds. I needed to make sure I was headed for the Air Force afterward. Not everyone who graduated would get into the branch of the armed forces they wanted. It was highly competitive. We were graded in every subject every day we spent at the academy, and those with the highest grades got first choice. My chances were good. I was near the top of the class; I graduated fifth out of 633 in our class of 1954.

When my time came for the career advice interview I was clear about what I wanted. My adviser was Captain John W. Miley, a professional soldier of the highest caliber. He was very sincere and took great interest in my career.

When I went into his office and saluted, he came right to the point. "Have you decided which branch of service you would like to select?"

When I said "Air Force," he seemed amazed. "Air Force? Why d'ya want to go into the Air Force?"

"Well, sir, I want to fly," I said. "I want to be a pilot."

"A pilot?" he said. "Don't you think that's a little dangerous?"

Dangerous? This man was a Korean veteran. He had the Silver Star and several Purple Hearts. He had been in battle after battle. He limped. He had really been through it. He was with the airborne infantry, pride of the Army.

"Well, no, sir," I said, "I think it's all right."

As I got up to leave he had a last piece of advice. "Mr. Scott, I want you to remember one thing. There'll always be a place in the infantry for you."

"Thank you very much, sir," I said.

What I really wanted to say was "Captain Miley, I like to sleep in a warm bed at night, not some cold foxhole." But I just saluted and left. I got my place with the Air Force. But I must say, had I not been able to go into the Air Force, I would have taken the route Captain Miley recommended; I had a great respect for him and his colleagues.

When I reported for duty at Marana Air Base in Tucson, Arizona in late summer 1954 I was at last where I really wanted to be, flying. I was in Tucson for six months' primary flight training and I got it in style. I had a great flight instructor, Chauncey P. "Chick" Logan—"Mr. Logan, sir" to me. He had a mustache, a ruddy complexion, rough skin. He was one of the original barnstormers, who flew from barn to barn in the countryside in First World War planes they'd fixed up, selling rides for a dollar. Chick Logan was the man who taught me how to fly and he taught me well.

The real joy of flying has never been better expressed, I think, than in the sonnet "High Flight," written by a nineteen-year-old pilot with the Canadian Air Force, John Gillespie Magee, Jr, who was killed when his Spitfire collided with another aircraft in cloud in December 1941. Shortly before he died he had mailed his parents the poem.

> Oh! I have slipped the surly bonds of Earth
> And danced the skies on laughter silvered wings;
> Sunward I've climbed, and joined the tumbling mirth
> Of sun-split clouds,—and done a hundred things
> You have not dreamed of—wheeled and soared and swung
> High in the sunlit silence. Hov'ring there,
> I've chased the shouting wind along, and flung
> My eager craft through footless halls of air . . .
> Up, up the long, delirious, burning blue
> I've topped the windswept heights with easy grace
> Where never lark, nor even eagle flew –
> And, while with silent, lifting mind I've trod
> The high untrespassed sanctity of space,
> Put out my hand, and touched the face of God.

Those stanzas sum up the enormous exhilaration I felt at achieving the ambition I had nurtured since I was a small boy.

* * *

After Marana I transferred to Webb Air Force Base near Big Spring, Texas, in the middle of almost nowhere, to learn how to fly a jet. Jets were still pretty new back then. There was a relatively high washout rate—people who did not make it through the program. But after six months at Webb I got my wings. In fact, I got my wings from my dad. He was a colonel then. He pinned his own original wings on my uniform. That was quite a ceremony.

Webb was followed by a brief period of training in gunnery—learning to use a jet fighter as a weapon—at air bases in Texas and Arizona. That was the first time I flew jets designed as fighter-bombers. One of the things I learned was how to deliver a nuclear weapon with a single-seat fighter. It was an interesting concept, called low-altitude bombing system or LABS. It was just ten years after Hiroshima and Nagasaki, yet here we were able to deliver a bomb with far greater capacity than either of the bombs dropped on those cities. This was where I began to get the picture of the seriousness of the Cold War. We were fast approaching the days of MAD—mutually assured destruction—when both the United States and the Soviet Union knew that if either one started a nuclear war we were all finished. The massive nuclear capabilities of both sides meant the world would be destroyed, yet both sides were playing every card they had in terms of lining up nuclear weapons against each other.

It made you feel very aggressive. If the other guy was going to do it, you were going to have to do it. You never thought about it in terms of people. If you were called upon to fight a war, as a soldier you would fight that war. Instead of a club, or a shield and a sword, you had a nuke.

By the time I finished at Luke Air Force Base, Phoenix, Arizona, I was really on my way. In the spring of 1956 I was posted to Holland to join a fighter squadron—the 32nd Fighter Day Squadron—at Soesterberg, near Utrecht. It was the best deal a fighter pilot could wish for. It was exactly what I wanted, aerial combat. This was flying at its most exhilarating: being pushed by another pilot to the limits of your ability, trying to outwit him with your skill and intuition.

You can be taught to become a good pilot, but you cannot be taught to become the best pilot. This depends to a great extent on your coordination, how quickly you react to situations and, perhaps most importantly of all, how sensitive you are to the slightest movement of the plane you are flying. When you are flying and are tightly coupled with a plane, that machine becomes an extension of yourself. Part of the beauty of flying is that feeling of becoming part of a beautiful machine which makes you more powerful, able to transcend the limitations of the human body.

In the often terrible weather conditions in Europe such skills were tested to their limits. We were flying the F-86 Saber, and then the F-100 Super Sabers on simulated combat missions in all weather conditions. We regularly flew to North Africa—Libya and Morocco—for training periods of several weeks, because of the clear skies there. Most difficult of all, though, was flying in the bad weather of northern Europe. Everything was still pretty crude back then. Navigational aids were not good. All we had was a dinky little needle pointing toward a 30-watt beacon to help guide us back to our base through thick cloud. But if you could fly under those conditions you could fly anywhere.

Little over six months after arriving in Europe, however, I was faced with a potential baptism of fire.

Alexei Leonov

As my schooling drew to a close I was determined I wanted to go to art college. At that time there were only two art colleges of real standing in the Soviet Union. One was in Leningrad, the other, a little closer though still around 600 km away, in Riga, Latvia. In the spring of 1953 I hitched a ride in an open lorry to Riga to apply for a place to study later that year.

I borrowed my father's best clothes, cut in the style of a Mao Zedong utility suit. I felt very smart. When I arrived I stuck out like a sore thumb. Still, I walked straight into the offices of the director of the art academy, convinced I would be enrolled without a hitch. I had taken some of my work with me and started chatting in such a loud voice with the director's pretty secretary that he came out to see what all the noise was about.

"I want to study here," I told him bluntly, when he asked what I was doing there.

"You'll have to come back and sit our entrance exam later," he replied.

But then he noticed I had a bunch of drawings under my arm and asked me to show him my work. As he browsed through my pencil drawings, watercolors and copied portraits of Chopin, Gorky and Peter the Great he gave his assurance that I would be given the opportunity to study at the institute after graduating from school.

When I left his office I stopped a young art student who was walking along the corridor and asked him where and how he lived. He told me it cost 500 rubles a month to rent a room in Riga, and my dream of becoming an artist seemed to crumble. I knew I would never be able to afford such a sum. My father earned only 600 rubles a month and still had a family to support, and I had no other male relatives left alive who could help me.

I returned home despondent. But I soon resolved I would follow, instead, my other ambition, to become a pilot. So I applied for military training. My parents supported my decision. One of my cousins, a professor, tried to talk me out of it, believing I should become an academic instead, but my mind was made up. I applied for a place at the Kremenchug Pilots' College in the Ukraine. The competition was fierce. But in September of 1953 I was offered a place.

It was considered such a great honor that not only my family but also all my schoolmates came to the station to see me off. When my mother found she did not have enough money to purchase a platform ticket, all my friends clubbed together to buy her one and then climbed over the fence to get on to the platform themselves. When my mother started to cry as the train pulled out, I climbed into the luggage rack and I cried bitterly, too, that I was leaving my family behind.

The next two years I spent learning to fly propeller planes at Kremenchug. I was then transferred to a Higher Military Pilots' School at Chuguyev, also in the Ukraine, for a further two years. At Chuguyev I started to learn to fly military jets. I felt proud that I had achieved this and I was still only twenty. Within a short time I was appointed assistant commander of our team of military cadets, with the rank of sergeant, and so started to learn the responsibilities of command and to take responsibility for myself at a very young age.

Throughout this period I continued to paint and draw in my spare time. The more I thought about it, the more I realized this was the ideal solution: I could be a pilot and continue my art as a serious hobby. Yet I would never have been able to be a professional artist and a fighter pilot, too.

By the time I got my wings my father had retired. He had taken a part time job cutting the grass at a military aerodrome in Kaliningrad. He made a point of chatting to pilots there, trying to pick up military jargon with which to impress me. When I returned to visit my parents he would casually let slip some of the sayings, one of which I took very much to heart.

"Don't wait until the end of the runway to apply the brakes," my father told me, smiling. "And don't leave love affairs too late in life to enjoy."

CHAPTER 2

Cold War Warriors

1956–61

Lieutenant David Scott

32nd Fighter Day Squadron, Soesterberg, Holland

Soviet tanks rolled into Budapest in October 1956 and we suited up for war. The rise of Nikita Khrushchev to power in the Soviet Union following Stalin's death had created expectations of change in the Eastern bloc. But the Soviet reaction to internal dissent was brutal. Demands by Hungarian workers for more freedom met with a bloody crackdown; within weeks 3,000 were killed.

As a front-line fighting unit our squadron was placed on highest alert. The whole squadron could be airborne and in combat within an hour. Four airplanes, in rotation, were on a five-minute alert, their pilots sitting in the cockpit on the runway for hours at a time. Another eight were placed on thirty-minute alert and the rest of the squadron on forty-five-minute alert. We knew the Russians had far more planes, something like four or five to one. They had excellent pilots, too, so our chances of getting into a fight and coming out of it OK were not good.

We had been trained in techniques of escape and evasion—how to evade capture by the enemy in the event of being shot down. We had been dropped, on one occasion, in Germany's Black Forest. Our mission: to evade Allied Special Forces deployed to capture us.

If we were captured we knew we would be thrown in jail and interrogated—Korean style. We would be forced to stand in front of bright lights for hours on end, have cold water squirted at us, be made to do calisthenics all night, be subjected to mind games and generally humiliated. If interrogators got too much out of you, your security clearance was taken away and you were grounded. Out of fifty men in our class, only three got through the course without being captured. Toward the end of the last day I was captured, tied up and thrown face down in the mud.

The most likely area of combat for our squadron, in the event of war, was considered to be the skies over Czechoslovakia. So we were issued with Czech maps and currency. We were also supposed to be issued with sidearms to defend ourselves if we were shot down. But when Hungary blew up all we had were signal guns, small pistols—not much good for personal combat. We had to act fast, so two guys hopped in a plane and flew down to Italy, where they went to the Beretta manufacturing plant and bought up as many Beretta pistols as they could load into the plane. When they got back they laid all these pistols out on a long table and everybody bought at least three—we had to buy them ourselves—together with holsters and ammunition, the whole nine yards. In those days we did not have a lot of sophisticated gear. All we had were coveralls and G-suits in mismatched colors which fitted like chaps on our legs. It was equipment left over from the Second World War and Korea. We stuffed the pistols wherever we could, tucking them under our arms and into our chaps along with as much ammunition as we could carry. We looked like something out of a Hollywood movie, a comedy scene from some tale about the Wild West.

But this was not playtime; it was a deadly serious business. The Army was on alert too with their tanks at the ready. This was about World War Three. We really thought the US would defend the Hungarian freedom fighters, since here was a group of people finally trying to break out of communism. There they were down on the streets, fighting. We thought the US would support them the way the US had been saying it would. But it did not.

It was beyond my political scope at the time to understand why we did not respond to the raw aggression of the Soviets. It was an ugly chapter of the Cold War. We let the Hungarians get had. I was too far down the ranks to know what was going on behind the scenes. All I knew was that after a few weeks on high alert our squadron was stood down.

Lieutenant Alexei Leonov

Higher Military Pilots' School, Chuguyev, Ukraine

Our newspapers were full of shocking photographs at the start of the Hungarian uprising. They showed the maimed bodies of supporters of the communist regime with the star symbol of the Soviet Union carved onto their corpses. I felt both disgusted and indignant. Had those involved in the revolt not been so cruel, they might have evoked a certain sympathy among the Soviet people. There were reasons for their discontent. Hungary's communist regime had brought the country to its knees. But the rebels behaved in a very cruel way. They shot, hanged and burned alive communist supporters and all those who disagreed.

I realize now that we did not know about all the events happening at the time. But we remembered very well that the Hungarians had been allied with Hitler during the Second World War and that a great number of Soviet soldiers had died fighting to liberate that country from the fascists. The prevailing feeling was that we had the moral right to have troops in Hungary and that Soviet garrisons had the right to defend their positions.

At the time all this was happening I had just completed my training with YK11 propeller planes at Chuguyev and was beginning to learn to fly MiG-15s. Our squadron was put on full alert. We prepared for combat. Looking back, I realize it is unlikely we would have been sent to fight. We were young, still in training. We were the future of the Air Force. But we felt prepared to go to war. I was twenty-one, ready for anything. Most of us believed the West would not be prepared to get involved in Hungary and face the might of the Soviet military; Europe had had its share of fighting. We were right. It did not. The rebellion was quickly put down.

Throughout the years of studies starting at Kremenchug Pilots'

College and Chuguyev Air Force School I always scored the highest marks in my exams. Then, at the very last, I took a wrong step. I dared to question an order to decorate a room in commemoration of Marshal Gyorgy Zhukov, the Soviet Union's military leader during the Second World War. I was too concerned with preparing for my final exams and said I would decorate the room once they were finished. This upset one of my tutors, and he marked me down a point in his subject, Marxist-Leninist theory. It upset me so much I vowed to re-sit the exam. But the commanding officer of my squadron said it would do no good: "Don't mess with those rogues. They'll always find a way to grind you down." However, he comforted me with the promise that I would be given first choice of where I wanted to be posted next.

He kept his word. I chose to serve with the elite 10th Guard Division, stationed in Vienna. I was not to know that by the time I was ready to take up my new post, the division would be relocated—to Kremenchug, the aerodrome where I had started out as a cadet four years before.

While we were making final preparations for our graduation ceremony, the astounding news came through that on 4 October 1957 the Soviet Union had launched Sputnik 1 into space. Everyone broke into spontaneous applause. We clapped each other on the back, congratulated ourselves on being the first country in the world to launch an artificial satellite. We were very proud of what Sputnik demonstrated to the world about the Soviet Union's advanced level of technology. We had no idea how spaceflight would progress from there—we thought it would be many years before a man would be launched into Earth orbit—but I was enthralled.

Within hours of Sputnik being launched the Soviet poet Nikolai Krivanchikov wrote a short verse which caught my imagination.

The day will come when we will embark on interstellar flight.
Who can prevent us from dreaming such dreams,
When it was Lenin who taught us how to dream?
All planets in the universe are waiting to be discovered.
We will chart the fifth ocean of space.
Chkalov set us on this path.

Valery Chkalov, one of the Soviet Union's most famous early
aviators, was a hero of mine. This poem has had great significance
for me all through my life.

David Scott

I was in a beer hall in Cologne when the Soviets launched Sputnik
1 into space. It was Oktoberfest and I loved to go to Germany to
drink the beer. So there we were, a bunch of bachelors taking it
easy. The Germans had strung a model of the Sputnik from the
ceiling of the beer hall and there it was as we drank our beer, above
our heads, making its distinctive "beep, beep, beep." I remember
people looking at us and could almost hear what they must have
been thinking: "What's wrong with you Americans? The Russians
have the Sputnik up there." They all thought the Sputnik was
great.

"Mmm, wonder why we don't have a Sputnik?" we asked
ourselves, with some confusion and embarrassment. The
psychological effect of that "beep, beep, beep" overhead for the
three weeks that the capsule's radio lasted was very significant. It
amplified our perception of the Soviet Union's technical capability.
At the time I viewed it very much in military terms. It was almost
as if the Soviets had developed a new cannon that could shoot
further. It enhanced their position considerably, no doubt about
that. For them to put a satellite into space before the United States
was a sound demonstration of their technological strength. It also
underscored the fact that we had a very strong foe on the other side
of the Iron Curtain and that they seemed to have better guns and
ammunition than we did.

Since the satellite was unmanned, I didn't view it as being highly
significant in terms of space exploration. It wasn't clear then that
the 184-pound shiny orbiting sphere was just the opening salvo of
a determined Soviet campaign to conquer space. As far as we were
concerned it might turn out to be a one-off Soviet stunt. Who knew
then if either the Soviets or the Americans would ever put a man
into space? Even when the Soviets launched Sputnik 2 a month later

with a dog on board, the full significance of what they were attempting to do did not really sink in.

Through reading the military newspaper *Stars and Stripes*, I was vaguely aware that our own attempts at launching a space satellite were not going well. But the paper did not convey the serious national angst being felt back home in the United States about the fact that the Russians were beating us into space. When the US Navy's Vanguard rocket attempted to launch a satellite on 6 December 1957, and it exploded only a few feet into the air— eliciting such headlines in the mainstream press as "Flopnik!" and "Kaputnik!"—we were not aware of the consternation it caused. We did not have too much contact with our friends and family at home; long-distance communications were not too good back then.

Even if I had been aware of the public reaction to such failures, the new domain of satellites and missiles seemed very far removed from my own. That was the space world and I was still much more concerned about airplanes, bombers and nukes. It was of far greater interest to me, for instance, to know how many MiGs the Russians had on the other side of the Iron Curtain. It wasn't clear that we would ever get a rocket to work. In so far as this might affect the military balance, I knew it was going to be significant, but I had absolutely no idea that this would mark the beginning of one of the most bitterly fought battles of the Cold War—or that I would soon become a warrior on its front line.

Alexei Leonov

Back at Kremenchug I spent the next two years flying modified MiG-15s which had been adapted for take-off and landing on soil airstrips both at night and during the day: risky maneuvers. Toward the end of those two years, I had a serious mechanical problem which almost cost me my life.

I was flying in heavy cloud with poor visibility, when one of the pipes in the plane's hydraulic system snapped. It was found later that the pipe had leaked oil on to one of the blocks of electrical engines, which caused a simultaneous failure not only of the navigational devices but also of the radio. Warning lights then indicated I had a fire on board, and an alarm was triggered. In such a situation the only solution was to eject. But that was impossible: I was too low. All I could do was cut the fuel supply to the engine and perform an emergency landing.

It turned out that there had been no fire; the signal was a false alarm. But the way I had handled the emergency drew the attention of a mysterious recruiting team who arrived at our base from Moscow shortly afterward and asked if I would be willing to join a school of test pilots.

The colonel in charge of the team told me that training of candidates for this school would involve testing entirely new types of military hardware and aircraft: rocket planes, traveling at huge speeds. If I was willing to accept the challenge, the colonel said, I would be invited to Moscow for a series of tests. If I passed, he said, I would have a very complicated but very interesting life.

I was twenty-five. I was flying the most modern military jets. I was doing well, enjoying life. So I asked him a question.

"There's this girl I really like," I said. (There was indeed a woman I wanted to marry, although she did not yet know it.) "Tell me frankly. If I take up the challenge you are proposing will it preclude me from having a family?"

The colonel broke into a broad smile. "If she is a good woman, there will be no problem. I would even approve of you marrying her."

Several days later I was invited to present myself at a military hospital in Moscow. It was 4 October 1959. I was one of forty candidates selected from three thousand pilots interviewed. All of us had had experience flying the most modern aircraft, MiG-15s and MiG-17s, under all conditions. No minimum number of flying hours had been specified, although by that time I had clocked up around 350 hours. We all had a high IQ and were below thirty years of age. At the end of a month of exhaustive tests only eight of us would be selected.

On the first day we were stripped of our uniforms and issued with hospital pajamas. I was then shown to a ward where another pilot sat in a corner reading a book. It was pretty hot that day and he was stripped to the waist. When he raised his head I saw he had a very handsome Russian face; large, sparkling blue eyes and a big grin. I looked more closely and saw that the corners of his mouth were turned up so that he looked as if he was always smiling.

"I am Senior Lieutenant Yuri Gagarin," he said. "My air base is in the north of the country. I fly MiG-15s."

"My air base is in the south of the country. I fly MiG-15s, too," I replied.

Yuri Gagarin turned out to be very talkative. Within half an hour I seemed to know his life story. He told me he had a wife and baby daughter. He said he was eager to get home quickly because the polar nights were about to start where he was stationed, near Murmansk on the Kola Peninsula. He said he was keen to add flying under the winter conditions of that unique and beautiful region north of the Arctic Circle to his record so that he would qualify as a pilot of the first class.

From that moment we became firm friends. We had both had hard childhoods—we had both begun working at an early age—and our careers had been similar. We liked the same films and books. The book Yuri was reading that day we met was one I did not know, *The Old Man and the Sea*, by Ernest Hemingway. It was not an easy book to get hold of, but when I read it I came to love it, too, for the way it portrays man's utter determination in pursuit of a goal.

Several years later, on my first trip to Cuba in the summer of 1965, I met Hemingway and I took the opportunity to tell him that his novel had been Yuri Gagarin's favorite book. He was touched. Hemingway was not a well-known writer in the Soviet Union then, because his books were so hard to find. But I found him a great character who showed special interest in our space program.

During that first month Yuri, the other candidates and I underwent tests at the hospital in Moscow, and doctors checked every conceivable aspect of our physical and mental condition. We were put in a silent chamber and set a series of complex tasks while blinking lights, music and noise were played to distract us. We were given mathematical problems to solve while a voice was piped into the chamber giving us the wrong answers. We were put in a pressure chamber with very little oxygen in extreme temperatures to see how long we could withstand it. We were put in a centrifuge machine and spun at high speed until we lost consciousness.

We understood perfectly well that whatever we were being tested for was beyond the scope of the test-pilot program. There were test pilots undergoing examination in the same hospital and the procedures we were subjected to far outstripped theirs in severity. We were instructed not to talk to them, informed that what we were being put through was top secret. We had our suspicions that the program would involve space flight. We were excited.

Finally eight of us were summoned by a senior air marshal. He addressed us in a fatherly way. He told us we had a choice to make. Either we could continue our careers as fighter pilots in the Air Force, or we could accept a new challenge: space. We had to think it over.

We left the room briefly and talked among ourselves in the corridor outside. Five minutes later we filed back into his office. We wanted to master new horizons, we said. We chose space.

We were told to return to our bases and await further orders. Before leaving for Moscow I had already been told I was to be posted to East Germany. There was no time to lose. When I returned to Kremenchug I knew that I had to talk to the woman I wanted to marry.

We had spent very little time together, but she had caught my eye the very first time I arrived in Kremenchug as a young cadet. She was still at school then. I remember seeing her walking along with her friends, big ribbons in her hair, smooth, dark complexion. When I was stationed in Kremenchug for the second time, I bumped into her in the street again. It was my birthday. I offered to take her to the Officer's Club—an invitation no girl was supposed to be able to resist. She thanked me but told me she had already been invited. When I saw her at the club later with another officer I vowed I would fight for her. I discovered she was called Svetlana and found out where she worked. After that I often walked her home. I did not want to lose her and, as I was about to be stationed abroad, I was very direct.

"Svetlana, I have very little time. I love you very much and I need you for my whole life. The day after tomorrow I want us to be married," I told her. "If not now, when? The day after that I am leaving for East Germany."

"I agree, Lyosha," was her simple reply, all I needed to hear to make me happy. Under normal circumstances, in Soviet times, a marriage could be registered three months after an application was made. But this delay could be shortened at the discretion of the local mayor and these were exceptional circumstances. We applied for a special license that very day.

We had only two days in which to prepare for the wedding. More than a hundred guests were invited, three thousand chrysanthemums ordered. My bride arrived late for the ceremony, with threads still hanging from her hand-sewn white gown. The next day I flew to East Germany. It was to be four months before she could join me.

At Altenburg, in East Germany, I was attached to a reconnaissance unit making flights over Hungary and East Germany to update military maps of these areas. The base was just 20 km from the border with West Germany, and often, as I flew near the air corridor separating East from West I saw American jet fighters on the other side. I felt no animosity toward the American pilots. Sometimes, as a mark of mutual respect, we tipped our wings in recognition of a fellow pilot.

But it was undeniably tense. The prevailing political climate put enormous pressure on us. All the time we were flying the control tower monitored our movements closely. If we flew too near the air corridor we were warned to change our flight path immediately. Pilots who strayed into the buffer zone were severely reprimanded—grounded.

David Scott

Late at night or early in the morning unmarked F-100s were deployed from bases in West Germany to fly reconnaissance missions over Eastern Europe. "Slick chicks" was the code name given to these spy planes. Specially equipped with photographic systems, they taxied out to the runway, flew at low level straight into East Germany, took photos and flew back as quickly as they could. Before we had satellites capable of taking pictures, this was our only way of knowing what the other guys were up to. The "slick chicks" were apparently flown mostly by CIA pilots; we didn't know much about them.

We were designated squadron fighter pilots and were issued with strict rules of engagement any time we flew near the air corridor between East and West—a no-fly buffer zone a hundred or so miles wide known as the Air Defense Identification Zone, or ADIZ. I flew close to the zone a lot while I was stationed in Holland, but to enter it was considered a serious violation, an aggressive move, which might be interpreted as an indication of war.

The rule on our side was that you never entered the corridor unless you were specifically authorized or in a combat situation. But every time our radar picked up a Russian airplane approaching the zone someone was scrambled to intercept it if necessary, in case it was some form of aggressive combat maneuver. Sometimes, to test the other side, ground control had us fly directly toward the ADIZ. When we got there we were instructed to turn. We could see the contrails of the Russian planes doing the same. The trick was to see who would turn first. It was like playing chicken, except it was deadly serious.

If a Russian plane entered the buffer zone we were authorized to enter, too, for inspection or interception. If it made an aggressive move, we made an aggressive move. If it shot at us, we'd shoot it down. If the plane seemed lost we pulled up alongside, got its number, and tried to contact it and tell it to go home. If it did not comply we pulled up in front and rocked our wings. According to international law, it should then follow us down. If it did not, our wingman would be lined up right behind it, his finger on the trigger. It was cat and mouse.

I strayed into the zone, once. It was the middle of the night, and my flight leader had taken us in there by mistake—his radio was screwed up. Radar in those days was pretty meager. Ground control couldn't tell him to get out of there. It was just the two of us. I was flying on the wing next to him, in tight formation. Suddenly a very bright light illuminated both of us. It was an interceptor—on our side. He'd been sent up to bring us home. We got down pretty quickly. My flight leader got into trouble over that.

While I was stationed in Holland I had three different squadron commanders. By far the best, head and shoulders above the others, was the third, a living legend by the name of Frederick C. "Boots" Blesse, or "Major Blesse, sir!" He was a brilliant pilot, a fighter-pilot in the Second World War and a "double ace" in Korea, having brought down ten planes. He wrote the book *No Guts, No Glory*, the early bible of aerial combat tactics. He was at the top of the fighter-pilot pyramid and I wanted to be just like him. He was full of life, full of vigor. When he took command of our squadron he made it clear he was going to fly one-on-one with every pilot in the squadron to see how good we were. That made for a few sweaty palms among the pilots, I can tell you.

When it was my turn to go up there, we really wrung it out. This was real eyeball-to-eyeball material: we flew straight at each other, at a given speed and given altitude, and when we passed we broke out and tried to see which of us could get on the other guy's tail, shoot the other guy down with gun cameras. Boots Blesse got more on my tail than I did on his, that's for sure. Next time up Boots would tell me to try and stay on his wing, and he put me through some pretty violent maneuvers to see if he could shake me off. It was a question of heart and mind—courage, nerve. Some people folded under the pressure; they simply didn't have enough confidence. But I did OK and when we got back down, as we took off our flying suits, Boots slapped me on the back.

"Well, Scotty, you hung in there pretty well. That was pretty good going," he said. "I don't mind flying with you some more." Six months later he chose me to fly on his squadron gunnery team. That was a pretty good feeling.

I had more than a couple of close shaves with aircraft problems in those early years, and they were valuable lessons in how to deal with emergencies. They used to say having a tour of duty in Europe made you a much better pilot than if you'd flown only in the United States. There was a lot more fast-moving bad weather to deal with, for a start. It was not unusual to get diverted, be low on fuel and have to put in at an alternate destination. I remember flying in Holland once, and being told to bring the plane down immediately because a thick fog bank was moving in; seconds later it was impossible to see the ground. We had to bring our planes down in Germany. We were very low on fuel and got there by only a whisker.

The weather wasn't the only problem. Once I was flying an F-100 in Holland when, just after take off, the throttle became disconnected from the engine. I was flying at full power. There was no emergency procedure for this. I had two choices: bail out or try to get the bird on the ground. The situation was so serious that ground control confirmed I had to decide fast whether to ditch the plane off the Dutch coast or try to bring her in. It's the toughest call for any aviator; I knew a lot of guys had died because they didn't get out when they could. But bailing out isn't without risk—if the ejection seat jams, or the canopy doesn't open, or the parachute fails to deploy, you're a goner. No pilot likes to do it, though with an uncontrollable plane you have no choice. With enough control most pilots will attempt to bring their plane home. I still had enough, I reckoned, to be able to land. I knew it would be tough, but just how hair-raising I had no idea.

To reduce speed I had to cut the fuel supply and shut the engine down, before gliding the plane in to land. This was less of a problem on a clear day, because you could glide from an altitude high enough to judge a proper landing from a circular approach. But the weather that day was bad. There was low cloud cover, which meant that someone on the ground had to calculate how long it would take for me to land from a low approach to the runway once I cut the fuel supply. It was a one-shot deal. I'd get only one try. Once the engine is shut down you are committed to land— somewhere! The person on the ground got his math wrong. He calculated that I should cut my engine 13 seconds from the runway. That was too late.

It was a short runway, about 8,000 feet. For an F-86, which landed at 120–130 knots, that was fine. But for an F-100, which under normal conditions landed at 170–190 knots, it left little margin for error, and I was going faster than normal. I had deployed the speed brake—the big plate that folds out from the belly of the plane—but the version of F-100 I was flying did not have wing flaps. I was still smoking when I came to the edge of the runway, doing about 200 knots. I reached up to my left and pulled the yellow handle that released the drag chute to slow the plane. But the chute was old and weathered; its panels tore and the chute collapsed into a streamer.

The F-100 was the first plane fitted with anti-skid brakes. But they were faulty, too. When I applied the brakes the anti-skid mechanism kept grabbing and then releasing the wheels too much, making the nose of the plane pitch back and forth like a porpoise moving through water. Finally, fast approaching the end of the runway, I engaged the "barrier" designed to catch and slow airplanes at this stage—the strut of the wheel at the front of the plane tripped a wire bringing up a thicker cable laced with heavy chain links. There were so many of these 40 odd-lb links that no plane was expected to drag them all. My plane not only dragged them all but dragged them with such force that the end link whipped up violently, tore loose, was hurled through the air and crashed into the roof of a house nearby.

The plane eventually slowed just as I came to the end of the runway and rolled forward into the mud. There were fire trucks racing along the runway toward me as I lifted the canopy and jumped to the ground, scrambling away as fast as I could in case what was left of the fuel exploded. Incidents and accidents like these are classed as minor or major according to how much damage is done to a plane. What happened that day was classed as minor, because my F-100 was virtually unscathed. It was a very close call all the same. That night in the bar most of the squadron came up to slap me on the back and buy me a drink.

Another time I was out over the North Sea "dragging the rag"— towing a banner used as a target for aerial gunnery. I had a new pilot with me in the back seat on orientation, a young second lieutenant who had just checked into the squadron. It was his first time flying outside the United States. After the target flight completed their practice, we were heading back to the Dutch coast dragging the target when the engine quit. So I called my friend Denny Kilroy who was leading the flight formation.

"Hey, Denny, I just had a flame-out."

"Aw, you're kidding me," he called back. "I'll see you at the bar."

"Nope, not kidding," I said.

"Uh-oh, then you gotta problem," Denny radioed back.

I looked in my mirror and I could see the young lieutenant in the back seat, his eyes like saucers.

"We're going to have to go through the bail-out procedures," I told him. It was winter. The North Sea was pretty darned cold. However, I managed to glide the plane down to reach a Dutch base on the coastline, just in time. But we barely made it.

Others weren't so lucky in emergency situations. I lost some good friends during my time in Europe. In my first six months with the 32nd Squadron we lost six planes and two pilots. In particular, we had a lot of problems with the F-100 Super Saber jets, which were being brought in to replace the F-86 Sabers. I had much affection for the F-86. That airplane had enabled me to join an exclusive club known as Machbusters—pilots who had broken the sound barrier at 750 mph—designated Mach 1 after Swiss mathematician Ernst Mach.

There was a technique in an F-86. If you applied the controls properly in a dive and let the plane roll the way it wanted to the drag would be reduced, whereby you could get her up to Mach 1 speed. Once you did that you would get a Mach 1 pin from North American, the manufacturer of the F-86. We all wanted that pin: it was a very big deal for a young fighter pilot. I got mine a few weeks after arriving in Holland.

By contrast the very early F-100s had a very poor reputation. They caught fire, or the engines quit; big problems. When we first started flying them our first squadron leader got everyone together in the officers' club, including the wives.

"We're going to go down to Africa to start learning to fly the F-100. It's really not such a bad airplane," he started matter-of-factly. Then he knocked us all out with his next comment, addressed to the wives. "I will guarantee to all you ladies that I will bring back as many of the men as I can."

I was still a bachelor when I arrived in Europe. By the time I finished my tour of duty I was a married man. I had met my wife-to-be, Lurton, before I left the United States, while I was visiting my parents in Texas. My parents knew Lurton's parents. Her father was a general, as was mine. They had both been stationed in England during the war. When I went back to Texas the following Christmas we spent more time together. Then, one summer, Lurton came to Europe on a tour and we decided to tie the knot. In the winter of 1958 she moved to Holland and we got a little flat.

Fighter pilots didn't earn a lot of money, but our standard of living was quite high. The exchange rate was good. I remember driving down to Stuttgart to buy a convertible Mercedes 190 SL for cash. After that we traveled around quite a lot. We took off for the weekends to Paris, drove to Denmark, or Sweden or Spain. There was a very good social life in the squadron; a lot of mixing and camaraderie, a lot of parties. Lots of flying, too. I was looking forward to spending another thirty years flying fighters. Life was good. I was having a ball.

Alexei Leonov

Twenty-four hours after my wife flew to join me at Altenburg I was summoned back to Moscow. I had been an officer for only four months; but I had been chosen. I was to start my training for the space program in March 1960.

The beginning was inauspicious. Even though the pilots chosen for the space program were considered the cream of the Air Force, the practicalities of everyday life were tough. When Svetlana and I arrived in Moscow I had only ten rubles in my pocket with which to take a taxi to the Institute of Aviation and Space Medicine. As the meter clicked over the ten-ruble mark I started to get nervous. Laughing, I told the driver he would have to reverse, that I could pay him ten rubles and not a kopeck more. He opened the door and left us there on the spot. We had to walk the remaining blocks in thick snow. When we arrived I had to warm my wife's frozen feet with my woolen aviator gloves.

Initially we were given temporary accommodation at a gymnasium attached to Moscow's Central Aerodrome; bunk beds in the corner of a volleyball court. We had to drape newspapers over the net in order to get some privacy, because another pilot and his wife were sleeping at the other end of the court. Later we were given a nicely furnished three-room apartment in the settlement of Chkalovskoye near the aerodrome. Slowly we were afforded some privileges regular pilots were not, such as access to a particularly well stocked grocery shop where we did not have to queue as most of the population did.

We had little opportunity to travel in those early days, certainly not abroad. Whenever we did travel as a group we were accompanied by an officer from the KGB. By the time we started training our initial group of eight pilots had grown to twenty, but still no more than four of us could fly together in one aircraft, in case there was an accident. Spare time was limited, too, but we did enjoy picnics in the countryside occasionally at weekends and a mandatory three or four weeks a year at special spa resorts.

We knew American pilots were also being vetted for that country's space program. We read press reports about the selection of seven astronauts for the Mercury program. We realized they were more experienced pilots than we were—older by an average of ten or fifteen years. We comforted ourselves with the thought that, though they were more experienced, they would never again be young and full of vigor as we were and that we had another fifteen years ahead of us.

David Scott

When NASA selected its first group of astronauts for the Mercury program, in April 1959, I remember reading their names and details of their Air Force and Navy credentials and turning to one of my flying buddies, amazed that these men had decided to join the space program.

"Oh boy. There go their careers. I wonder what those guys think they're going to do," was the general feeling among most of us military pilots at the time.

At that very early stage of America's space program there was no actual flying involved. Essentially, those guys were being recruited to become passengers in capsules which would be carried into space by the sort of rockets we had read regularly blew up. The idea of a seasoned pilot being made to ride on the back of an unreliable rocket over which he would have no control seemed not only highly risky but incredibly frustrating. At least with an airplane you were in control and if something went wrong you stood some chance of bringing it in to land. Then, when they put a couple of monkeys in capsules on top of the rockets and launched them into space, amid fears about the effects of zero gravity on living organisms, we snickered at the prospects facing America's first astronauts.

"Would you want to follow in the footsteps of a couple of monkeys?" we would ask each other and our answer was always the same: "I don't think so."

But we were still pretty cut off from the outside world. We did not have a television set—not that we would have had much time to watch one if we had—and most radio stations were in Dutch. We had no idea of the immense excitement back home that surrounded the selection of these first astronauts—the "Original Seven," as they later became known. The media hysteria that surrounded their every move from the moment of their first press conference passed us by. Our main source of news was still *Stars and Stripes*, so I guess we had a rather narrow view of the outside world.

On Saturday nights we watched RKO newsreels in the movie hall before the feature films started. I do remember seeing footage of Khrushchev's so-called "kitchen debate" with Vice President Richard Nixon on the relative merits of American and Soviet washing machines. It was at a Moscow exhibition on the American way of life, in the summer of 1959, and was a brief attempt by both men to ease tensions between the USSR and the USA in the deepening Cold War.

The following year, however, the Soviets shot down one of our high-altitude U-2 spy planes, captured its CIA pilot, Gary Powers, and put him on trial. It wasn't clear at first who he was. Far less was known about the operations of the CIA back then. Even we as military pilots stationed in Europe didn't know a lot about the U-2s. We had just been told that if we did see a strange-looking plane flying in our sectors we were to pay no attention to it—that it was classified. After serving two years of his ten-year sentence for spying, Powers was freed in exchange for the convicted Soviet spy Rudolf Abel. But the whole embarrassing affair caused a major international diplomatic row. All hope of easing tensions between the world's two superpowers faded once more.

But even the Gary Powers incident didn't feature much on my personal radar. I was much more concerned with making sure I took the next step up the flying ladder. That next step was very clear. It had to be test-pilot school and, as far as I was concerned, the best place to be was Edwards Air Force Base in California, home to the greatest test pilot of them all, the legendary Chuck Yeager.

This was way before Yeager had been immortalized by Tom Wolfe in the classic book *The Right Stuff*, but he was already a legend. Yeager was the test pilot at the top of the flying pyramid: he was where we all wanted to be. He had secured his place in history by becoming the first pilot to break the sound barrier, Mach 1, in an X-1 rocket airplane in 1947. Until his plane's supersonic boom echoed across the high Californian desert that day, there had been a fear that a plane would disintegrate under the pressure and stresses caused by excessive drag at supersonic speed. Test pilots had died trying to push their planes beyond Mach 1. Until Yeager beat it, Mach 1 was regarded as an insuperable barrier, which no human being could cross and emerge alive. Yeager opened up a whole new era of supersonic flight when he broke this taboo. Long before the Russians had put their little silvery Sputnik into space we were flying supersonic. As far as I was concerned, that was a more significant opening salvo in the technological race of the Cold War.

The best way to reach test-pilot school, I was counseled by the Air Force, was to get a graduate degree in aeronautics. So, while still stationed in Europe, I applied to, and was accepted by, the best engineering school in the country, Massachusetts Institute of Technology (MIT).

By the time Lurton and I got back to the United States, it was 1960 and John F. Kennedy was running for president. The country was on the brink of enormous change and social upheaval. The

Civil Rights movement was gathering pace in the South. Technology was evolving fast, too; most people, it seemed, had television sets in their homes when we got back, which was not the case when I left for Europe. Stereophonic sound had recently been invented. The music scene was changing fast, too. The hippie period was just round the corner.

However, the transition from military to civilian life and the world of academia had a greater impact than all the social changes going on around me. Our daughter Tracy was born in the spring of the following year, so I was caught up in my own world of the family and studying. And, man, oh man, was I unprepared for graduate school.

From the moment I arrived it was like trying to drink water from a high-pressure fire hydrant. I had to work flat out every day for two years. Compared to the hard grind of MIT, the five or six years I had spent flying fighter jets felt like playing. While the curriculum was extremely demanding, the teaching system was much more relaxed than I had ever experienced before. At West Point we had been graded every day in every subject, but at MIT there was no requirement to go to class and essentially only one grade, the final exam. Some of the professors could not have been more different from my teachers at West Point or flying school. Some of them had long hair. At first I thought they were weirdos. I had one math professor who wore shorts, carried a backpack and used to run down the corridor to class. Turned out he was brilliant, and I came to like him a lot.

Not only were the academics intense, but the course had recently been changed to include "astronautics." So instead of just aeronautics my degree turned into one in aeronautics and astronautics. Though I was familiar with the practical movement of an airplane through the atmosphere, I knew little of the theory of aeronautics and absolutely nothing of astronautics. In contrast to the long-established field of aeronautics, the course material for astronautics was only then being developed by some of the leaders in this new field.

It was my first exposure to space and it was both fascinating and exciting. Having stepped, temporarily, out of the military domain,

I gradually began to understand, if not yet share, the enthusiasm others felt about the future of space exploration. It was still very early days, but slowly my interest in the universe beyond our own planet began to grow. It was the most exhilarating time to be at MIT. The team of professors and teachers there at that time was quite outstanding. They included Charles S. Draper, Walter Rigley and Richard Battin, all of whom wrote bedrock textbooks on the three main areas of focus in astronautics: guidance, navigation and control. (MIT was later chosen to build the guidance and navigation system for Apollo spacecraft.) They were pioneers in their field, long remembered for their groundbreaking work.

The difference between aeronautics and astronautics at the most basic level, I soon discovered, was that, while aeronautics deals with the movement of an object through the Earth's atmosphere, astronautics deals with the movement of an object through the vacuum of space, 100 miles above the Earth's surface and beyond, where there is no "air." In each case, the major force on an object is gravity, but in aeronautics objects also have to contend with the effect of the Earth's atmosphere and its "air" particles of oxygen, nitrogen and water, which provide "lift" or drag to moving objects.

The way objects move between the two domains of the Earth's atmosphere and space was an important subject. It encompassed the study of booster rockets to launch objects into space and, on their return, how objects could be made to withstand the intense heating caused by the friction of the Earth's atmosphere as they slowed from the high velocity of spaceflight. Another important subject was orbital mechanics: how objects have to move at very high velocities in space to counter the force of gravity, which tends to pull them back to Earth, and how they "control" their movements using rocket engines.

My specialization and the subject on which I wrote my thesis, however, was the mathematical application of guidance techniques and celestial navigation—how many star fixes you would have to take, and when, to go to the Moon, or Mars. The framework for celestial navigation in Apollo consists of identifying and locating thirty-six stars roughly equally spaced in the celestial sphere located in different constellations—including well-known stars such as

Sirius, the brightest star visible to the naked eye (after the Sun) in the constellation of Canis Major, and Aldebaran in the constellation of Taurus—and using them to provide information for movement from one location to another by means of "inertial guidance." Little did I know then how important to me this would later be.

When I was at MIT three of those thirty-six stars were referenced, but unnamed. Years later they were named after three men whom I worked with closely, the three Apollo astronauts: Ed White, Gus Grissom and Roger Chaffee. The stars were named Regor (Roger Chaffee's first name spelled backward), Navi (Ivan backward, Ivan being Gus Grissom's middle name) and Dnoces (second backward) after Ed White II. All three men were to become my good friends. But that lay in the future, as did the tragedy of their deaths.

Alexei Leonov

Long before we met him, one man dominated much of our conversation in the early days of our training; Sergei Pavlovich Korolev, the mastermind behind the Soviet space program. He was only ever referred to by the initials of his first two names, SP, or by the mysterious title of "Chief Designer," or simply "Chief." For those on the space program there was no authority higher. Korolev had the reputation of being a man of the highest integrity, but also of being extremely demanding. Everyone around him was on tenterhooks, afraid of making a wrong move and invoking his wrath. He was treated like a god.

It was six months before we were permitted to meet Korolev. We spent most of our time training at the newly built Cosmonaut Training Center, a special military facility under construction on the outskirts of Moscow. Later it became known as Zvyozdny Gorodok (Star City). We were also flown regularly to Engels Aerodrome near the Volga river for intensive training in parachute jumping. During my training as a pilot I had been required to make only one jump a year. In the first month of training I made seventy. Within a short time I became an instructor of parachuting.

The reason there was such emphasis on parachute training soon became clear. The earliest Soviet spacecraft were designed in such a way that those within the craft were ejected from the module at an altitude of approximately 400 meters, as it returned to Earth, and from that height they had to descend by parachute. All Soviet spacecraft touched down on land, rather than in the sea as American versions did. But whereas later models incorporated small retro-rockets to slow the final stage of the capsule's descent, softening the impact on landing, early models did not.

The man in charge of our training on a day-to-day basis was General Nikolai Kamanin, assistant to the commander-in-chief of the Soviet Air Force. Kamanin was a legendary war veteran, a Hero of the Soviet Union. He came to know our families and us well, and was very approachable. He was a keen sportsman and particularly loved tennis. I used regularly to play doubles with him and his wife.

The Chief Designer was a far more shadowy figure. Yet it was he who ultimately had the greatest influence on our lives. At the end of the first six months we were summoned to the Institute of Aviation and Space Medicine in Moscow and told we were to meet the Chief. We were excited. I remember the first time I caught sight of him. I was looking out of the window when he arrived, stepping out of a black Zis 110 limousine. He was taller than average; I could not see his face, but he had a short neck and large head. He wore the collar of his dark-blue overcoat turned up and the brim of his hat pulled down.

"Sit down, my little eagles," he said as he strode into the room where we were waiting. He glanced down a list of our names and called on us in alphabetical order to introduce ourselves briefly and talk about our flying careers.

There had been lengthy discussions at the very beginning about who should be chosen to train for the space program. Eventually it was decided that fighter pilots were best suited to working in extreme conditions, because of their ability to react instantly in dangerous situations and their multifaceted skills—not only flying but also acting as navigator, radio-communicator, gunner and engineer. Sergei Pavlovich Korolev had been a test pilot in the 1930s. We knew that about him, but little more.

When Korolev came to Yuri Gagarin's name on the list he seemed to warm to him straight away. Yuri turned red in the face and got slightly flustered. Korolev seemed amused: crinkles started to appear at the corner of his piercing dark-brown eyes. He put aside his list, as if he had forgotten about the rest of us, and asked Yuri to talk about his childhood and career as a pilot. They spoke for maybe fifteen minutes. Eventually the meeting continued as before. When he came to me I said I had six sisters and three brothers. "Your mother and father did a very good job," he said, smiling. He seemed to like me, too.

After the Chief left we all gathered round Yuri. "You are the chosen one," I told my friend. Years later I discovered that I had been right. Korolev had left us, met other designers in the space program and confided, "This morning I met a handsome Russian man with lively blue eyes, sturdy and strong, an excellent pilot. He is the man we should send into space first."

It was not until the spring of the following year that Korolev was able to realize this ambition for his protégé. Before that, disaster struck the Soviet space program. Not once, but twice.

The first tragedy cast a heavy pall over the early days of the program. It happened on 23 October 1960 at the Baikonur launch complex, in the remote, barren steppes of the southern republic of Kazakhstan, where snow lies thick on the ground throughout the long winter and in summer intense heat and fierce sandstorms make it almost unbearable. The site was a restricted military zone, a top-secret facility. (The American CIA pilot Gary Powers was believed to have been attempting to take surveillance photographs of Baikonur when his U-2 spy plane was shot down over the Ural Mountains and he was put on trial in Moscow in May 1960. His mission was regarded as a great insult to the Soviet Union, provocation by the United States, a sign that the Cold War was accelerating.)

The first we knew of this early tragedy came when many planes were scrambled from our aerodrome in Moscow carrying emergency supplies of blood bound for Baikonur. General Kamanin quickly gathered us together and told us what had happened. A new R-16 rocket was being tested when it exploded. A hundred and sixty-five people were killed, including Marshal Mitrofan Nedelin, head of Soviet Missile Deployment. The intensity of the explosion meant the dead were unrecognizable; Nedelin's remains were identifiable only by the academy headquarters badge he was wearing. The accident had no direct effect on our training—the rocket was being developed for military purposes—but it was a tragic event.

More shocking for us personally was the death of one of our own—the youngest member of our team, Valentin Bondarenko, several months later—on 5 March 1961. Preparations were almost complete for the first manned spaceflight scheduled for 12 April 1961. Yuri Gagarin and Gherman Titov had already been dispatched to Baikonur in preparation.

It was 23 March 1961 and the rest of us were continuing with a series of isolation-chamber tests at Star City. These were to last up to fifteen days to determine our physical and mental state under widely varying conditions. Medical sensor pads were attached to our bodies so that our heart rate, brain activity and blood pressure could be monitored. In Valentin's case the pressure in the chamber was lowered to simulate the Earth's atmosphere at a height of 5 km. To compensate, the oxygen content in the chamber was increased to a partial pressure of 430 mm. Like all of us, Valentin had a small electrical heating element in the chamber so that he could make himself tea and something to eat. He had been rubbing his skin with a pad soaked in alcohol, before reattaching a medical sensor, when he accidentally dropped the pad on to the heating element.

The resulting spark turned into an instant fireball in the oxygen-rich environment. Poor Valentin did not even have time to blink. His woolen clothes, soaked in oxygen, turned into an explosive device; he lived for four hours after he was dragged from the chamber. He was deeply mourned, for we were all close friends by then. Following Valentin's death all experiments with heightened partial pressure of oxygen were stopped. Some heads rolled.

The tragedy caused a major rethink of the space program. The system of using an oxygen-rich environment in our spacecraft was abandoned in favor of regenerating oxygen through a system of filters—a much safer but more bulky process. We knew the

Americans also used an oxygen-rich environment in their spacecraft, and it eventually cost three US astronauts their lives, too; even after those deaths the Americans did not change their system.

The Soviet Union did not alert those in charge of the US space program to our tragedy. In those days there was only a limited exchange of information between our two programs via international forums and congresses, and no bilateral mechanism for exchanging information of this sort. What had happened to Bondarenko was considered an internal matter, in any case; not something we wanted openly discussed. Like nearly every aspect of our space program, it became a closely guarded secret.

CHAPTER 3

Red Star, White Star

1961–5

Captain Alexei Leonov

Petropavlovsk-Kamchatsky, Kamchatka, Soviet Far East

Trained dogs were strapped into Sputnik capsules many times before it was considered safe to send a man into space. First Laika in Sputnik 2, then Strelka and Belka in Sputnik 5, Chernushka in Sputnik 9, and Zvezdochka in Sputnik 10. All were launched into orbit to prove it was possible for living beings to survive in zero gravity.

Engineers often had difficulty coaxing the dogs into the space modules. Sputnik 9's original dog, which had been trained to push buttons with its muzzle while in flight to obtain food and water, ran away. It had to be replaced at short notice, so the engineers used a stray they found wandering around the Baikonur complex. In fact, Chernushka, the stray, seemed to fare far better than some of the trained dogs. Whereas some of them got sick and disoriented in flight, Chernushka returned barking loudly and was so frisky that she ran off as soon as she was released from her harness, and had to be chased by technicians who wanted to test her heart and blood pressure.

In the final weeks before we were scheduled to put the first man into space a further test was conducted which led to wild speculation in the West. A dummy dubbed Ivan Ivanovich—the Russian equivalent of John Doe—was launched into orbit aboard a spacecraft of the same Vostok design as would be used for the manned flight. During the course of its short flight the dummy was programmed to activate a prerecorded radio broadcast. The broadcast, which bizarrely included, among other things, a recipe for borscht (Russian beetroot soup), was picked up by Western listening posts. Since no announcement of a flight had been released by our state news agency, TASS, rumors spread like wildfire that a manned space flight had gone wrong and been covered up.

But on 12 April 1961, the West was left in no doubt about the success of another Vostok flight, that of Vostok 1.

A week before the flight I had been dispatched to Petropavlovsk-Kamchatsky on the remote Kamchatka Peninsula in the Soviet Far East, where one of five UHF radio stations which would monitor the flight of Vostok 1 was located. During that early period the Mission Control Center at Yevpatoriya in the Crimea, crucial during later missions, was still under construction. So communication and control from the ground during this flight were performed from the radio stations above which the spacecraft flew.

When I left for Kamchatka I was still not certain who would be selected to make the flight and, if successful become the first man in space. Would it be my close friend Yuri Gagarin? Or Gherman Titov? Both had been chosen from a group of the six—all under 1.7 meters tall—who had undergone training for this flight. Vostok's interior was so cramped that it could not accommodate a taller man like me. Both had been sent to Baikonur launch station. Not until a few hours before the flight would one of them be informed he was to make history.

I was under strict orders not to initiate communication with the spacecraft unless instructed to do so by central command in Moscow. Communication was still rudimentary in those early days. Radio contact utilized an ultra-short-wave frequency, an open channel which could be monitored by the West. Every effort was to be made to keep the technical details of the flight secret. We did not know then that President Kennedy had been informed the night before that a Soviet manned space mission was imminent, though

I realize now it would have been a serious failure of intelligence if the Americans had not been aware of our plans.

Even after Vostok 1 was launched successfully, we were not told who was on board. I knew it would take approximately fifteen minutes for the spacecraft to come within range of our radio transmitters; even then we would be able to maintain radio contact for only a very short time. But before this happened grainy images appeared on a television monitor in our station. The quality was poor but I could just make out the shape of a man strapped inside Vostok 1. I could not make out his facial features but I could tell from the way he moved that it was Yuri. I immediately felt enormously proud. We had become very close friends. Within a few minutes I heard his voice broadcasting my radio call sign.

"Zaria 3, Zaria 3 [Dusk 3, Dusk 3], what is my flight path?" said Yuri. For reasons of security he did not want the term "space trajectory" to be used on an open radio link.

The radio operator by my side did not realize his finger was depressing the button that opened the radio link with the module when he turned to speak to me, "What shall I tell him, Commander?"

"Tell him everything is going fine," I replied.

Yuri overheard me speaking. "OK, message received. Give my regards to Blondie," he said, using the nickname I was given because of my fair hair—although I had lost much of it.

As he passed out of radio contact I became very concerned that I had disobeyed orders unintentionally by communicating with the spacecraft without prior permission. A short while later my fears were allayed when a telex clattered into the radio station. It congratulated me for "displaying the correct initiative."

An hour and forty-eight minutes from the beginning of its flight, news came through that Yuri had ejected from the Vostok capsule and landed safely by parachute near the Volga river in the Saratov region. My mission was over. Another telex came through later that day requesting me to return to Moscow immediately, but I decided to stay on for an extra day. The Kamchatka Peninsula, known as "the Land of Fire and Ice," is famous for its volcanic activity, lava fields, geysers and hot springs. It is a spectacularly beautiful part of the country, which I had always longed to see. So I spent the next day sightseeing.

I did not know that the Kremlin had organized a spectacular celebration of the flight and a lavish reception for all those who had contributed to its success. On the morning of 14 April Yuri and his family were invited to stand alongside Khrushchev and other leading party members on top of Lenin's mausoleum in Red Square. Helicopters buzzed overhead, and a massive military parade marched past.

At the time none of us expected this first spaceflight would receive so much attention, not only in the Soviet Union but all over the world. Even Khrushchev, we later discovered, was surprised. When Korolev phoned him the night before the launch to tell him that everything was ready, Khrushchev had seemed almost exasperated. "Go ahead," he said. "You are a professional. You know what you are doing. Don't bother me."

Due to my sightseeing exploits I did not get back to Moscow until the afternoon of 14 April. I flew back on a regular commercial flight and then took a bus into the center of Moscow. When I got out of the bus I could not believe my eyes. The streets were full of jubilant crowds setting off fireworks. When I asked a couple of lads what was going on, they looked at me in disbelief.

"Are you mad?" one shouted above the noise. "Don't you know? Yuri Gagarin has flown into outer space!"

Captain David Scott

Massachusetts Institute of Technology, Boston

It was roughly midway through my course at MIT that Yuri Gagarin chalked up the first manned spaceflight. For many of those on my astronautics course it was a major landmark along the road of space exploration. The buzz in the astronautics department when it was announced in April 1961 that the Soviets had put a man in space was palpable. Across the country it was front-page news. This time I was well aware of the electrifying public reaction in the United States to the Soviets getting ahead of us again.

From a military point of view it was another demonstration that the Soviet Union had a technological capability that we did not. Yet I did not immediately view it in that way. I had departed the potential theater of war in Europe and shifted gears into academia, so I didn't think about it in a Cold War context. My temporary separation from the military allowed me to view it more objectively. Gagarin's mission was a very spectacular and exciting achievement. From the perspective of the astronautics course I was in the midst of, however, Gagarin's Earth orbit mission, while a clear indication that astronautics had come of age, was a long way from the sort of interplanetary navigation I was studying.

Spacecraft in those days were still very limited. The earliest space missions still did not allow any actual "flying"—by which I mean control of the spacecraft. Even when Al Shepard went into space the following month, chalking up America's first sub-orbital flight, I didn't take a lot of notice of the parades. Following man's very first steps into space held less appeal for me than studying hard to make sure I would achieve my ambition of going to test-pilot school.

But Gagarin's flight and the rapturous public response to Al Shepard's mission, even though it amounted to only five minutes of weightlessness, galvanized the Kennedy administration. America wasn't going to stand for continually being upstaged by the Russians. Kennedy was adamant that the United States had to take the lead. His speech before Congress on 25 May 1961 went down in the history books.

"I believe this nation should commit itself to achieving the goal, before this decade is out, of landing a man on the Moon and returning him safely to Earth." In just one sentence he committed the United States to a Herculean task. Most people, including the top brass at NASA at the time, so I discovered later, had no idea how they would achieve that task. It was not clear at that very early stage whether Congress would back Kennedy anyway and agree to fund such a huge space program. There were more pressing political concerns, from the Bay of Pigs fiasco, to the Berlin wall going up and, later, to the Cuban missile crisis. These were very tense times, and we were close to major open conflict again. If a nuclear war started, we were all finished.

The "Tsar" H-bomb detonated by the Soviets on 30 October 1961 in northern Siberia further raised the specter of such nightmarish scenarios. The size of the bomb was almost beyond comprehension; 58 megatons—more than 2,000 times the size of the 20-kiloton H-bomb dropped on Hiroshima in August 1945. This was three times more powerful than anything the US had ever tested.

When John Glenn became the first American to go into orbit on 20 February 1962, my mind was totally focused on passing my final exams. After that I couldn't wait to get back to flying again. I had already been invited to go to Edwards Air Force Base in California for an interview and a flight evaluation for test-pilot school. My visit there seemed to go pretty well. I felt good about it. My final exams at MIT went well, too. I was sure I was in good shape, on my way to Edwards.

Shortly before I was due to finish at MIT, I received my official Air Force orders telling me where I would be reporting for duty when I graduated. When I opened the envelope containing my instructions I could not believe what I was reading. I checked the envelope. It had my name on it, my serial number. But I was totally confused.

The instructions were that I was to report for duty that July as a professor of aeronautics and astronautics at the newly opened Air Force Academy in Colorado Springs. A professor. Colorado Springs. "You have to be kidding," I thought.

That was most definitely not the world I wanted. I had worked so hard. I had done everything by the book. I had graduated near the top of my class at West Point. I had cut my flying teeth with Chauncey Logan. I had been posted to Europe and flown front-line missions along the ADIZ border with the Iron Curtain countries. I had nearly been killed more than once flying missions in appalling weather. I had been chosen by Boots Blesse, the man at the top of the flying pyramid, to fly on his team. I had sweated every waking hour for two years, studying the complexities of aeronautics and astronautics, because I had been told this would give me the best chance of getting into test-pilot school. And now they wanted me to be a professor!

This was such a blow. There was no way I was going to give up the dream I had cherished since I was three years old. That dream was to fly, not to stand in a classroom talking about flying. I was absolutely determined that I was not going to give in without a fight. It meant challenging orders, and in the military service you are not encouraged to challenge orders; you say "Yes, sir!" and get on with it. But as far as I was concerned the stakes could not have been higher. The only ambition I had ever had was on the line. So I made up my mind. I would try to persuade someone in the Pentagon to change my orders.

A short while later I flew down to Washington. I didn't know where I was going or whom I was going to see. The Pentagon is a maze. It took me all day to find my way around the system. Eventually I was granted an interview with a colonel. I was just a young captain then, and I sat outside his office for maybe an hour waiting until I was finally admitted.

"Why are you here, Captain?" the colonel barked.

"Well, sir," I said. "I want to discuss my orders."

The colonel looked at me with absolute astonishment. "What d'you mean, 'discuss your orders?' What are your orders, Captain?"

"Sir, I've been assigned to the Air Force Academy as a professor."

"Oh, wonderful," he said. "Congratulations."

"But, sir," I had the temerity to say, "I'd like to go to test-pilot school."

Oh, man. Here was this salty old colonel, in a job he obviously hated, and I'd rubbed him up the wrong way. He read me the riot act for coming to the Pentagon, taking his valuable time.

"Don't you know you never challenge orders? Why are you challenging orders?" he demanded.

I thought, boy, I shouldn't have done this. I've done a stupid thing challenging the system. My career is over. But I summoned up the strict bearing I had been taught at West Point, where you learn to give only one of three answers to a question: "Yes sir," "No, sir" or "No excuse, sir." You do not give excuses or explanations for anything, and give a reason for a particular course of action only if invited to do so. So there was only one answer to his question.

"No excuse, sir."

With that he seemed to soften slightly. He leaned back, put his feet on the table and pulled out a cigar. In that moment I realized he was going to loosen up a little and ask me to explain why I was seeking his help and advice. So I told him that I had been counseled by the Air Force that if I wanted to go to test-pilot school I should go to graduate school.

"Sit down, son," he said. I sat, and he chewed on his cigar for a while, looking at me and, eventually, breaking into a broad smile, "I like your initiative," he said. "I like that you came down here. I'll see what I can do."

He turned out to be such a nice guy. He realized, I think, that I had been given the wrong advice by the Air Force, or, perhaps, that procedure for getting into test-pilot school had changed, but that I had followed the advice I was given. Several days later I got a new set of orders. I was to report for duty at the US Air Force Experimental Test Pilot School at Edwards Air Force Base, California. Edwards! Eddy! It was the most advanced experimental test-pilot school there was. For fighter pilots it was the Holy Grail.

Alexei Leonov

From very early on I was convinced that one of the best ways for the Soviet Union and the United States to cooperate, to build bridges, was through their respective space programs. I remember once drawing a sketch of an American and a Russian standing side by side on the Moon holding their respective flags. As I was finishing the drawing I decided this might not go down too well with some of my superiors, so I replaced the national flags with a more neutral beacon. I still have that drawing. I wish now I had left it as it was. But I could not have realized then what significance that image would have for me many years later.

By the early 1960s I had come to the conclusion, privately, that our troops should no longer be stationed in Eastern Europe, that their presence there compromised the Soviet Army. When Kennedy was elected president, I remember feeling hope that relations between our two countries would get better. He seemed a decent man, a gifted politician. But the Soviet Union had given too many reasons to be viewed with fear and suspicion by the West. The political system in our country was that of a totalitarian regime. The sad truth was that ordinary people in the United States and elsewhere knew far more about how our regime operated than we did ourselves.

Any hope of thawing relations between our two countries seemed to evaporate in the autumn of 1962 over the deployment of Soviet nuclear missiles in Cuba. Again the Soviet Union and the United States came to the very brink of war.

I thought then that I would have to abandon the space program and return to regular duty as a fighter pilot. It seemed certain the conflict would end badly. I had nightmares about an imminent nuclear holocaust. But Kennedy was wise beyond his years. He managed to defuse the situation. No shots were fired. The missiles were withdrawn.

As the crisis subsided we became focused once more on our program. Most of us did not take that close an interest in politics. We had not been vetted according to our political beliefs before joining the space program, and it was not a precondition that we be members of the Communist Party before being recruited. Most of us were party members, but some, like Gherman Titov, Valery Bykovsky and Konstantin Feoktistov were not. Titov and Bykovsky joined subsequently, but Feoktistov never did.

Our training was intensive, a punishing regime which pushed us beyond what we thought we were physically capable of. Every day started with a 5 km run, followed by a swim before we even began our individual programs. Every aspect of our daily routine was carefully monitored by a team of doctors and nutritionists. Our diets were carefully prepared. We were allocated special rations not widely available to the rest of the population, such as 50 g of chocolate a day, to keep up our energy. I remember being amazed later when I discovered that American astronauts were allowed to eat virtually whatever they wanted. We were expected to be in peak physical shape.

In addition to the exhaustive physical training, we were enrolled at the Zhukovsky Higher Military Engineering Academy (which some called the Soviet West Point) to gain further academic qualifications. The curriculum covered a wide spectrum of disciplines including physics, mathematics, chemistry, metallurgy, aircraft construction and technical drawing. Zhukovsky made you work like a horse. But there was one man in particular who inspired in us such confidence, respect and loyalty that we always approached our work with great enthusiasm. He was Professor Sergei Mikhailovich Belotserkovsky, a doctor of engineering sciences with a high military rank and a State Prize laureate.

The first time we attended his class, in September 1961, we all filed into his lecture hall in our military tunics with briefcases packed with books tucked under our arms, like first-formers not knowing what to expect. We stiffened with astonishment when we saw a giant wind tunnel in the auditorium, together with various model airplanes. As we gathered around, examining each in turn, Yuri Gagarin saw Belotserkovsky enter the room and quickly called us to attention.

"Comrade officers!" he said, his loud voice echoing around the hall, and then, addressing Belotserkovsky, "Comrade Colonel! Cadets from the cosmonaut preparation center in attendance."

"Comrade officers, please take your seats," came the professor's reply and we sat down to take our first close look at the man who would guide us through the most complex courses over the years to come. His tunic was without a wrinkle and his tie was elegantly knotted, a sign of his precision and accuracy. His wavy gray hair was in some disarray, fitting perfectly with his scholarly image. His eyes wandered from face to face with an intent stare—not as a demonstration of his superiority, but rather as a sign that he was pleased to meet each one of us.

"As I see, you are all high-class fighter pilots of great experience in flying the latest planes," he began. "Now let's try to comprehend in depth why these vehicles that are heavier than air are able to fly . . . Then we'll start discussing and, who knows, we may even jointly design a new airplane type spacecraft," which, in later years, we did. Then he outlined the course ahead of us, and from that day onward we never had any problem understanding each other. We were always amazed by his ability to draw on the blackboard, using colored chalks and with no drawing instruments, diagrams that looked as if they had come straight from the pages of a textbook. Belotserkovsky treated each one of us as a creative individual and encouraged us to excel. Step by step the topics became more and more complicated, but whenever he was about to introduce us to a complex new subject he addressed us in the same way.

"Good geese," he would say, "let's see what comes next."

Once, we invited him to join us at an important football match in the Upper League. It was partly boot-licking, a way, we hoped, of encouraging him to be more lenient in exams. But it was also to give him a chance to relax. Yet even this he was later able to turn into a lesson on aerodynamics. When one of the players scored a fantastic goal, the crowd erupted, some ecstatic, others yelling in agony. To an untrained eye the ball should not have landed between the goalposts. We stood there perplexed, too. But the professor just mused quietly, "An amazing problem . . . In our next lab work you'll have to find the solution," which we did with a model football in the wind tunnel. We were able to demonstrate exactly how, if the ball was spun, its trajectory altered and then went on to apply this to the landing trajectories of our spherical Vostock and Voskhod spacecraft.

Fascinating as they were, however, these studies at Zhukovsky had to be fitted into our little spare time during our training program. At first I felt ashamed. I was falling behind the other students, most of whom were studying full time. They tried to put me in a class of female students at the beginning. When I refused I had to study on my own. That took up most of my weekends and many evenings over the next seven years. In the end, I finished the course six months ahead of some of those full-time students with whom I had started and gained first a higher engineering degree and later a doctorate of technical sciences.

At the end of our first year, having passed a series of complex exams, we were given the official title "cosmonaut"—which in Russian means one who voyages within the range of our cosmos or within the solar system.

We regarded the American term "astronaut"—implying, as it does, one who flies between stars—as rather boastful. Privately, we used to joke that we were mere "cosmonauts" not "astronauts."

David Scott

Out there in the high desert, a hundred miles northeast of Los Angeles, Edwards was a pretty rough place. It was remote and the housing was basic, at best. But that was OK. When I arrived there in July 1962, I was exactly where I wanted to be. The sun would shine almost every day in warm, clear skies—for flying a very welcome change from the almost constant gray slate of a sky and wet, cold, grim flying weather in Europe.

Most importantly, Eddy was home to Chuck Yeager. Aside from his status as a living legend, I'd heard a lot about Yeager from my former roommate in flying school. He had been assigned to Yeager's squadron when Yeager was a squadron commander in France in the early 1960s. This guy enthused endlessly about the great esprit de corps Yeager engendered, and talked with passion about what a great pilot he was. When I arrived at Edwards in July 1962, Yeager had just taken over as Commandant of the Test Pilot School (TPS). During the two months before my TPS class began, I got to know him quite well, though never as Chuck Yeager. I was just a captain. To me he was "Colonel Yeager, sir!"

I got in some flying time with him early on. He was a great pilot—the smoothest stick I ever knew. To Yeager, and any really good pilot, this meant that the plane he was flying was more like a physical extension of himself than a machine he was controlling. A "smooth stick" can feel the motion of the plane with every muscle in his body; every sense is attuned to the plane's balance and movement, so finely that it feels almost like gliding. There are never any awkward, jerky movements. Being in a plane with Yeager was like taking part in a graceful dance—an aerial ballet, but with a deadly purpose.

The Cold War meant the technology race was on. We were really pushing the envelope when it came to developing faster, better planes than the Soviets. A lot of guys were dying. Within ten years nearly half of the men in my class would be dead.

We knew the risks but, as far as we were concerned, helping to develop a new aircraft was the ultimate. There was nothing more exciting for us than to get a new plane and put it through loops, rolls and dives to make sure it was OK. That was why we were all there. We were exploring new frontiers.

I thought I knew how to fly before I went to Edwards. But at Eddy I really learned. The difference between being a fighter pilot and a test pilot is precision and accuracy. Everything you do as a test pilot has to be precisely on the mark. You have to fly at exact speeds, at exact altitudes in set traffic patterns—always on the mark. You also learn how to fly planes at the very edge of the envelope. You push the machine into uncontrolled maneuvers to learn ways of pulling it back from the brink of disaster.

I flew with a lot of really good guys. But the best was always Chuck Yeager. Not only was he always the best, but he was also a lot of fun. He was a loose enough guy to want to do what all fighter pilots want to do: show how a plane could really perform, "shine" ourselves in front of the boys.

One time I was with him in New Mexico on a clear sunny Sunday in a T-38, a sleek, white, pretty thing, a plane you could do some fancy footwork in. I was flying, and Yeager was in the back seat. On take-off he told me to pull the nose of the plane straight up instead of flying the regular, slow, climbing flight path. Only the highest-performance plane could manage a maneuver like that. We all liked to do it—but not normally with the boss in the back seat. It was not an approved procedure. But there was the Yeager, giving me a little rein.

It was a fun feeling, a bit like riding the incline of a steep fairground ride—right on up into the blue. I kept expecting Yeager to tell me to pull her round. But he just wanted me to fly straight up. All I could hear from the back seat was his voice keeping on telling me to "Pull her up. More, more."

The ultimate accolade came when I graduated from test pilot school. I won the award for best pilot in my class. No one knew in advance who would win. I remember when it came to the ceremony we were all lined up in the front row. My name was called out and I stepped up to the podium, where Yeager presented me with the award. I felt pretty good about that. That was really exciting. To win the best pilot award at such a hard school to get into, and from the best pilot in the world, was a more cherished achievement than anything else I had ever done. Where was I to go from there? I remember wondering at the time.

My next move was quickly apparent. At the end of our course, some of us were chosen to graduate up to the newly opened Aerospace Research Pilot School (ARPS), also at Edwards. There we would get the opportunity to train as "space pilots"—something like the guys who were flying at speeds beyond Mach 5, up through the atmosphere above an altitude of 50 miles, the generally accepted lower boundary between the Earth's atmosphere and space. In terms of manned spaceflight, the Air Force had a lot more opportunities at the time than the new space side of the "National Advisory Committee for Aeronautics" (NACA) which had recently been reorganized and changed its name to the "National Aeronautics and Space Administration" (NASA). Along with the original NACA people the Air Force was jointly developing all these booming and zooming airplanes, such as the X-15 rocket airplane and the X-20 spaceplane or Dynasoar, short for dynamic soaring. This was the exciting stuff.

During the early weeks of our assignment to ARPS we learned about how the body functions in space. It was a short, intensive course in physiology and medicine at the School of Aviation Medicine, Brooks Air Force Base in Texas. Total immersion. We did everything from learning about the body's internal organs to observing an autopsy. The idea was that we would be able to monitor our own reactions while flying at very high altitudes or in space and, if necessary, be able to describe accurately what was going on so that a doctor on the ground could help us try to deal with any problems.

In the very early days of spaceflight some thought there might be a need, in dire emergencies, for space pilots to perform minor operations during long spaceflights—appendectomies and the like. So we were expected to learn as much about the human body as we could during the couple of weeks that we were at Brooks. The doctors there were just great, and some of what we learned was pretty light—the effect of placebos, how to deal with broken bones or troublesome toenails. But parts of it were pretty rough—particularly the autopsy right at the very end.

During the first week of the course we were shown two pairs of lungs—a light-colored healthy-looking pair which had belonged to a nonsmoker, and a dark oily-looking pair that had been those of a two-packets-a-day man. I didn't smoke then and that ensured I never would. After that we were presented with a case of open-heart surgery on a dog and were invited to massage the dog's heart to learn how its rhythm could be kept going until it was repaired by surgery. The next morning, we traveled to the nearby Air Force base hospital for our "final exam." After a thorough briefing by the presiding pathologist, we were ushered into a very stark, but quite small, clinical room and were all standing around chatting and joking when a corpse covered with a sheet was wheeled in.

Up to that point each man on the course had been convinced he was a tough pilot, not easily shocked. But when the sheet was pulled away from the cadaver each one of us shrank back against the walls to get as far away as possible from what we were about to witness. First of all it was explained that the man had died of liver cancer, and we were reminded of the details of human anatomy we had learned earlier in the course. These were about to be most graphically displayed. After making a series of incisions, the surgeon began removing organs and placing them in steel basins for later inspection. It was almost as if he was taking the pieces of a car apart, except that the odor of decaying flesh became almost overwhelming.

It was a small room with little ventilation. Half the men left the room very swiftly. Of those who remained, several lit cigars to try to mask the stench. I stood there trying to take as shallow breaths as I could. In spite of this experience, or perhaps because of it, the course was excellent. It was more intensive medical tuition than we would ever be given subsequently at NASA.

As initial training for flying on the fringes of space, we then started flying F-104 Starfighters, the first fighters designed to fly at speeds greater than Mach 2; all fighters before that barely broke the sound barrier. The F-104 was a high-speed, high-altitude operational fighter which looked more like a rocket than an airplane. Its long, tube-like body had short, stubby wings, so razor-sharp that their edges had to be capped with a slim plastic casing to prevent accidental injury to maintenance crews.

Our task was to learn how to control these aircraft in the much thinner atmosphere at heights of up to 90,000–100,000 feet where airplanes perform in a very different manner. It is a very delicate business, but exhilarating. The F-104 is a fighter, so can maneuver well, drop bombs and fire cannon; it just travels extremely fast and lands fast, too. Our task would be to fly a high-altitude parabolic arc—called a "zoom climb"—to such an altitude that we had to don a pressure suit to support our body in the vacuum of "space" in case the aircraft's cabin pressure failed. These profiles really pushed pilots to the limit. Even Chuck Yeager had had to bail out of a "rocket-boosted" version of one of these aircraft when it went out of control as he was taking it over the top of a "zoom." But the thrill of testing both yourself and your aircraft to the edge of your ability was almost indescribable.

Zoom climbs involved flying to a height of around 35,000 feet, then pushing the airplane as hard as possible to a speed of slightly more than Mach 2. When that speed was reached, you pulled the nose of the plane up at about 3.5 G until you reached a 45-degree angle, and then continued climbing at 45 degrees. At around 60,000 feet the temperature of the engine reached over 600° C, at which point you had to cut the engine to prevent it from overheating. The plane would then continue climbing like a bullet until it started losing energy and the speed bled off to around 120 knots—at a height of around 90,000 feet. You then let the plane drift out of the climb to level off before beginning the descent, holding the controls very, very lightly.

In the thin atmosphere at the top of the climb, the sky was dark blue and you experienced a short period of weightlessness, though the only indication was the feeling of your butt lifting away slightly from the seat as your body pulled against the tight straps. After gliding down to a height of about 35,000 feet you lit the engine, then lowered the landing gear, the wing flaps and speed brake in order to reduce the speed before bringing the plane in to land on the concrete runway at Edwards. If the engine failed to relight, as it sometimes did because of the difficulty of the maneuver, you had to bring the plane in to land on the dry lake bed at Edwards—the world's longest runway.

In addition to zoom climbs, we also practiced what were called "low-lift-to-drag ratio landings." They involved landing a plane without much lift, which comes in like a rock and has a very high "sink rate" as it approaches landing—such as the X-15 rocket plane. For this we configured the F-104 with the same lift-to-drag ratio as the X-15 by firing the engine up to medium power and deploying all the landing gear and other braking mechanisms at around 25,000 feet. It was on one of those runs that I had perhaps the closest shave of all—one of my more exciting experiences.

I was with my best friend at that time, Mike Adams, flying a two-seater Starfighter when our engine failed during a steep X-15-type landing approach. We had started our circular approach to landing at 25,000 feet and were coming in at a speed of 300 knots and an angle of about 20 degrees; passenger airliners, by comparison, land at an angle of about 3 degrees. After lowering the landing gear and wing flaps to slow us down, as we reached a height of around 6,000 feet, close to the end of the runway, both Mike, who was sitting behind me observing, and I realized we were sinking much too fast. We could see the ground looming up toward us too quickly, the runway was getting rapidly wider. We did not know it, but part of the engine had failed. We were sinking like a rock.

In that split second, with no time for discussion, we both realized we were in deep trouble. We would be on the ground in a matter of seconds. In that instant we made different decisions about how to get out of the situation. Mike's instinct was to eject, and he probably thought I was going to bail out, too. Mine was to get us down, try to get us on the runway. I pulled the nose of our F-104 up to "flare" the plane—make it rear up as a duck does before landing—to slow it down further. Still it was a very hard hit; the landing gear snapped off, and the plane slid away from the runway on its belly—shaving off the bottom of the fuselage and careering out into the desert. When I looked down I could see the hard-packed desert crust through a hole beneath my feet and the orange glow of flames beginning to lick away at the twisted metal around what was left of the belly of the plane. I knew I had to get out quick and opened the canopy. But I couldn't move. There was something pinning me in my seat.

The F-104 was the first plane to incorporate stirrups and cables to pull a pilot's legs back toward his seat before ejecting; it was a safety mechanism designed to stop his feet flailing when bailing out at high speed. For some reason the stirrups had been activated and my feet were locked in place. Scrabbling quickly for the cable-cutter handle behind my seat, I sliced through the cables to release the stirrups and managed to hurl myself free of the cockpit. With fire engines hurtling toward the burning plane, I glanced back to where Mike had sat and saw he was not there. Part of the jet's engine had pushed its way through to where his seat had been, destroying the rear compartment. In that instant I feared the worst.

But, looking down the runway in the shimmering heat, I caught sight of a figure waving energetically at me. It was Mike, a parachute trailing in his wake. He had pulled his ejection handle just before we landed. It turned out the decisions we made in those few seconds were absolutely the right decisions for each of us. Had Mike not ejected, the engine would have crushed him. Had I tried to eject, I would have been killed: my seat, it was discovered later, had buckled and would have blown up in the cockpit. We lucked out. As was usual following an accident like that, we were taken to hospital for overnight observation, even though we had only a few minor cuts and bruises.

"I guess you got out of that one OK," Yeager said when he strode into the hospital to see us afterward. Then, with a wry smile, he indulged a dark sense of humor by handing me a piece of melted aluminium from the wreck as a souvenir. He was still by my bedside, the two of us looking through a whole bunch of photographs of the wreckage, when Lurton came to visit me. Some time afterward she told me she had never seen two men move as fast as we did to conceal those photos so that she would not see how close I had come to "buying it."

The classic way of preventing fear building up was to go right back out there. So two days later I flew the same profile, several times—no problem. The day before I had taken an F-104 out with an instructor, and Mike had, too, and we flew every conceivable profile to try to find out what had happened. We couldn't figure it out. An accident investigation board later discovered the nozzle of the engine had been faulty and opened mid-flight, causing the plane, almost fatally, to lose power.

Alexei Leonov

The first time we were permitted to see the spacecraft under construction at OKB-1 (Sergei Korolev's Experimental Design Center), I was taken aback. Being an artist I had imagined a totally different type of space capsule, something more sleek and aerodynamic. Yet here they were, neatly lined up along a large conveyor belt in the construction hall, looking like an underground station in the Moscow Metro—no windows, but brightly lit. Then large silver balls. Each craft was in a different stage of assembly. Some seemed little more than empty spheres. Others were nearer completion.

How small they seemed. I remember wondering, "Could they not manage to give us a little more room?" It was clear that I would never be able to fly in such a craft. I was 1.74 meters tall, and the craft had been designed for people under 1.70 meters. They were the first generation of Vostok space capsules, like the one in which Yuri Gagarin had orbited the Earth. I felt disappointed.

But I was quickly assured that scientists had already begun work on a second generation of spacecraft. It was for a mission in this type of craft that I would train. This second generation became known as Voskhod, which means "sunrise." Their mission would be much more ambitious. Just how ambitious was not clear until the end of the following year, 1962.

In the meantime, three further Vostok flights carried cosmonauts into space for longer duration. First was Gherman Titov, who orbited the Earth seventeen times. Dizziness and problems he experienced with his eyesight led to an adjustment of Vostok's life-support systems. Then came Andrian Nikolayev, who made sixty-four orbits, followed by Pavel Popovich, who launched a day after Nikolayev and flew a parallel mission orbiting the Earth forty-eight times. Valery Bykovsky and Valentina Tereshkova, the first woman in space, undertook the last two parallel Vostok flights in June the following year. Tereshkova's achievement was officially hailed as a triumph for equality. Khrushchev congratulated her on live television and declared her mission had been a sign that "men and women are treated equally in the Soviet Union."

Most Russians believed that women should not meddle in what was regarded as men's work. When Tereshkova returned to Earth and spoke about how simple the mission had been, our hearts sank. The prestige of the cosmonaut corps sank a little, too. Several decades later it was revealed that her mission had been anything but a simple affair. She was courageous and completed the tasks assigned, but she did suffer serious problems with her balance. The Soviet policy of strictest secrecy meant that none of this was revealed at the time, however. Korolev simply concluded that his spacecraft were unsuitable for women.

When we were summoned for our first viewing of the second generation of Voskhod vehicles in early 1963, which I was scheduled to fly two years later, we were issued with unusual instructions. We were told to arrive at Korolev's design bureau, OKB-1, in civilian clothes. Up to that point no one had taken a step out of uniform; few of us even possessed an overcoat. We had only one day to buy ourselves new clothes. We all wanted an Italian suit and coat. We went to different shops in search of the best we could find, but there was very little choice—in fact, there was only one model on sale. When we gathered the next day to catch the bus that would take us to our meeting we burst out laughing. We had simply exchanged one uniform for another. We were all wearing the same brick-colored overcoat, tailored by expensive Italian designers.

When we entered the bureau we were all told to take off our coats, anyway, and issued with white protective jackets and covers for our feet. Again we were shown a row of space vessels. Most looked very similar to the Vostok-type capsules we were used to.

But at the end of the line two stood out. They had a more interesting design. They had a transparent airlock attached, with a movie camera installed.

Korolev entered as we were studying the craft. "OK, gentlemen," he said, "take a good look." The explanation he then gave of the purpose of the airlock was astonishing. "Every sailor on an ocean liner has to know how to swim. Likewise a cosmonaut must know how to swim in open space," he said confidently. "Not only swim, but accomplish assembly and disassembly work outside the vehicle. Man must learn how to do this." He took a good look at all of us.

Then he addressed me directly. "Well, my little eagle," he said, "please put on that spacesuit and go and sit inside the space vessel. Try to exit the vehicle through the airlock. You have until twelve o'clock to prepare to make a presentation to the engineers of how you find the vessel." With that, he abruptly left.

I started to feel very hot under the collar. I wondered if Korolev had singled me out by chance. But I knew he had been following my progress closely. He knew my technical abilities, my excellent physical condition and that I excelled at sports. Yuri Gagarin turned to me almost immediately and congratulated me. "Now it is your turn, Lyosha," he said, slapping me on the back. "Now it is you who has been selected."

I took a good look around the cabin and realized there were quite a few differences between the Vostok capsule with which I was familiar and this Voskhod model. Both spacecraft had the same landing module. But unlike Vostok—from which the cosmonaut ejected shortly before touchdown, descending by parachute—it was clear from the design of the seats that the Voskhod was designed to bring its occupants back to Earth with a "soft" landing. There were also two television cameras fixed at certain points of the capsule, and there was an independent navigation system and switches to control all environmental parameters within the spacecraft.

Some of the control panels I was familiar with from Vostok had been shifted to different positions. The optical orientation system had been moved 90 degrees to the left. The switchboard controlling the airlock chamber and the spacesuit was directly above my head as I sat in one of the two seats. The position of the airlock switchboard was particularly awkward—I was to bump my head on it regularly over the months of training until I got used to its position. My first impression was that the cabin was very cramped. I later found that in zero gravity Voskhod took on a more spacious feel, and could even become a quite comfortable and reliable temporary home.

When Korolev returned later with the engineers, however, I started firing off some of my initial observations. He told me to slow down, not be so hasty. There was still time to perfect the design of the craft, he said. That was the moment when I began training in earnest for my mission aboard Voskhod 2. Every week after that I returned to the spacecraft as its design was modified, to familiarize myself with every inch of the vessel. I knew every nut and bolt in that spacecraft. It became my home. I used to sit in the cabin regularly in my spacesuit without turning the ventilation system on, to test my stamina.

Korolev's power, influence and responsibilities at that time are almost impossible to comprehend. Not only was he in charge of all space-related issues, but he was also in charge of some of the design of rockets for military purposes. He oversaw the design and testing of communications and surveillance satellites, too. Although he delegated responsibility for each program to trusted designers in separate engineering bureaus, his workload was enormous. The rough Soviet equivalent of NASA, the Ministry for Medium Machine Building, was established in 1962. But in truth it was Korolev who was the mastermind behind all these programs. I marveled at his ability. He was a very profound thinker. But he was tough. He did not suffer fools gladly. He had the ability to silence a person with the smallest gesture of a hand.

That Korolev's single-handed mastery in all these different fields was not publicly acknowledged never seemed to bother him. I never had the impression that anonymity weighed heavily on Sergei Pavlovich. He was not bothered by such petty concerns. If he had been known publicly, that might have interfered with his work. He appreciated more than most the Kremlin's fears that an attempt might be made on his life by enemies of the Soviet Union if his identity and importance to our country's space program were ever revealed.

Wherever Korolev traveled, a so-called secretary—a bodyguard—accompanied him. When he went on leave to a Black Sea sanatorium it was sealed off for his exclusive use. What bitter irony he must have seen in the state's sudden concern for his welfare. His life had been worth nothing under Stalin's brutal regime. He had been left to rot in a Siberian gulag.

David Scott

By the spring of 1963 NASA had moved beyond the stage of "spam in a can"—missions flown by astronauts with little control over the spacecraft. After John Glenn became the first American in space, Scott Carpenter and Wally Schirra had flown two more Mercury missions in May and October 1962, and completed three and six Earth orbits respectively before splashing down in the Pacific Ocean amid great fanfare. In September that year NASA was beginning to look at real flying. The space agency had recruited a second batch of nine astronauts to fly its Gemini and Apollo missions.

Six months later, in March 1963, the word went out that NASA was looking to recruit a third batch of astronauts and I put in my application, principally because it was what you were expected to do. I knew three of the men who had been chosen in the second batch selected, Frank Borman, Jim McDivitt and Ed White. They were all Air Force, and they'd all been at Edwards. They were really good pilots. If the space program was good enough for them, I reasoned, it must be pretty good.

And yet NASA's budget was still terrible at that time. Shortly after I put my application in, *Time* magazine ran an article with a photograph of Borman and White sitting around in an office in Houston, alongside an article detailing the lack of funding for the Gemini program. The implication was that they were wondering whether they had made the right choice in giving up their Air Force careers.

Congress was still reluctant to approve all funding requested for the space program, even though Kennedy had made another big speech about America's commitment to space exploration in the autumn of 1962. After touring NASA's new launch facility at Cape Canaveral, Florida, he had visited the Marshall Space Flight Center in Huntsville, Alabama, where Wernher von Braun was building the Saturn rockets that would later be used in both early Apollo tests and, ultimately, lunar landings. Then Kennedy had stood in the stifling heat of a packed campus football stadium at Rice University in Houston, Texas, and outlined his vision of America's place in the "space race" to a wildly cheering crowd.

"We mean to be part of it," he had promised. "We mean to lead it . . . For the eyes of the world now look into space, to the Moon and to the planets beyond, and we have vowed that we shall not see it governed by a hostile flag of conquest, but by a banner of freedom and peace.

"We choose to go to the Moon in this decade and to do the other things, not because they are easy, but because they are hard," he had continued as applause began to grow. "That challenge is one that we are willing to accept, one we are unwilling to postpone, and one which we intend to win." The crowd had erupted.

This rousing speech signaled the start of the space race. It captured the nation's imagination. When JFK was elected two years earlier, my reaction had been "Too bad, wonder what this guy Kennedy's going to do? A Liberal Democrat. Looks like the country's going down the tubes." I had voted for Nixon. I was a conservative because of my upbringing, a Republican. But this speech, apart from demonstrating Kennedy's great oratory skills, was a clear demonstration of his visionary leadership.

Still, no one had any idea then how easy or hard it would be to get to the Moon. I certainly did not. How much more complicated could it be, I thought, than flying a rocket plane? I just assumed NASA would build the craft needed to get to the Moon and someone would go fly it. Looking back, it is clear Kennedy would have been very well briefed on the area in which American technology was ahead of Soviet. Crucially, for instance, our inertial guidance systems were much lighter and more accurate than the Soviets'. So this was the arena he chose to demonstrate the benefits of free society over totalitarianism. This was one aspect of the Cold War he was confident he could win.

It was another year before Kennedy returned to Texas with the intention of rousing the nation with another landmark speech on space. The tragedy that awaited him in Dallas in November 1963—before he was able to deliver that speech—was unthinkable. In the spring of that year we were all still full of optimism.

Yet going to the Moon still seemed a very remote concept to most of us. All I had in mind at that time—if I was offered a place on the space program and if I accepted it—was maybe flying in space once or twice, and then resuming my career with the Air Force.

Shortly after sending in our applications to NASA, a whole bunch of us were invited to undergo a week-long physical exam as part of the selection procedure. Luckily, we were not subjected to the indignities endured by the Original Seven. When they were recruited the survival of human beings in space was such a big unknown that their physical tests included having every bodily orifice probed and checked. By the time we went through the selection process, at least they knew the human body could survive spaceflight.

Nevertheless, in addition to the usual cardiograms, treadmill tests and electroencephalograms, we were given hypoxia tests in which we were starved of oxygen to see how we reacted. We were also spun in a chair in a dark room to test our tolerance of motion sickness and given a quart of glucose to drink before having our blood taken every hour to monitor our blood-sugar levels. We were also subjected to the pretty unpleasant experience of having ice-water poured in one ear to see how our inner ears reacted to the imbalance of the inner canals in one ear being warm, while those in the other ear were so cold. It was very disorienting. The brain didn't know how to cope with these mixed messages. It made your eyeballs spin pretty wildly. It was a test of the workings of our inner ears to some extent. But it wasn't much fun.

After this we were put in an isolation chamber for several hours at a time and our reactions were monitored. Part of this time in the chamber was spent working with what we called a "busy box," a very cleverly designed, multifaceted electronic box on which there were a lot of dials and switches. A group of doctors watched us closely as we were given cues to perform increasingly complex tasks with the box. In the end a number of tasks became so complex that they were impossible to complete, so it was like a game that ran until you lost. It was a way of seeing how you responded to stress. Some guys took it very seriously and got very intense. But it did not worry me too much.

Then there were the sessions with the psychologists and psychiatrists. They really had us trying to second-guess what we thought they wanted to hear. One of the favorite tests was to present each of us in turn with a blank sheet of paper. Mike Collins, who had also graduated from the test-pilot school and ARPS at Edwards and had a great sense of humor, had gone through NASA's selection rituals once before. He had not been chosen then and spent some time warning us that perhaps we should take this test a little more seriously than he had done the first time round. When asked what he saw on the white sheet of paper that time, he had talked about a bunch of polar bears fornicating in a snow bag. The shrinks' faces had "tightened with displeasure," he said. Another smart response—which I thought wiser to keep to myself—was to study the blank piece of paper and then inform my inquisitor that it appeared to be upside down.

No matter how hard you tried with these guys, though, it wasn't hard to make a wrong move. When I was queried by another psychiatrist about my background and mentioned that I had been at MIT he asked me how I had liked Boston.

"Beautiful country," I said. "But I come from Texas where people are really open and friendly and I found the people in New England a little hard to get to know—rather cold." There was a long pause. I knew from years of experience as a pilot that you could never afford to get on the wrong side of doctors or psychologists; they had the power to ground you in a flash.

"I sure hope you're not from Boston," I added a little nervously.

"Born and raised," came his grave reply.

It didn't seem to do me too much harm, however. Following the tests a shortlist of thirty of us were invited to go to Houston for a series of interviews and written examinations. If we were selected from that group, we were told, we would be getting a phone call from Deke Slayton, then director of the Flight Crew Operations Division at NASA, which was responsible for all astronaut evaluation, training and crew selection. The call would give us the opportunity to accept or decline the offer.

Nobody knew how many people would be selected or when. It was all very tenuous. I remember spending most of the long drive back to California from Houston with my buddy Mike Adams, who was also on the NASA shortlist and who I was sure would be offered a place, endlessly discussing whether joining the space program was really such a good idea. The time I had spent studying astronautics at MIT had certainly stimulated my interest in space. But at the time the Air Force had its own ambitions for a space program with the X-15 rocket plane and the X-20 spaceplane, or Dynasoar, program and even the planned Manned Orbiting Laboratory. With Congress quibbling over funding for NASA, the Air Force might be a safer bet. Shortly after I returned to Edwards I was offered a place as an instructor at the ARPS, so staying on at Eddy seemed pretty good to me.

Late that summer a group of us from Edwards, including a few other guys who'd been shortlisted by NASA, went on a tour of test-pilot schools in Germany and France. We were staying in a hotel in Germany when word went out that Slayton was making his phone calls. That evening I got to thinking seriously again about whether I really wanted to join NASA if I was offered a place. I decided to talk it over with one of my senior officers and late that evening went to the commander of the ARPS, to talk it over.

He was in no doubt. "If you're offered it, take it," he said. "Take it."

"Mmm, OK," I thought. Then I decided to go have a word with the flight surgeon, a bright young doctor, a man I also really respected. I shared with him a very private thought.

"I've got a problem," I admitted. "I don't know what to do. I really don't know what to do."

"Well," he said, "how do you feel about it?"

"I think I want to stay at Edwards."

"Well, then, you ought to stay at Edwards."

Oh, man. Two differing opinions. Totally opposite advice from two men I greatly respected. I agonized over it, really agonized. I sat on my bed in that hotel room into the early hours of the morning wondering what I would say if the phone call came. If I left Edwards I would be leaving behind the world of Chuck Yeager, the world I had dreamed of since I was a small boy. I was torn between two great opportunities. The bottom line was that the future looked good. It was like a chess game. Which piece should I move in which direction?

As it turned out no phone call ever came from Deke Slayton. The next morning when I went down to the front desk of the hotel, one of the other guys, Ted Freeman, was standing with a big smile on his face.

"We've got our orders," he said.

"What d'you mean?"

"Ask the guy behind the desk," he replied and, sure enough, there was a letter waiting for me. I quickly tore it open. The instruction was brief, but clear: "Report to Houston in three days. You've been assigned to NASA."

So, I never did get to make the decision to become an astronaut. It was made for me.

I can't deny I was pleased, very pleased, to be one of the fourteen men selected in the third batch of astronauts. I never could understand why my good friend Mike Adams had not been selected, too. But he stayed on at Edwards and he went on to a great career, including flying the X-15.

Crucially, I never did know what I would have said that night if I'd been asked outright if I wanted to join NASA. I'd have probably said yes. As it was, I just got ready to pack my bags for Houston.

Alexei Leonov

Intriguing news. *Life* magazine revealed it had paid half a million dollars over three years to America's first group of seven astronauts, selected in 1959, for the privilege of printing their personal stories. Our salaries at that time were around 50 or 60 rubles a month, and that was before the deduction of Communist Party dues. There was no official exchange rate, as it was strictly illegal to possess dollars, but on the black market a dollar fetched around 60 kopecks; so our salaries would have been the equivalent of about $100 a month.

I followed the progress of the American astronaut corps closely. There were many detailed articles in American scientific journals about the country's space program, and we were given summaries of them in regular intelligence briefings. This gave us some advance warning of who was scheduled to fly next and what the purpose of their mission would be.

On the lighter side, I was fascinated to read about the lives of the astronauts in the many articles that appeared in *Life*: first the original seven Mercury astronauts, then the nine men selected in 1962 and, the following year, the group of fourteen. I knew all their names and how many children they had. There were endless photographs of them with their families relaxing in the comfort of their spacious condominiums and smart houses.

We felt little envy. The cosmonaut corps and our families were well looked after. Most of us had young children by this time. My daughter, Viktoriya, had been born a few weeks after Yuri Gagarin's first spaceflight. Yuri had a small daughter, too, and our families spent a lot of time together. The cosmonaut corps was a very tight-knit group. Decisions were taken collectively.

A long time was spent discussing what form the accommodation to be built for us at Star City should take. The option of living in individual houses was ruled out. We opted instead for Soviet-style collective living in two interconnected high-rise blocks of flats with a communal hall for parties and other celebrations. It was as if we thought we would be young forever and wanted to live side by side, almost like college students.

Decisions on matters of corps discipline were also taken among ourselves. We held regular meetings and very early on took the unanimous decision that, if a member of the corps violated the strict discipline needed to achieve the objectives of the program, that member should resign. This resolution was soon put to the test.

Late one night four cosmonauts, Grigory Nelyubov, Mars Rafikov, Valentin Filatyev and Ivan Anikayev were sitting together, drinking heavily, in a small restaurant at the railway station at Chkalovskoye. This was strictly prohibited. The more they drank the more argumentative and abusive they became. Eventually they picked a fight with one of the restaurant's waiters and the stationmaster called the local military base.

"OK, men, it's time to pack up and go home," said the officer sent from the base to sort the situation out.

Nelyubov retorted, "Shut up. It's none of your business." He had chosen the wrong person to treat with such disrespect: the officer turned out to be a senior commander.

"If you don't apologize to me by nine o'clock tomorrow morning, you are in big trouble," the officer warned him.

The next morning the four were summoned to see the chief commander of the space center. When we found out what had happened we immediately called a meeting for that evening. It was a short meeting. We took a vote; all those who believed the four should resign were asked to put up their hands. The decision was unanimous: even the four themselves raised their hands. It was a sad day. They were all highly qualified pilots, and they had made it through the extraordinarily tough selection process. They all went back to the Air Force.

The fate of two of them was rather strange and sad. Anikayev was at a party one night when someone stole his car keys from his pocket. They took his car and, in their haste to get away, ran someone over, killing them. The thief then put the keys back in Anikayev's pocket. Anikayev was arrested and sent to prison for a year until it was discovered he was innocent. But after he was released he never flew again.

Nelyubov's was an even sorrier tale. He was stationed as a test pilot in the village of Cherniskova in the Soviet Far East after he left the cosmonaut corps. He carried on disobeying disciplinary rules. One night he was walking along a railway track in a blizzard with the collar of his jacket pulled tight around his ears to protect him from the snow. He did not hear a cargo train approaching from behind. A plank protruding from one of the carriages hit him on the head, and he died instantly. Twice I went to visit his grave in the remote village. His death was a tragedy.

But when they resigned, we had little time for regrets. We were all under immense pressure. Our schedule was hectic, and there was no room for those who could not keep up.

During a short break on one of these days of intensive training at Star City, in November 1963, I switched on the radio. Top of the news bulletin: President John F. Kennedy had been assassinated. Later that day I watched on television the terrible scenes of him being shot in the head. Many Soviet people mourned his passing, many laid wreaths at the entrance to the American embassy in Moscow. I did so myself.

Kennedy was respected as a great leader. I remember vaguely thinking then that the American space program might be stepped up in the wake of his death; that the Americans would be determined to pay tribute to their dead leader by realizing his ambition of landing a man on the Moon by the end of the decade.

My overwhelming feeling at that time, however, was that America must be a lawless state. If such a killing could take place in broad daylight with all the protection the superpower afforded its president, the country must be overrun with gangsters. How could I have known then that several years later I would come within inches of losing my own life in an assassination attempt on our own head of state?

David Scott

I was in Seattle visiting the Boeing plant that manufactured the X-20 Dynasoar space glider on 22 November 1963, during my final few weeks at Edwards before joining NASA, when, late in the day, we were informed that President Kennedy had been assassinated. My immediate reaction was immense shock. I thought the president was protected. Nothing like this had happened in living memory in our country, though it was to be just the first of a series of assassinations in those years.

After the initial shock there was a feeling of lost hope and promise. In the three short years of his presidency Kennedy had been a dynamic, popular president. Even though I had not voted for him, I had come to admire him and respect what he had done for the country. Those years had been exhilarating. America was moving forward very fast, both socially and technologically. There was a real feeling of progress: the country was going places. We were on a steep upward curve in terms of prosperity and technology—satellites, launch vehicles, spacecraft. More than anyone else Kennedy had created an aura of romance around space. He had been the greatest single supporter of NASA. More than anyone else, he had captured the public imagination. His enthusiasm and absolute determination convinced many Americans that we could win the race to the Moon.

But those shots fired in Dallas brought the country to a juddering halt. I didn't give too much thought at the time to the impact this would have on the space program, which by then had taken on a life of its own. Congress had at last thrown its weight behind NASA and approved the funding needed for its program. However, the funding was annual and we had to ensure we maintained the support of Congress and the president. Vice President Lyndon B. Johnson assumed power immediately and fortunately he was a strong supporter of NASA—he had put together a lot of Kennedy's policies on space. The program would go on. But that day in November the clock stopped, temporarily, for me as it did for everyone in America. We terminated all our briefings, packed up and went back to Edwards.

Like everyone, I watched Kennedy's funeral on TV. It was an overwhelming event. When Johnson changed the name of NASA's launch facility to the Kennedy Space Center and Cape Canaveral to Cape Kennedy it was very moving. The speculation and conspiracy theories that started circulating shortly afterward, however, created still greater mistrust of the Soviet Union. (Even after the Warren Commission eventually concluded that Lee Harvey Oswald had acted alone, the fact that he had lived briefly in the USSR led many to believe that dark forces in Moscow were somehow involved with Kennedy's assassination.)

It was only a matter of weeks later that I moved with my family to Texas to start training with NASA. The nation was still subdued by the tragedy of Kennedy's death, but when we arrived in Houston in January 1964 the excitement in the air was nevertheless palpable.

Everything was new. Houston's vast Manned Spacecraft Center complex was still under construction in a bayou area about twenty-five miles south of Houston. It was very near a Gulf inlet lake called Clear Lake, which was actually more muddy than clear. The whole area had been hit hard by Hurricane Carla several years before and was pretty depressed and run down. There were a number of big oak trees left, but apart from them it was pretty much scrubland. Yet the MSC, as it became known, had the feel of a new college campus—well groomed, with modern buildings and good roads. New residential communities were springing up in addition to the housing developments at Timber Cove and El Lago where many of the Original Seven and second group of nine astronauts lived. One of these, built on spec, called Clear Lake City, was where most of us rented houses when we arrived.

From the moment we arrived in Houston, those of us who had come straight from the military had to pack away our uniforms and work in civilian clothes. This was quite a departure for those who, like me, had been totally focused up to that point on a military career. Yet there was a great feeling of excitement at having made it on to the space program and we were keen to get to know each other. The first time we actually got together as a group was when NASA presented us at a press conference in Houston, as their third batch of new astronaut recruits. We all talked about our backgrounds a little. Some of the other guys, like Dick Gordon, had come from the Navy. C. C. Williams was a marine. Others, like Rusty Schweickart and Walt Cunningham, had military flight backgrounds, but had left the armed services and were following civilian careers when they were recruited.

Shortly after that we all got together one evening for a big party at Dick Gordon's home in Clear Lake City. It was a great evening, a big buffet was laid out and we all got the opportunity to relax together for the first time. Our wives were there, too, and few of them had met each other before. While they talked about how our family lives would take shape in this new community, we spent a lot of time listening to what Dick had to say about what we could expect of our new careers as astronauts.

Dick, who was all smiles most of the time, was the informal "leader" of our group in those early days. He was a close friend of Pete Conrad from the second group, who had arrived a year earlier, and so was able to share what he had learned about the program from Pete. We discussed what we would be doing in the coming months of early training and how the Gemini program was expected to get off the ground—the first manned Gemini mission was still over a year away and the Apollo program existed only on paper. We were all surprised that so many of us had been selected; after all, the Gemini missions had two-man crews and Apollo seemed far in the distance. The general atmosphere that evening was very friendly. We knew we would be competing with each other to fly missions, but it seemed right from the start that we would be good friends, too.

There was some friendly rivalry between the Air Force and Navy pilots on the program. We kidded the Navy pilots—used to coming in slow and landing their jets hard on the deck of an aircraft carrier—that they couldn't bring planes down without a thump. They kidded us Air Force guys—used to landing jets smooth and fast—that we needed too much runway to get a bird on the ground. But we became a pretty close-knit bunch during that first year in Houston. We did a lot of socializing.

From the moment our names became known, everyone in the community in Houston fell over themselves to accommodate us. They gave us almost more support than we could handle. Although the fanfare that surrounded the selection of the original Mercury seven had died down, we were still treated as celebrities.

The banks all wanted to have our accounts. The car dealers wanted to sell us cars. For the Original Seven, having a hot car had been quite a big deal, and a lot of them charged around in sleek Chevrolet Corvettes. But I had shipped my Mercedes 190 SL over from Europe. Lurton and I had two children by then—our son, Douglas, had been born in October 1963—so we also bought a more sedate Chevrolet station wagon.

Our first major task, however, was to find a permanent place to live. I spent some time discussing this with Ted Freeman, an instructor at Edwards, who was also in our third group of astronauts, and with Ed White, from the second group of astronauts, whom we both knew from Edwards. Ed lived in a lovely house in El Lago. He told us about a very new area just south of the MSC, on a small bayou running out of Clear Lake, called Nassau Bay. They were clearing the woods at Nassau Bay, and were getting ready to build homes. There were no roads, but Ted and I took a drive down there and we both thought it looked pretty nice. We each chose a lot and employed an architect to design a house. Lurton and I decided on a single-story L-shaped, ranch-style home, with three bedrooms, a big, high-ceilinged family room and nice back yard. It was a great improvement on the rudimentary accommodation at Edwards.

There wasn't much else in Nassau Bay—maybe one grocery store—and it was too far to go into Houston regularly. There was also, in the beginning, limited schooling for the kids, of which there were many in the group—Dick had six. But slowly the area started to develop and eventually even a little yacht club opened, with a swimming pool. Many of the people who moved into the area worked for NASA, though it was not a NASA neighborhood. Other families who had nothing to do with the space program lived there, too, which gave it a nice mixed feeling. Some of our neighbors worked for IBM and at the university. A lot of families had kids, and it was a great, very sociable environment. Few of us had owned a home before and many had not lived in civilian neighborhoods, so it was quite a change from what we had been used to.

Almost as soon as we arrived in Houston *Life* magazine signed us up for exclusive rights. Since the late 1950s the magazine had had an exclusive contract for the astronauts' personal stories. The Original Seven had been given extremely lucrative contracts but by the time we came along the offer was significantly less, around $10,000 a year, as I recall. However, this was still a lot of money. I was on an Air Force captain's salary when I joined NASA and throughout my time with the space program continued to be paid by the Air Force. We were all paid according to our rank, or, in the case of the civilians, according to a civil service grading they were accorded when they were recruited. This meant our salaries varied quite a bit.

Apart from supplementing our salaries, the *Life* contract also had the major benefit of limiting media access to our personal lives: our wives and children were bothered less by inquisitive reporters from other publications. Over the years they had to get used to the fact that one day everyone was interested in their dad and the next day he seemed to be old news. In the immediate run-up to a mission, for instance, it was quite common for a reporter and photographer to follow the children of the assigned crew to the bus stop and elsewhere. As soon as the mission was over the reporter would ignore them and concentrate on the children of the next crew due to fly.

We soon discovered that, as the "new kids on the block"—those of us in the third batch—we were in line behind the first two groups when it came to just about every privilege; including flying time in the limited number of T-33 military aircraft at NASA's disposal. But as some of the Original Seven began to leave NASA after completing their Mercury programs, we moved up the pecking order. John Glenn was one of the first to leave. After running for public office in 1970, he won a Senate seat in his native Ohio in 1974.

One of the first courses we undertook was survival training; learning how to survive if we had to bail out from a spacecraft in an unintended environment. The trajectory of all Gemini and Apollo missions—launching toward the east from Cape Kennedy at a latitude of 28° N with planned landings in the Pacific Ocean close to Hawaii—meant we did not have to concern ourselves with survival in cold climates. Our training concentrated on jungle, desert and ocean survival tactics.

Jungle survival training took us to the Air Force tropical survival school in Panama, where we learned how to build a shelter and how to chop down palm trees to retrieve the hearts of palm to eat; we soon got tired of hearts of palm, I can tell you. We paired off for several days and we learned to catch and cook iguana, too. It took me three days to catch an iguana, so that tasted pretty good. The Air Force survival manual noted that "anything that creeps, crawls, swims or flies is a possible source of food." I stuck to iguana and hearts of palm, as I recall.

The water survival course was closer to home. It mainly involved bobbing about in a life raft in the Gulf of Mexico and learning how to operate a solar still, by which sea water could be condensed into pure drinking water using solar energy. Desert survival training took us to Nevada, where we were dumped out in the desert somewhere near Reno Air Force Base, in daytime temperatures approaching 130° F. We spent most of our time during the day digging a hole, keeping in the shade, reading novels and conserving water.

Those nights in the desert when it was cool were, by contrast, really quite beautiful. It was so still and quiet and the view of the stars was magnificent. I did my desert survival training with Rusty Schweickart, who had also studied at MIT, and was something of an expert on astronomy. He was quite a wit, too. Some time later, when we were training together, we developed a mnemonic to help us remember the zodiacal constellations from Aries through Pisces: All Tall Girls Can Lead Virtuous Lives; Short Sisters Casually Accept Propositions.

Out there in the desert at night we spent hours picking out the various constellations and talking about life in general, the program, our families and kids. Rusty is a highly cultured man, very liberal, widely read. I really enjoyed his company. Our kids, who are about the same age, got along particularly well, too. It made us smile when, a few years later, my daughter Tracy took a shine for a while to one of Rusty's twin boys. They must have been about nine at the time.

After completing the basic survival courses we spent some time touring the country to visit the design and construction facilities of the many different government contractors responsible for various parts of the space program. The prime contractor for Mercury and Gemini spacecraft was McDonnell-Douglas—or Macdac, as we called it—in St. Louis, Missouri. We went to the Marshall Space Flight Center in Huntsville, Alabama, to observe the development and tests of the launch vehicles and to the Cape to look at the launch facilities.

Back in Houston we had engineers come in and teach us how the different spacecraft systems operated. In addition, each of us was assigned to represent the astronaut corps in one particular area of the program; my academic background at MIT led me to be assigned to guidance and navigation. We also spent a fair number of hours in the classroom, studying everything from astronomy to rocket propulsion, digital computers and even geology—a box of rocks was waiting on our desks at each class to introduce us to this branch of science. Later, inspired by an outstanding teacher, I developed a love of the subject, but in those early days I sat at the back of the classroom with Charlie Bassett and we discussed when we would next get some flying time.

We kept on flying high-performance jets throughout our time on the space program, because we were required to keep up our flight proficiency and piloting skills. It was still one of my greatest joys and I tried to get in as much flying as I could.

In addition to four or five T-33 trainer jets loaned by the Air Force, NASA had at its disposal a couple of F-102 delta-wing fighter jets, which we all wanted to fly. They were beautiful machines with what was known as a "coke-bottle" design, because the fuselage of the plane tapered in the middle to give them less drag so they could go supersonic.

The F-102 single-seater interceptor was a front-line fighter jet, and only those of us who had previous experience of this type of aircraft got the chance to fly them. While stationed in Europe and at Edwards I had flown every one of the aircraft in the Century series: the F-100, F-101, F-102, F-104 and F-106. So I was put on the F-102 flying list at NASA. So was Charlie Bassett. Charlie became my best friend, and we had a great time flying together.

Like most astronauts from military backgrounds, I had joined NASA on a three-year tour of duty. This was military policy. I assumed I would go fly in space, then go back to regular duty in the Air Force, but the policy changed shortly after we arrived. As the complexity of the space program became clearer, we were all assigned to NASA indefinitely. Indeed, NASA finally had to recruit another two groups of astronauts for Apollo, as it took a lot more people to run the program than originally thought.

By the time we were told our assignment to NASA was indefinite I felt few pangs of nostalgia for Edwards. From the moment I arrived in Houston I was totally caught up with space.

CHAPTER 4

A Fair Solar Wind

1965

Major Alexei Leonov

Baikonur Cosmodrome, Tyura-Tam, Kazakhstan

Conditions at the Baikonur Cosmodrome, in the barren steppes of Kazakhstan in Central Asia, were enough to test the endurance of any human being in the early days of spaceflight. The desert, where the 2,000-square-mile complex was located, swarmed with scorpions, snakes and poisonous spiders. I once witnessed a technician, a young captain, being bitten on the neck by a spider. He collapsed and died within minutes. There was nothing we could do.

Few allowances were made for the brutal extremes of weather when accommodation was built for cosmonauts in the lead-up to their missions and for the many engineers and designers permanently stationed there. The brick apartment blocks and small houses were constructed according to Moscow specifications. Years later the chief developer of the complex was awarded with a high honor by the state. I would have had him severely punished.

Hurricane-strength winds reduced temperatures to below –40° C in winter. Once the snow melted, the incessant high winds hurled sand against the buildings with such force that, although we jammed towels against doors and windows, fine dust and sand got everywhere: in our clothes, eyes and food. At the height of summer the heat was so intense—soaring to 40–50° C—that we had to wrap ourselves in wet sheets at night to reduce our body temperature. The moisture attracted insects into our rooms.

Slowly, facilities at Baikonur improved. But for the first generation of cosmonauts, the time spent there was grueling. The decision to locate the cosmodrome there—1,300 miles southeast of Moscow, 100 miles east of the Aral Sea and 500 miles west of Tashkent—was logical. Proximity to the equator gave rockets maximum impetus at launch, since the Earth rotates west–east fastest at the equator and a vehicle launching at this point benefits from the added velocity.

It is not in the town of Baikonur, though this was the name by which it has always been known. The cosmodrome was in fact 200 miles southwest of the town, nearer a small railway town called Tyura-Tam (the geographical discrepancy was intended to confuse Western intelligence sources). But American spy planes, such as that flown by Gary Powers, were known to circle the area, attempting to pinpoint the exact location of its rocket launchers.

The site's remoteness and the sparseness of the area's population also meant that Baikonur was considered the safest location should accidents occur. And they did. The explosion of an R-16 rocket in the autumn of 1960, which cost 165 people their lives, would have been an even greater tragedy had it happened in a more heavily populated area.

Four years after that disaster our space program had not only fully recovered but gone from strength to strength. In the wake of Gagarin's flight and the five record-breaking Vostok missions that followed, we had successfully launched the first of the newly designed Voskhod spacecraft in October 1964. This spacecraft completed a flight of sixteen orbits with three cosmonauts aboard, prompting both envy and admiration in the West that we had developed a multi-seat spacecraft.

By that stage, the United States had only managed to launch one astronaut at a time into orbit, aboard their first generation of Mercury spacecraft. The Voskhod capsule was originally designed to accommodate two men. By sending three men into orbit in the adapted version, we had moved one step further ahead in the space race.

My mission in the spring of 1965, in another Voskhod capsule, was to chalk up another first for the history books: man's first walk in open space.

I trained intensively for two years, together with my commander on the mission, Pavel Belyayev. Pasha was my closest friend in the cosmonaut corps besides Yuri Gagarin. There were those within the program who had wanted Yevgeny Khrunov to command the mission, and the doctors opposed Belyayev because he had broken his leg in a parachute accident years before. But I lobbied hard for Pasha, whom I thought more capable than Khrunov. I had worked with him more; I trusted him. In the end they agreed, though it caused some rancor in Khrunov.

Apart from intensive preparation for piloting and navigating the spacecraft, our program of training for the Voskhod 2 mission concentrated heavily on simulating as closely as possible every moment immediately before, during and after my space walk. The space walk itself was scheduled to last between ten and fifteen minutes. But the whole process of getting into the airlock, before exiting into open space, and then reentering the spacecraft would take over an hour. Every second of the exercise had to be practiced in weightlessness, which was a time-consuming and exhausting task.

There was no way of simulating pure weightlessness in any laboratory on the ground, so we had to train in a modified Tupolev aircraft, a TU-104, flown at a speed of about 1,000 kph in a series of parabolic arcs. Under such conditions the interior of the aircraft could simulate the weightlessness of space for approximately 30 seconds at a time when the plane swooped sharply downward—it was rather like an extreme exaggeration of the effect of going over a hump-back bridge. But such limited, disconnected periods of weightlessness meant that the hour and fifteen minutes of exiting the spacecraft, performing the space walk and reentering the airlock had to be practiced over the course of over two hundred TU-104 steep climbs, during which zero gravity was achieved.

We also had to practice for every emergency conceivable,

including the possibility that I would lose consciousness while in open space. It was a very real fear since a space walk had never been tried before and no one knew what to expect. If it happened Pasha would have to exit the spacecraft to rescue me. The physical strain of training for an emergency like this was so great that the commander of the back-up crew suffered a minor heart attack during training and had to be replaced.

There had recently been a major shift in the political climate of our country. On 13 October 1964, the day the Voskhod 1 crew returned from orbit, they received a telephone call from Nikita Khrushchev, who was at his retreat in the Crimea, to congratulate them on their successful mission. Within a few days Khrushchev was summoned back to Moscow and dismissed. Deposed from power, he was shunted into retirement and replaced by Leonid Brezhnev as First Secretary of the Communist Party. The change in political leadership had little effect on the space program. Though Khrushchev had shown little enthusiasm for space initially, he had come to appreciate the enormous political capital to be gained from our early superiority in space exploration. He had given the program his full backing. Brezhnev was to do the same.

Despite intense pressure to push ahead with the space program, it was policy to conduct tests of unmanned spacecraft prior to manned missions which were considered particularly risky. Voskhod 2 was such a mission. So three weeks before we were due to launch in mid-March, an unmanned prototype was sent into orbit from Baikonur. It exploded.

The explosion happened when confused signals sent from the ground triggered the spacecraft's self-destruct device. The device was incorporated into unmanned craft in case they went out of control and threatened to land in populated areas. By that time Pasha and I, our back-up crew, and the most senior designers and engineers on the program—including, of course, the Chief Designer, Sergei Korolev—were already at Baikonur.

Korolev came to see Pasha and me in the hotel just hours after the explosion. He looked exhausted and strained. He had not been well: he had been suffering from a high fever as a result of a lung inflammation. But nothing would deter him from consulting with us and ensuring our safety. Late that evening he sat down with us and presented us with a stark choice.

"All the data from the unmanned mission has been lost," he told us. "We have only one Voskhod vehicle remaining which is ready for immediate launch. That is your vehicle. If we use this vehicle for another unmanned launch to test the equipment for your space walk, Lyosha, your mission will be delayed by a year until a replacement spacecraft can be built. What is your opinion?

"It is up to you. I cannot tell you what you should do," he continued. "There is no textbook answer. Nothing can prepare you for exactly for what you will experience on your mission. There are risks—there is no doubt that there are risks. It is your decision."

Then, very cannily, he added that he believed the Americans were preparing their astronaut Ed White to make a space walk in May. He knew how to get our competitive juices flowing. He must have known what we would say. We didn't want to lose a year. We were at the height of our preparations. That night in late February 1965 we were full of self-confidence. We felt we were invincible. Despite the risks, we said, we were ready to fly.

Although it had long been clear that I was to perform the space walk and that Pasha would pilot Voskhod 2, we were not confirmed as the prime crew for the mission until a week before the due launch date. A formal selection committee made up of senior political figures and military officers met in Moscow to consider the results of tests we had been subjected to over the previous few days. When they confirmed that we would be undertaking this historic mission I felt very proud. But I also felt a huge responsibility. I knew I could not have trained any harder or been better prepared for this mission, but I kept asking myself if I would be strong and calm enough to complete it as I was supposed to.

I knew my wife, Svetlana, would worry about me if she knew the full details of my mission, so I did not tell her that I would be performing a space walk. We cosmonauts did not discuss the details of our work with our families, anyway. All she knew was that ours was to be a particularly complex and challenging mission. We spoke on the phone the day before our launch. I knew Svetlana and our small daughter, Vika, would have the support of other cosmonauts' wives and families as she followed the mission at home.

Wives were not allowed to travel with us to the cosmodrome. Nor were they allowed to enter the control center, initially at Yevpatoriya, in the Crimea and later in Moscow. This was both for security reasons and because we cosmonauts considered it bad luck to see any woman who was not directly involved in the program immediately before a mission. This superstition is common in our military; submariners believe, for instance, that it is bad luck to have women aboard a submarine. Our children had so little idea of the nature of our work that Vika, when small and asked by friends in her kindergarten what sort of spacecraft her father flew, replied "Well, a bus, of course." All she knew was that a military bus picked us up outside our apartment each morning and she assumed that was the vehicle that took us into space.

By the evening of 17 March 1965, we were ready. The few personal belongings we were allowed to take with us on the flight had already been loaded into the spacecraft: in my case a small drawing pad and set of colored crayons. I intended to sketch what I saw in space during any spare moment I had.

Our food had also been loaded, in sealed plastic pouches and metal tubes; small cubes of cheese, borscht and blackcurrant juice. In addition I had requested a small portion of kharcho, a spicy Georgian soup of rice, meat, onion and garlic. But at the last moment I ordered most of this food to be replaced with extra ammunition for my pistol. What use would we have for food in a mission expected to last only twenty-four hours? Much better to carry more cartridges for self-defense, in case our spacecraft landed in an area with wild animals.

Then Pasha and I observed a custom which had come into being after Yuri Gagarin became the first man in space four years before. This tradition dictated that cosmonauts setting off on a mission should transfer from the rooms where they had been staying at Baikonur to the small house where Yuri and his back-up, Gherman Titov, had stayed the night before that historic flight. That night I lay down to sleep in the bed in which Yuri had slept. Pasha took the bed used by Titov.

Throughout the night we were both monitored by doctors to see how well we slept. It was intrusive. Who can sleep well when they know they are being watched? The next morning I felt as if I had hardly slept at all. But the doctors told me I had slept deeply. After a short medical Pasha and I were pronounced fit to fly. We then sat down to a breakfast of boiled eggs, bread and butter, mashed potatoes and tea.

After breakfast a small group of us—including Yuri, who would be monitoring our mission throughout our flight, and Korolev—opened a bottle of champagne, as was also customary. Each of us took only a sip. Then we signed the label of the half-empty bottle and Yuri put it to one side with the promise, "We will finish this when you both return to share this bottle with us." We observed another tradition, as most Russians do before embarking on a long journey, which was to sit briefly to compose ourselves: "Friends, let us all take a seat," Yuri said. We all sat. Then he sprang up. "OK," he said. "Let's go."

Our small group made its way to the bus that would take us to our spacecraft. On the way we performed an act that both was a necessity and also had become one final tradition before a flight. We all stood in a small circle and pissed on the wheel of the bus.

Once sealed inside our Voskhod capsule, ready for launch, there was very little sound. Less than we were used to as military pilots when the cockpit of a fighter jet would reverberate with the roar of engines revving up before take-off. Sealed inside our space capsule, all we could hear was the gentle purring of electrical devices and the sound of engineers' voices coming through our radio headsets. More distinctive was the smell inside the capsule. It was the smell of fresh paint and glue No. 88, a special adhesive containing surgical spirit, which I found strangely pleasant.

After so many hours of training in the spherical capsule it did not seem cramped. But it was small: just over 2 meters in diameter. The seats into which Pasha and I were strapped, lying on our backs with our legs bent, resembled small metal cradles. They were suspended at either end on shock absorbers to soften the vibration we would feel on take-off and landing. As the engine of the rocket beneath us ignited, we felt a light vibration start to build. Lifting away from the launchpad, we were pushed back into our seats. Now we felt the full force of the rocket propelling us upward through the Earth's atmosphere. It felt as if we were being lifted vertically by a speeding train. From this moment on we were required to report constantly on how we felt.

"Diamond One," Pasha reported, using his call sign. "I'm feeling calm." "Diamond Two," I followed up. "I'm feeling perfect."

For the first eighteen seconds after lift-off, if anything had gone wrong with the rocket we would not have survived. The design of our spacecraft did not allow for escape—by ejection seat or parachute—in these first vital seconds. It was extremely hazardous. Korolev had explained this aspect of the capsule's design as being no more risky than that of a passenger plane, which also, at that time, stood little chance of landing safely if anything went wrong during the first twenty seconds after take-off. We had no time to dwell on such risks, anyway. We were too busy monitoring all aspects of the spacecraft.

Our vision out of the small circular windows in front of our seats was obscured at first by a protective aerodynamic shield within which the capsule was cocooned. But 80 km above the surface of the Earth the shield was discarded. I looked out of my small porthole and caught my first glimpse of the Earth. I was disappointed.

As a military pilot I had often seen the Earth from a height of around 15 km, and the view from my porthole was not that different. I had expected to see the curvature of the horizon against a dark sky, but we were not yet high enough for that. Ten minutes into the flight, at an altitude of almost 500 km, our capsule separated from the rocket with a loud flap. We were flying far beyond the confines of the Earth's atmosphere. When we separated from the booster and the roar of its engine stopped, we reached a state of weightlessness. Our first orbit of the Earth had begun.

Small loose items began to float around inside the capsule. Without the sound of the rocket beneath us it was suddenly very quiet. So quiet that we could hear the clock ticking on the control panel and the gentle clicking on of our orbital operational devices.

As we moved out of sunlight we flicked on the lamp to illuminate the capsule. At that point we had little sensation of flying in space. It felt almost as if we were back in one of our training simulators.

For two or three minutes it was extremely uncomfortable. I had the feeling that I was suspended upside down, which is a well-documented phenomenon: once the force of gravity ceases, the senses become confused. But we quickly got used to it and started going through a complex series of checks to verify that all systems inside the capsule were operating correctly.

"Diamond One confirms that all systems are OK," Pasha reported to Mission Control. "Diamond One and Diamond Two are feeling perfect."

After we entered orbit Pasha requested permission to extend the airlock chamber in preparation for my space walk. Permission granted, he activated the devices, which started to pump air into the small rubber tubes that ran the length of the hollow canvas chamber. From a coiled length of 70 cm the airlock quickly extended to its full length of 2 meters. In the meantime I had strapped the bulky breathing apparatus, containing metal tanks with ninety minutes' worth of oxygen, on to my back. I was ready to clamber inside the airlock and undergo decompression before exiting into open space.

With a sharp slap on my back Pasha gave me the command to enter the airlock. "Poshel," he said. "Go."

Once inside the airlock I closed the hatch and waited for the nitrogen to be purged from my blood. To avoid suffering from what divers call the bends, I had to maintain the same partial pressure of oxygen in my blood once I emerged into space. With the pressure inside the airlock finally equal to the zero pressure outside the spacecraft, I reported I was ready to exit.

Ground control had to carry out careful checks of all my systems before granting permission to open the outer hatch. When it opened I was lying on my back. There wasn't enough space to turn or move much. But I craned my neck backward to catch my first, unobstructed, glimpse of the Earth. This time I was not disappointed.

What I saw as the hatch opened took my breath away. Night was turning to day. The small portion of the Earth's surface I could see as I leaned back was deep blue. The sky beyond the curving horizon was dark, illuminated with bright stars as I looked due south toward the South Pole. I craned my neck back until it hurt. I wanted to see more. Moving at 18,000 mph, the scene below me rapidly began to change. Very soon the outline of the African continent came into view.

All the time I was waiting for the moment when I would be given permission to ease myself free of the spacecraft. It seemed a long wait. Then my earpiece crackled into life.

"Diamond Two," Mission Control radioed, "we can see you very well. It is time for you to start your mission."

My heart was racing. I realized the moment I had waited for so long had arrived. It took me just a few seconds to push my upper body out of the airlock. I brought my feet to its rim, holding on to a special rail and took a final look around before I let myself go.

By now we were passing over the Mediterranean. When I raised my head I could see a vast panorama. It was as if I was looking at a gigantic, colorful map. I could see the whole of the Black Sea. To my left were Greece and Italy, ahead was the Crimea, and to my right lay the snow-capped Caucasus Mountains and the Volga river. When I raised my head I could see the Baltic Sea.

Lenin once said the universe is endless in time and space. It is the best description of what I saw in those moments. Then I pulled my thoughts back and responded to Mission Control.

"I'm feeling perfect," I reported as I pulled the breathing tubes connected to my life support system out of the airlock. With a small kick, as if pushing away from the side of a swimming pool, I stepped away from the rim of the airlock.

I was walking in space. The first man ever to do so.

Nothing will ever compare to the exhilaration I felt in that moment. No matter how much time has passed, I can still remember quite clearly my conflicting emotions.

I felt almost insignificant, like a tiny ant, compared to the immensity of the universe. At the same time I felt enormously powerful. High above the surface of the Earth, I felt the power of the human intellect that had placed me there. I felt like a representative of the human race. I was overwhelmed by these feelings.

When my four-year-old daughter, Vika, saw me take my first steps in space, I later learned, she hid her face in her hands and cried.

"What is he doing? What is he doing?" she wailed. "Tell Daddy to get back inside. Please tell him to get back inside."

My elderly father, too, was distressed. Not understanding that the purpose of my mission was to show that man could survive in open space, he remonstrated with journalists who had gathered at my parents' home.

"Why is he acting like a juvenile delinquent?" he shouted in frustration. "Everyone else can complete their mission, properly, inside the spacecraft. What is he doing clambering about outside? Somebody must tell him to get back inside immediately. He must be punished for this."

His anger soon gave way to pride when he heard a live broadcast of Brezhnev's message of congratulations beamed up to me from the Kremlin via Mission Control.

"We, all members of the Politburo, are here sitting and watching what you are doing. We are proud of you," Brezhnev said. "We wish you success. Take care. We await your safe arrival on Earth."

Televised pictures of me pushing away from the airlock and floating free in space had been transmitted via Mission Control, with only a few minutes' delay, into the homes of millions of my compatriots.

Floating free of the spacecraft, I had been facing directly toward the Sun. Even with the gold filter across my visor cutting out nearly all ultraviolet light it was like being somewhere in the south, Georgia maybe, without sunglasses on a summer's day. For a moment I slightly eased open the filter to see what it was like looking down at the Earth through the clear glass of my helmet.

A vast area of the Earth's surface, over 2000 square miles, lay spread below me. I could recognize everything, as if I were in a geography class. I took in scenes, almost subconsciously, to paint when I returned home. But it was too bright. I had to pull down the filter again.

The heat was incredible, and I had great difficulty trying to activate the Swiss camera incorporated in the breast of my suit. The switch to turn it on was sewn into my upper trouser leg, and it was just 1 cm out of reach. I kept brushing the top of my leg with my glove. It looked very strange when I saw televised footage of my space walk later. But at least the cine camera mounted on top of the airlock seemed to be working, as were the two television cameras attached to the exterior of the space capsule.

It was important that this mission be captured on film. There must be no dispute that, besides Gagarin's historic flight, we had also achieved the first space walk too. I knew the Americans were planning to send one of their astronauts on a space walk in a couple of months' time. Ed White was his name. I knew about the entire astronaut corps. I had spent time studying them. They were an interesting bunch, different from our cosmonaut corps in some ways, but similar, too.

(I did not, of course, know then that the Americans would dispute what we had achieved, anyway, despite the film footage. But it did not surprise me. The race between our two countries for superiority in space was intense. Personally I did not believe in all this boasting about who did what first, the Soviet Union or the United States. If you did it, you did it. The very fact of what you had achieved should have been enough. But I knew politicians did not feel that way. For those in the Kremlin and in Washington, space was a fierce battleground for superiority not just in technology but ideology. I had no time or desire to contemplate such conflict. As far as I was concerned I was there to prove to my fellow man what human beings were capable of.)

The tranquillity of space was profound—far greater than anything you might imagine on Earth from diving deep beneath the surface of an ocean. And I felt a huge desire to disturb this motionless environment by moving my body, my arms and my legs, as much as I could, given the restrictions of my spacesuit. I felt like a seagull with its wings outstretched, soaring high above the Earth. It was this desire to test my movements that led me to push myself away from the side of the spacecraft, making me tumble uncontrollably until my communication and breathing cords yanked me to a halt. I did not panic or feel scared; I was only too aware how fear could affect the ability to think and to take whatever action was needed. But the concern in Pasha's voice as he lost sight of me in the television monitor was evident.

"Where are you? Can you hear me? What are you doing?" he asked. Then, realizing I was OK, he admonished me a little. "Be careful!"

As I pulled myself back toward the spacecraft I was struck by how fragile and vulnerable it looked in the vastness of the universe. The brightness of the sunlight reflecting off its rounded hull gave the Voskhod such a deep golden sheen that it truly earned its name, "Sunrise." For many years afterward I tried to recapture in paintings the extraordinary shade of gold of our spacecraft that day, but I could never quite replicate the intense tone. My senses were so heightened that everything I saw in that short time left such deep impressions in my memory that they will last my entire lifetime.

As I pulled myself back toward the airlock I heard Pasha talking to me once more. "It's time to come back in." I realized I had been floating free in space for over ten minutes. In that moment my mind flickered back for a second to my childhood, to my mother opening the window at home and calling to me as I played outside with my friends, "Lyosha, it's time to come inside now."

With some reluctance I acknowledged that it was time to reenter the spacecraft. Our orbit would soon take us away from the Sun into darkness. It was then I realized how deformed my stiff spacesuit had become, owing to the lack of atmospheric pressure, and that my feet had pulled away from my boots and my fingers from the gloves attached to my sleeves, making it impossible to re-enter the airlock feet first.

I had to find another way of getting back inside quickly and the only way I could see of doing this was by pulling myself into the airlock gradually, head first. Even to do this, I would carefully have to bleed off some of the high-pressure oxygen in my suit, via a valve in its lining. I knew this might expose me to the risk of oxygen starvation, but I had no choice. If I did not reenter the craft, within the next forty minutes my life support would anyway be spent.

The only solution was to reduce the pressure in my suit by opening the pressure valve and letting out a little oxygen at a time as I tried to inch inside the airlock. At first I thought of reporting what I planned to do to Mission Control, but I decided against it. I did not want to create nervousness on the ground. And anyway, I was the only one who could bring the situation under control.

But I could feel my temperature rising dangerously high, starting with a rush of heat from my feet traveling up my legs and arms, due to the immense physical exertion all this maneuvering involved. It was all taking far longer than it was supposed to. Even when I at last managed to pull myself entirely into the airlock I had to perform another almost impossible maneuver. I had to curl my body round in order to reach the hatch to close the airlock, so that Pasha could activate the mechanism to equalize pressure between it and the spacecraft.

Once Pasha was sure the hatch was closed and the pressure had equalized, he triggered the inner hatch open and I scrambled back into the spacecraft, drenched with sweat, my heart racing.

My serious problems when reentering the spacecraft were, thankfully, not televised. My family was spared the worry and anxiety they would have had to endure had they known how close I came to being stranded alone in space. They were also spared the trauma they would have suffered had they known the grave danger that Pasha and I faced in the hours that followed. For the difficulties I experienced reentering the spacecraft were just the start of a series of dire emergencies which almost cost us our lives.

From the moment our mission looked to be in jeopardy, transmissions from our spacecraft, which had been broadcast on both radio and television, were suddenly suspended without explanation. In their place Mozart's Requiem was played again and again on state radio. The custom in the Soviet Union at that time was for such solemn music to be played after a senior political figure had died, but before an official announcement of their death was made.

* * *

We knew nothing of this news blackout. Nothing could have been further from our minds. As I tried to catch my breath after struggling so hard to reenter the spacecraft, I knew that if my physical training had not been so intensive I would never have been able to perform the complicated maneuvers that had saved my life.

Pasha also realized how close I had come to being stranded outside the spacecraft, how very nearly the highlight of this mission had turned into disaster. But he was calm. As soon as I opened my helmet to wipe the sweat from my eyes I was drenched again. Pasha told me to rest a little before I wrote up my report of the space walk in our logbook. We had another full Earth orbit before we needed to perform our next task, ejecting the airlock. That meant I had an hour and a half in which to write my report and rest.

But my adrenalin was pumping so hard there was no chance of my sleeping. I reached for my sketch pad and colored pencils and sat quietly drawing my first impressions of the panorama I had seen while floating free in space. I tried to capture the different shades of charcoal rings that make up the Earth's atmosphere, the sunrise or air glow over the Earth's horizon, the blue belt covering the Earth's crust and the spectrum of colors I had observed looking down at the globe.

I had time for four sketches before we made our preparations to eject the airlock by detonating small explosive devices between it and the main body of the spacecraft. Almost immediately after these devices were activated, the spacecraft started to roll and roll. We were still in daylight, though about to enter the night zone, and the light flashing in through our windows was dizzying.

We had only enough fuel to perform one orientation correction, and that correction had to be our final adjustment for reentry. But there were another twenty-two hours of our mission to go. I could not believe that we would have to endure this roll of 17 degrees per second—ten times stronger than expected—for that long, but when we reported our problem to Mission Control they remained silent.

As we entered the night zone the rotation bothered us less. It was peaceful, quiet. But as soon as we moved back into daytime the sunlight would be streaming through our window, disorienting and distracting us. There seemed nothing we could do. Exhausted from the rigors of my space walk, I felt I could bear the discomfort of this 17 degrees per second rotation.

Then another, much more serious, problem developed. As I was going through routine instrument checks I noticed that the oxygen pressure inside the cabin was steadily increasing. My instruments should have shown an oxygen pressure of 160 mm, but they showed the pressure creeping up first to 200 mm, then 300, 400 . . . 430 . . . 460. The spacecraft had dozens of power engines and electrical circuits on board and Pasha and I both knew that, if a single spark occurred in any of the electrical circuitry, the atmosphere of the spacecraft at such high oxygen pressure would trigger a massive explosion. It was similar to the situation that had killed Valentin Bondarenko during the ground test four years before.

We immediately reported the emergency to Mission Control. They advised us to lower the temperature and humidity in the spacecraft. This should have reduced the oxygen pressure to some extent, but, although the pressure stopped climbing, it remained perilously high. We were alarmed.

We tried to narrow down the causes of the problem. We knew there must have been a leak in the spacecraft, causing too much oxygen to be regenerated, more oxygen than we could consume. Perhaps the hatch or part of the body of the spacecraft had become deformed, we calculated, owing to the intense heat to which it had been subjected when Voskhod 2 had faced the Sun without rotating, while I performed my space walk. Such a deformation could have caused a slow leak. It was a critical problem. We could think of no solution. We were both deeply troubled.

I thought of the last words that Sergei Pavlovich Korolev had spoken to me before we mounted the gantry to Voskhod 2 that morning: "I won't give you a lot of advice and ask a lot of you, Lyosha. Just don't outsmart yourself. Just exit the ship and come back in. Keep in mind all the Russian sayings that have helped a Russian man during difficult times." Korolev's faith in me gave me great strength. But who could have foreseen such difficult times?

As I consulted our charts I realized we must be over Moscow. I gazed out of my window and looked down as the spacecraft rotated toward the Earth. Covered with a blue mist, Moscow looked like a giant lobster cut open by the veins of the rivers running through the city. I thought of Svetlana and Vika. They could not know I was so close to them, thinking of them in those moments.

Tired, cold and hungry, we slowly drifted into fitful sleep.

The sound of a hissing upper valve woke me several hours later. I immediately checked the oxygen indicator. The pressure, I realized, had been slowly diminishing and it had now fallen below the critical point of 460 mm at which it posed a danger.

I nudged Pasha. "Look, the pressure's dropping." We still had several hours left before we were due to reenter the Earth's atmosphere. Voskhod 2 was still rotating, but the situation seemed to have stabilized enough for us to continue with our set program until our orbit reached the designated time for our reentry procedure to commence. As soon as the automatic landing systems came into operation, the rolling stopped and we were able to enjoy a few delicious moments of tranquil flight. I was even able to shoot a short film of our activities.

But then the spacecraft started behaving strangely again. Just five minutes before our retro-engine was due to start, I checked our instruments and realized our automatic guidance system for reentry was not functioning correctly. The rolling had started again. We would have to switch off the automatic landing program. This meant we would have to orient the spacecraft before reentry manually, and would also have to select our landing point manually and decide on the exact timing and duration of the retro-rockets that would bring us back into the Earth's atmosphere. There was nothing for it but to deactivate the automatic system. Once we did this the craft's rotation slowed again and the feeling of stillness was quite beautiful. After all the difficulties we had encountered, I still felt as if I could continue orbiting in space for another hundred years.

Despite the critical situation we started quietly assessing the state of our on-board systems. Without a correction to its orbit, our spacecraft could continue on its current path, circling 500 km above the surface of the Earth, for another year. But, though our life support systems would keep us alive for another three days, we had sufficient fuel for only one or two orientations. We knew our landing would have to be performed during our next orbit and that, despite our best efforts, we would be coming down off-target— 1,500 km west of where we were supposed to land.

As our orbit brought us above the Crimea we received our first communication from ground control for some time.

"How are you, Blondie? Where did you land?" It was Yuri Gagarin; he always called me "Blondie." It was good to hear his voice. Even in such difficult circumstances he sounded full of warmth, even relaxed. Some of the other cosmonauts were serious all the time, but we shared a sense of humor. We had spent a lot of time together. Yuri was a kind, down-to-earth man, not full of himself at all, as you might imagine he would be after becoming so famous. Yuri and I shared another bond: Korolev had singled us both out as favorites. But from what he was saying it was clear Mission Control thought we had already landed.

Pasha clicked on his microphone. "We had to turn off the automatic landing system. It was nonoperable. We have only enough fuel to do one correction and, besides that, the indicator shows that the main engine for reentry is very low on fuel," Pasha reported in as steady a voice as he could. "We can make only one attempt at reentry. We are asking you therefore to go into emergency mode."

"Yes, we give you permission to employ the emergency procedure," Yuri came back almost immediately, showing no surprise that we were still in orbit.

It was my job, as navigator, to determine where we would land. Our orbit would take us right over Moscow: we could set down in Red Square. But we had to choose somewhere as sparsely populated as possible. I decided on an area close to the city of Perm, just west of the Ural Mountains, which form a natural divide between the Volga region and western Siberia. Even if I miscalculated and our orbit took us beyond Perm, we should still land in Soviet territory. We could not run the risk of overshooting so much that we came down in China; relations with the People's Republic were poor at the time. I advised Mission Control of my decision. But it was not clear whether they had heard. We received no confirmation.

There was no time to worry whether they had heard or not. Pasha began orienting the craft for reentry. This was no easy task, but Pasha performed it brilliantly. In order to use the optical device necessary for orientation he had to lean horizontally across both seats inside the spacecraft, while I held him steady in front of the orientation porthole. We then had to maneuver ourselves back into our correct positions in our seats very rapidly so that the spacecraft's center of gravity was correct and we could start the retro-engines to complete the reentry burn. As soon as Pasha turned on the engines we heard them roar and felt a strong jerk as they slowed our craft. We then simultaneously counted on separate second-counters the short time the rockets should be left to burn before our hands reached out in silence to switch them off. As the roar of the rockets died down, it became very quiet again. According to the flight schedule, our landing module would separate from the orbital module ten seconds after retro-fire. I counted the seconds down in my head. But something was very wrong.

It felt as if we were being dragged from behind, as if something was pulling us back. When we began to reenter the Earth's atmosphere we started to feel gravity pulling us in the opposite direction. The conflicting forces—my instruments indicated 10 Gs—were so strong that some of the small blood vessels in our eyes burst. Looking out of my window, I realized with horror what was happening. The communication cable connecting the landing module with the orbital module was still holding the capsule together. It was causing us to spin round its common center of gravity as we rapidly entered the denser Earth atmosphere.

The spinning eventually stopped at an altitude of about 100 km when the cable connecting the two modules burned through and our landing module slipped free. Then we felt a sharp jolt as first the drogue chute and then the landing chute deployed. Everything became very peaceful, very calm. We could hear and feel the wind whistling in the straps as the module swung gently on the landing chute.

Suddenly everything became dark. We had entered cloud cover. Then it grew even darker. I started to worry that we had dropped into a deep gorge. There was a roaring as our landing engine ignited just above the ground to break the speed of our descent. Finally we felt our spacecraft slumping to a halt. We had landed in 2 meters of thick snow.

We had no idea where we were. I should have turned off the electronic-mechanical orientation system once I had set it to our designated landing site, which would have allowed us to work out the exact latitude and longitude. But I realized that I had left the apparatus running. Now it was indicating that we had landed 2,000 km beyond Perm, in deepest Siberia.

"How soon do you think they'll pick us up?" Pasha turned to me, concerned, as the landing module juddered to a standstill.

I tried to make light of our situation. "In three months, maybe, they'll find us with dog sleighs."

First, we needed to assess our location. We had to get out of the spacecraft. We wanted to feel firm ground beneath our feet. That was not as easy as we thought. When we flicked the switch to open the landing hatch, the explosive bolts holding it shut were activated and a smell of gunpowder filled the cabin. But, though the hatch jerked, it failed to open. Something was blocking it from outside. Looking out of the hatch window we could see it was jammed against a big birch tree.

We had no alternative but to start rocking the hatch violently back and forth, trying to shift it clear of the tree. Then, using all his strength, Pasha managed to push the hatch away from the remains of the bolts, and it slid back and disappeared into the snow.

We took in a deep draft of fresh air and felt our lungs contract with the sudden blast of cold. After so many emergencies, the relief at drawing breath on Earth again was indescribable. We threw our arms round each other, slapping each other on the back as best we could in our bulky spacesuits.

Pasha maneuvered himself close to the hatch opening but I found I couldn't move, because my legs were stuck under the TV screen console. Pasha tried to dislodge them, pushing and pulling me until I at last managed to pull myself free; my boots remained stuck behind. We both squeezed out through the hatch. First Pasha and then I tumbled free of the spacecraft and sank up to our chins in snow. Looking up, we could see we were in the middle of a thick forest, a taiga of fir and birch. Our main chute fluttered in a tangle above us caught in the branches of trees that must have been 30 or 40 meters tall.

The heat of the landing module was rapidly melting the snow and ice beneath it and slowly, before our eyes, it sank on to firm ground. We knew we had to try and determine our whereabouts quickly and raise the alarm, let people know we were alive. I clambered back into the landing module to retrieve a sextant and emergency transmitter. The sun was shining and I tried to determine our approximate location by measuring its height above the horizon. But it soon disappeared behind clouds. The sky grew darker and it started to snow, so we sought shelter back inside the spacecraft.

Fortunately, Pasha and I were used to harsh climates. He had been born in the Vologda region, north of Moscow, and had spent much of his childhood hunting in the forest close to his home; his first ambition had been to become a hunter. I had spent my childhood in central Siberia and dreamed of becoming an artist. No one who has not been through the rigors of those times can begin to imagine the hardship and difficulty of surviving in such an extreme wartime environment. But it had made me tough; I felt I could withstand almost anything. Pasha and I both felt we had already been tested to our limits, though we knew there was no way of telling how long we would have to fend for our ourselves in this remote corner of our country.

It was vital to try and get a message to Mission Control to help them locate us. I started tapping out a signal in Morse code on our emergency transmitter.

"V - N - Vse Normalno, V - N . . . V - N." Again and again. "O - K. Everything's OK, OK . . . OK." There was no way of telling whether the signal was being received.

We were only too aware that the taiga where we had landed was the natural habitat of bears and wolves. It was spring, the mating season, when both animals are at their most aggressive. We had only one pistol aboard our spacecraft, the firearm I had stowed away at the last moment, but we had plenty of ammunition. As the sky darkened, the trees started cracking with the drop in temperature—a sound I was so familiar with from my childhood—and the wind began to howl.

Even though Mission Control lost contact with us on reentry and had no idea where we were or whether we had survived, our families were informed that we had landed safely and were resting in a secluded dacha before returning to Moscow. Our wives were advised to write us letters welcoming us home. The letters, they were told, would be passed on to us at the dacha.

We had no idea if our coded signal had been received. It turned out later that Moscow had not received it, because the vast expanse of forest in the Northern Urals where we had landed interfered with the radio waves. But it was picked up by listening posts as far away as Kamchatka, in the Soviet Far East, and even Bonn, in West Germany.

More importantly, a cargo plane flying close to our landing site had also picked it up. A search party had been dispatched, as a result, and all military and civil aircraft in the area where our signal originated were instructed to join the search.

Late in the afternoon, several hours after landing, we picked up the sound of a helicopter approaching. We plowed through the thick snow into a clearing and stood waving our arms. The pilot spotted us. But we soon realized it was a civil not a military aircraft. He and his crew would have no idea how to rescue us.

They saw it differently. Eager to help, they tossed a rope ladder down to us and signaled that we should grab it and clamber on board. It was impossible. We would have had to be circus acrobats. It was a flimsy, unreliable ladder and our spacesuits were too heavy and stiff to allow us the agility of scaling its rungs.

As news of our whereabouts was relayed from pilot to pilot in the area, more aircraft started to circle above us. There were so many at one point that we worried there would be a serious accident if one collided with another. But they meant well. A bottle of cognac was tossed out of one plane; it broke when it hit the snow. A blunt axe was thrown from another. Of far more use were two pairs of wolf-skin boots, thick pairs of trousers and jackets. The clothes got caught in branches, but we managed to retrieve the warm boots and pulled them on.

But the light was failing fast and we realized we would not be rescued that night. We would have to fend for ourselves as best we could. As it grew darker the temperature dropped rapidly. The sweat that had filled my spacesuit while I was trying to reenter the capsule after my space walk was sloshing around in my boots up to my knees. It was starting to chill me. I knew we would both risk frostbite if we did not get rid of the moisture in our suits.

We had to strip naked, take off our underwear and wring the moisture out of it. We then had to pour out what liquid had accumulated in our spacesuits, to separate the rigid part of the suit from its softer lining—nine layers of aluminium foil and a synthetic material called dederone—and then put the softer part of the suits back on over our underwear and pull our boots and gloves back on. Now we could move more easily.

We tried for a long time to pull the vast parachute out of the trees so that we could use it as extra insulation. It was exhausting work and we were forced to rest briefly in the snow. But, as it grew even darker, the temperature dropped further still, and it began to snow much more heavily. There was nothing for it but to return to the capsule and try to keep as warm as we could. We had nothing to cover the gaping hole left by the detached exit hatch, and we could feel our body heat dropping sharply as the temperature plummeted to below –30° C.

The next morning we woke to the sound of an airplane circling overhead. When we scrambled out of our spacecraft we saw an Ilyushin-14 flying very low above us and heard its pilot revving his engines loudly. We realized later he had been trying to frighten away wolves, which he must have spotted closing in on our landing site. Above the roar of the engines we could just hear voices in the distance. I took a signal gun and fired a flare to direct them to where we were.

Slowly, a small group of men on skis came into view. Led by local guides, the advance rescue party included two doctors, a fellow cosmonaut and a cameraman, who started filming as soon as he saw us.

It was to be another twenty-four hours before a further team of rescuers could chop down enough trees to make a clearing big enough for a helicopter to land close by. We would have to survive another night in the wild. But this second night was a great deal more comfortable than the first. The advance party chopped wood and with it built a small log cabin and an enormous fire. They heated water for us to wash in a large tank flown in especially by helicopter from Perm. And they laid out a supper of cheese, sausage and bread. It seemed like a feast after three days with little food.

By the next morning we were ready to ski 9 km to a clearing where a helicopter was standing by to fly us to Perm. From there we were flown to Baikonur, where we disembarked from the plane to find a large group waiting for us on the runway. At the head of the group were Korolev and Gagarin. They looked very serious: there were no smiles or looks of relief that we had survived. We were confused. Were they angry?

As it turned out, they were confused, too. The temperature in Baikonur then was 18° C, but we were swaddled in heavy jackets, polar hats and high wolf-skin boots: our rescue team had forgotten to bring us additional clothing. We looked quite exotic standing alongside the young pioneers dressed in white shirts with red ties.

As we approached the group, their faces suddenly broke into broad smiles. They strode to meet us and slapped us on the back. We hugged each other, laughed and joked. We were then driven in an open-topped jeep to the town of Leninsk at the Tyura-Tam rail junction. A motorcade stretching for several kilometers followed behind. Flowers were tossed in our path. Young pioneers lined the route, saluting.

A government committee was awaiting our arrival, ready with many questions about our twenty-six-hour flight. We had to deliver reports on how our mission had gone. Mine was brief and to the point: "Provided with a special suit, man can survive and work in open space. Thank you for your attention."

*   *   *

There was no mention in subsequent statements to the press of any of the emergencies we had experienced during our mission. Nobody explained anything to the public. As far as they knew, our mission had passed without incident. As far as those in power were concerned, the crises were a matter for discussion and analysis by the relevant technical committees. They were not to be aired in public.

By the standards of the day we were rewarded generously for our efforts. I was immediately promoted to lieutenant-colonel and both Pasha and I were made Heroes of the Soviet Union. In addition we were each given 15,000 rubles and a Volga car. We were also given forty-five days' leave. This was followed by a series of official visits abroad to scientific symposiums and conferences and to meet a series of international leaders to commemorate our trip. We traveled widely in Eastern Europe—Czechoslovakia, Hungary, Yugoslavia, East Germany and Bulgaria—and to France, Austria, Greece and Cuba.

Our families accompanied us to Bulgaria and as we were driven through the streets of Sofia crowds tossed flowers in the air. These were not crowds who had been brought to the roadside by buses and instructed to cheer. They were ordinary people full of warmth and genuine goodwill. Vika was presented with a small lamb with a big bow round its neck. We didn't really know what to do with it. Svetlana tied it to a rosebush outside our hotel room, but it disappeared overnight. The press reported that it had wandered off.

When we dined with the Bulgarian president the next evening, he asked Vika why she had not taken better care of the lamb.

She put him straight. "He didn't run away," she announced. "Your men took him away because he was trampling on the roses." Everyone roared with laughter.

All the attention we received was hard on Svetlana. Both at home and abroad, when we were invited to parties in the evening many women came up to Pasha and me and asked to dance with us.

"Don't monopolize your husband. We all want some time with him," they'd say. Svetlana was young and sometimes became jealous. It took a while for her to become accustomed to all the attention we received.

The attention included a mountain of correspondence. After Yuri's flight, so many letters had poured in for him that a special department was set up at Star City to deal with them. In the first days after our flight many letters poured in for us, too—at least fifty a day. In the beginning I tried to answer them all, but then it became a bit of a nightmare and I had to leave the department to deal with them. Many were quite sad; people asking for help in getting flats or jobs or medical treatment. But some made me smile, one in particular.

"Hi, there, Lyosha and Pasha," it said. "Your work is risky. My work is risky, too. If something goes wrong in space, you die. If something goes wrong in my line of work, I go to prison. When I finish this letter I vow that I will quit my life of crime. Sending you my best wishes." It was signed "The Chicken"—who was a notorious burglar!

Some years later, I passed the letter to the minister for home affairs and asked him to check what had happened to the thief. A short while later he came back and said "The Chicken" had indeed been on the straight and narrow since writing to us. The letter has since been put on display in the ministry's internal museum.

Our trip to Cuba in the summer of 1965 coincided with celebrations for the anniversary of the Cuban revolution. We stood for nine hours with a crowd of over a million people listening to Fidel Castro deliver a victory speech. The crowd was rapturous, but it was very hot and humid and many people had to be taken to hospital with sunstroke. Castro often invited our delegation to his home for dinner during our stay. He talked at length about the agricultural achievements of his country, about pig farms and a cow, to which a monument had been erected, which produced over 80 liters of milk a day.

He discussed problems, too. He spoke of his anger at the Soviet Union for withdrawing its rockets from Cuba. At that time Cuba had closer relations with China, and many of the propaganda slogans of Cuba and China were the same. This was the time when China was talking about world revolution, which appealed to Castro. Few in the Soviet Union supported such rhetoric. We had had our fill of revolution. Castro talked with disdain of the Soviet

Union sometimes. We felt as if he was patronizing us. We pretended we did not notice. He would learn soon enough how much he needed the support of the Soviet Union.

Castro rarely took his cigar out of his mouth, even when he was talking. The pungent odor of those cigars was very unpleasant, especially for a nonsmoker like myself. I started smoking cigarettes on that trip, to try to mask the smell of the smoke from the endless cigars. It took me three years to kick the habit.

In addition to traveling abroad, we received a visit from a delegation from the United States. It was in April 1965, just a few weeks after we had returned from our mission. Our meetings with the delegation took place in the offices of a Soviet press agency. At first I thought the delegates were American journalists, but I soon realized that they were not, that they were specialists from NASA.

Pasha and I met them over the course of several days for many hours at a time. First, they were shown the film of my space walk, which had been broadcast live on Soviet television. Then they sat asking questions for hours and hours. They recorded the interviews, and filmed them with several cameras under special lighting. The fact that we were meeting was reported in the Soviet press.

I did not talk too much about the difficulties I had experienced in reentering the spacecraft. I did not want to present myself as a hero. The televised transmissions of my space walk had not shown the difficulties, and neither did a longer film about our mission broadcast two weeks after we returned. The film, which was later shown at the Cannes film festival, included footage of Brezhnev speaking to me in open space from the Kremlin. It also showed us skiing out of the forest to a waiting helicopter when we were rescued, though it did not say where that film was shot.

I do not believe that what NASA learned from us during those discussions was put in a drawer and forgotten. I believe the information we provided changed the course of America's next step in the space race.

CHAPTER 5

Death of a Visionary

1965–6

Major David Scott

Manned Spacecraft Center, Houston, Texas

Dick Gordon was the first to burst through the door with the news. "Hey, Dave, did you hear the Russians did an EVA? The press is calling it a 'space walk.' They say some guy named Leonov was outside for about ten minutes."

At first I didn't believe it. "You've got to be kidding," I said. But I could see Dick wasn't joking. Some of the other guys might pull a stunt like this, but Dick was serious. I had got to know him pretty well by then. We shared a small office on the astronaut floor of Building 4 at the Manned Spacecraft Center (MSC) in Houston. Dick was a Navy test pilot, chosen in the third group of astronauts with me, and he and his family lived near us in Nassau Bay. Dick was frowning, thinking hard.

My mind started spinning with questions. "How can they be that far ahead of us? What's our latest planning on going outside? We don't have an EVA planned 'til Gemini 4, do we? And even then it's just standing up in the cockpit and taking a look outside. It's pretty risky stuff. How could the Russians have got this far so quickly?"

"Yea, those guys are moving fast," said Dick. "They keep getting the jump on us."

I carried on firing off questions. "I wonder what suits the Russians are using? How is their spacecraft configured? What kind of spacecraft has that capability? How was he tethered? What did he do outside? Do we have a mission report on it?"

"I really don't know. MPAD [Mission Planning and Analysis Division] is trying to find out what they can," said Dick scrambling to keep up. "I wonder what the bosses are doing? They must be really choked up, because they keep telling us that the Russians aren't really ahead."

"Not really ahead?" I was on my feet pacing by this point. "If this EVA is real, they're not only ahead but pretty far ahead, at that. What proof do we have that this guy really went outside?"

The first grainy photographs released to the world press of Alexei Leonov floating in space sparked a heated debate in the West. Some claimed the photos were faked. They simply would not accept that the Russians had chalked up another first. They did not believe the Russians were capable of sending a man outside a spacecraft so early on in their program. Yet we knew nothing about how Soviet spacecraft operated. We had no idea what their operating systems were. Pretty soon it was generally accepted that Leonov had indeed taken man's first steps in space. It put a massive booster under NASA's plans for a similar mission that summer.

NASA's primary focus at this point was the Gemini program, a complex series of ten two-man space missions, each of which would demonstrate how space operations and techniques had to be successfully accomplished to prepare the way for the next program, Apollo. Apollo was the program that would realize Kennedy's dream of a Moon landing by the end of the decade, but first Gemini had to pave the way. It was uncharted territory. Test-pilot heaven.

Gemini had four key objectives. The first was to prove it was possible for two orbiting space vehicles to rendezvous and fly formation in space—"station-keep"—and not only rendezvous but link up as one vehicle, or "dock." For a pilot this would be a real adrenalin rush. Rendezvous and docking would be essential for the Apollo program. Second, EVA—extra-vehicular activity, or space walking—which the Russians had just proved was possible. Third, we had to conduct long-duration missions equivalent to a lunar landing. The Russians were ahead of us on that front, too. Their cosmonauts had completed 70 or 80 Earth orbits in five days, compared with Gordon Cooper's 22-orbit, 34-hour flight on the last of the Mercury missions nearly two years before. Fourthly, the program had to prove that a precision landing back on Earth by means of a guided reentry could be achieved successfully.

Two unmanned Gemini missions had already been launched successfully, and the first manned mission Gemini 3 was scheduled to launch on 3 March 1965, just five days after Leonov performed his EVA on Voskhod 2. Until Leonov stepped into the history books, NASA's ambition for the second manned Gemini flight scheduled for June had been conservative. An astronaut had been planning to open the hatch of his capsule, stand up in his seat and put his head and shoulders out into space—the first step toward a full EVA. But after March 1965 that changed, and plans for the mission became more ambitious. We had to go one step further than the Russians or at least equal their feat.

The crew of Gemini 4 had already been announced: Jim McDivitt and Ed White, both Air Force test pilots. I had known Ed for years; he had been two years ahead of me at West Point and had flown at Edwards. He was a track star and a natural leader, whom I looked up to as one of the "big guys." Shortly after Dick burst into our office on that spring morning, Ed began training in secret. At night, after most NASA staff had left the training facilities, Ed started practicing new EVA procedures. The details of his mission were kept strictly under wraps, however; we wanted to surprise the Russians.

Although most aspects of NASA's activities and mission planning were openly discussed, few details of the Soviet space program were known, and competition was really heating up. Starting before Kennedy's assassination in 1963 and lasting until several months after Brezhnev replaced Khrushchev in autumn 1964, there had been a series of meetings between American and Soviet officials to discuss cooperation in space. These were largely prompted by Kennedy's concern about the growing cost and risk of the Apollo program.

Despite declaring the Moon race a national goal early in his presidency and having reaffirmed his commitment to it very publicly at Rice University the following year, records show that continuing pressure from Congress led him to begin privately seeking ways of cutting the cost of the space program by discussing ways of cooperating in certain fields with the Russians. The idea was to save money and prevent duplication in certain space-related issues such as communications and weather forecasting. Once the American presidency passed to Lyndon Johnson—who as head of the National Aeronautics and Space Council under Kennedy had been a prime architect of the Apollo Moon decision—and Brezhnev's hard line took hold in the Soviet Union, however, such attempts at cooperation were abandoned.

Kennedy's brief attempt to unite the USA and USSR in the journey to the Moon had, in any case, been deeply resented by many within NASA, who were convinced that the Russians would be unlikely to share any information of great significance. Little of value had apparently been learned, for instance, as a result of a US delegation to Moscow to talk to Leonov and Pavel Belyayev after their Voskhod 2 EVA mission. At least, if it was, none of this information had been passed on to the astronaut corps. The difficulties Leonov had experienced reentering Voskhod 2 after his suit expanded during his space walk, and the way he dealt with the problem by bleeding the suit of oxygen, would have been of great interest not only to Ed White, who was about to perform our first EVA, but also to me—although of course I did not know it then.

At that time I had been assigned as the back-up capcom (capsule communicator) for the Gemini 4 mission at a remote site; the capcom was the person—the only person—who spoke directly to the crew. This was to assure consistency and continuity, a single point of contact with someone a crew knew. The capcom did not say anything to the crew unless the Flight Director approved it, and usually the capcom was an astronaut; pilots speak to each other in a very concise, clear way—"pilotese"—which can often speed up communication in vital situations. For instance, they call enemy planes or other close-flying aircraft "bogies" and give their position according to the hands of a clock. Translated into "space-speak" this would turn "a meteorite or unidentified object to the right of a spacecraft" into "a bogey at three o'clock." Being capcom was a coveted opportunity. It was widely regarded as taking you one step closer to being selected for a mission. Whatever we were doing we knew our performance was being judged—informally, but judged all the same.

Being capcom on Gemini 4 would take me to Carnarvon, western Australia, where one of a series of worldwide remote tracking stations was located. A team of two dozen or so permanent Australian staff ran the Carnarvon station; we would be there to represent NASA. It wasn't until we were on the plane bound for Australia, shortly before the launch of Gemini 4, that we were authorized to open the envelope containing revised final details of the mission.

"Well, take a look at this," said Ed Fendell, leader of the five-man NASA contingent, as he handed me pages for the flight plan marked "Confidential." As I skimmed the sheets I could see why his eyes had been wide with amazement. Ed White was going for a full EVA and he would be maneuvering himself about in space by using a little handgun, a "zip gun" which would help him move by firing spurts of pressurized oxygen. It turned out this was one of the things Ed had been practicing in secret.

"Well I'm proud to be an American," Fendell burst out. "We're gonna beat those Russians yet."

In the hours before the launch of Gemini 4 we took a few moments to set up arrangements for a touching tradition which had started with the first American manned space flight, that of John Glenn. As a tribute to the astronauts, the Australian city of Perth had taken to switching on as many lights as possible when a manned spacecraft was passing overhead during nighttime. The Australians loved the space program. The spirit of exploration was in their bones, I guess. Their location meant they played a vital part in keeping track of our space vehicles. The effect of lighting up the national grid of that sprawling city was, I later discovered, from outer space like a jewel sparkling in the dark.

Ed exited the spacecraft and walked in space as Gemini 4 orbited the Earth from Hawaii to the east coast of Florida. He enjoyed himself hugely and, owing to a poor communications link with

Houston, didn't immediately pick up on Houston's prompting that is was time to terminate the EVA, and ended up staying outside the spacecraft for twice as long as scheduled. He found his spacesuit rather more bulky and difficult to maneuver than anticipated. It led to some serious problems with closing the hatch after he reentered the capsule. Apart from that, and a brief difficulty due to a post-launch station-keeping with the Titan launch vehicle and a computer failure that required a rolling ballistic reentry, the four-day, 66-orbit mission was flawless.

My stint as capcom had gone well. There were no glitches at our end. The time we had spent before the mission in simulations, either locally or with Houston, running through every eventuality we could think of, had paid off. We had even practiced having to deal with medical emergencies aboard the spacecraft with a very genial Australian doctor assigned to support the mission. But there were no emergencies during the flight.

After Gemini 4's mission was over I spent a few days in Australia relaxing. Carnarvon was a great place, little more than a fishing village, really, with just one hotel. We spent most of our time out at the tracking station, which was at the end of a dirt road in poor, dry land, with the Australian "trackies." At the end of our trip we were invited to spend our last weekend at a sheep station—shearing sheep and eating kangaroo.

Lt. Col. Alexei Leonov

Sofia, Bulgaria

Ed White's space walk took me by surprise. I heard of it during our official visit to Bulgaria. It was much more ambitious than our intelligence reports had led us to believe was originally intended. We had understood White was planning only to open the hatch of his spacecraft, pull his upper body out of the capsule and take photographs before closing the hatch again. But the information the Americans had gleaned from us during their delegation's meetings with Pasha and me in the weeks following our mission had, I firmly believe, enabled White to undertake this more ambitious schedule.

White's mission was pretty risky, too. The Gemini spacecraft was very different from our Voskhod. Without an airlock, which our spacecraft used, the atmosphere inside Gemini had to be completely depressurized before the crew could open its hatch and White could exit. If there had been any problems with this, the mission might have ended in disaster. But it went perfectly, as far as I could tell. I personally felt little sense of competition with the Americans at that point, because our space program was so far ahead of theirs that they could not possibly outstrip us. I had done something that no one else could do. I had been the first man to walk in space.

But many press reports in the West later claimed that White had been the first to perform a space walk and that mine had been a fake; that the film of me outside Voskhod 2 had been staged in a laboratory. These reports were taken so seriously that the *Guinness Book of Records*, for instance, for some time recorded White as the first man to walk in space. NASA did nothing to contradict the false claims.

We have a saying in Russian: "The dogs may bark but the caravan continues on its way." It means that, no matter what anyone says, no matter how many people carp or complain, if you are right you have the confidence of a giant. I knew I was the first.

The Americans more actively cultivated competition, mistrust and paranoia, I believe. I experienced it at first hand just a few months later when Pasha and I traveled to Athens that August for the International Astronautical Congress. An American delegation was there, and it included Deke Slayton, then Director of Flight Crew Operations at NASA, and two astronauts, Pete Conrad and Gordon Cooper. Cooper and Conrad had just completed their eight-day, 128-orbit mission aboard Gemini 5 in August 1965.

It was the first time I had met any American astronauts. Despite reading a lot about them in *Life* magazine, I had no idea what to expect. Gherman Titov had been the first cosmonaut to visit the United States after his Vostok 2 mission four years before. He had been hosted by John Glenn, the first American in orbit, and they had got on well. But that was before the Cuban missile crisis, before Kennedy's assassination, before Khrushchev was overthrown and replaced by Brezhnev. Since then the Cold War had deepened. Paranoia was rife on both sides.

My first impression of the astronauts we met in Athens was bad. Our first planned meeting started out on a sour note. Pasha and I turned up at the agreed time and place, but the Americans did not. We waited for over half an hour, but they did not come: their excuse was that they had overslept. The Greek newspapers reported the next day how rude the Americans had been to snub us. We had a running joke for a long time after that. If the Americans and Russians were ever, in the far distant future, to agree on a joint space mission, we reckoned the venture would never get off the ground; the Americans would oversleep.

When we did eventually get to meet Conrad and Cooper, two days later, the situation was very awkward. They explained that they had not overslept at all. They did not turn up for our meeting, because they had not been given final clearance by their government to do so.

After they told us that, we all got on fine. While the Russian doctors and other officials met with their American counterparts to suck each other's blood and drag out details about our respective space programs, Pasha and I sat drinking cognac with Conrad, Cooper and Slayton for about four hours. I don't know how we managed to communicate because they did not speak a word of Russian and we knew very little English. But we gestured a lot, and, anyway, many of the words we needed to talk about space have the same root in both languages. Most importantly, we were in the same profession.

We had all four been fighter pilots, albeit on different sides of the Iron Curtain. Now we were all members of this tiny, privileged elite who had tasted life beyond the confines of our own planet. We felt a special bond. It was a very pure meeting. No amount of political manipulation could interfere with that. When we parted I remember thinking what nice guys they were.

After that meeting we were all invited to spend the day with Aristotle Onassis on his yacht. We swam, had dinner. By the time we parted, it felt as if we were all members of one crew.

It was on that trip to Athens that I first met one of the masterminds behind America's space program, the German rocket scientist Wernher von Braun. One evening Pasha and I had dinner with von Braun and his wife. There was one question I simply had to ask: "How is it possible, if America is—as you boast—so much more technologically advanced than we are, that the Soviet Union was the first to launch a Sputnik? The first to send a man into space? And the first to enable a man to perform a space walk?"

Von Braun's reply was more frank than I had expected. "Technologically we had the capability to do all those things. But, perhaps, we did not have the same determination as your Chief Designer," he said. The respect he showed Korolev was, I knew, mutual. I had heard Korolev speak with admiration for von Braun, who had very nearly been brought to the Soviet Union to work on our space program instead of America's. But, however good a scientist and designer von Braun was, he would never match the genius of our Sergei Pavlovich.

Not only was Sergei Pavlovich a brilliant designer and manager, but he had an iron will and an incredible determination born of immense hardship. How great that hardship had been, we were about to learn.

David Scott

Almost as soon as I arrived back in Houston from Australia I had temporarily to disengage myself from the excitement of being directly involved in a mission and step into a role that all astronauts, I think without exception, did not enjoy—that, essentially, of a traveling salesman. "Week in the barrel" we called it, and, with a few notable exceptions it was most of the time a real chore.

Weeks in the barrel consisted of public speaking tours, which all astronauts were roped into annoyingly often. For a week at a time every several months, we would be sent to small communities all over the United States. Criss-crossing the country from Louisiana to Indiana, from California to Michigan, we had to give speeches three times a day about the space program. NASA needed the support of Congress. The idea was that, if we did local congressmen a favor by supporting them, they would vote in favor of more funding for the program.

There were a few lighter moments. One I remember quite vividly involved a fine young lady voted "Miss Holiday in Dixie," whom I was asked to accompany to a ball while I was on a speaking tour in Louisiana. I remember getting off the bus taking us to this event at a nearby stadium, arm in arm with "Miss Holiday," when who should be standing right there but Chris Kraft—Christopher Columbus Kraft, Jr, NASA's legendary Flight Director—and his wife, Betty. Chris gave a big smile straight away.

"Glad to see you're working hard, Dave," he said, and he gave me a nod and exaggerated wink.

It was also fun to witness the enthusiasm people felt about space—rows of Boy Scouts or classrooms full of children barely able to keep in their seats as they stretched their hands in the air, bursting to ask about the life of an astronaut. But as far as I was concerned these weeks on the road, driving from one Holiday Inn to another, took me away from my main focus. And that was to show I was capable of flying a spacecraft as soon as possible.

The difference between being assigned to a mission and going through training while waiting to be assigned was huge—we could all see that as soon as we arrived at NASA. From the very moment you were assigned to the crew of a flight, your life underwent a dramatic transformation. People started to look to you to make decisions, since the crews had a lot of influence on the way missions and procedures were drawn up and planned. You got to run the show, in a sense. This added hugely to the excitement of your working life. As a member of an assigned crew you were also given priority with just about everything at the Manned Spaceflight Center in Houston. Everything you wanted was at your disposal: from simulators to T-33s allowing you to fly to wherever you needed to be for that particular day's training.

My background, starting with my swimming ambitions through the time I spent at West Point and MIT to my time at Edwards with Chuck Yeager, had demonstrated my fiercely competitive nature, first kindled, no doubt, by my father's exhortations to "Get out there and mix it up." At NASA we were all like that; we were all high-achievers. We all wanted to outshine the next man; we all wanted to be chosen for the next crew, and everything we did we did with that in mind.

Even when we were relaxing, playing sport, we wanted to come out on top. There never was any formal physical training at NASA, other than the annual physical exam. But very early in the space program Ed White had persuaded NASA to build a handball court around the back of the Manned Spacecraft Center, Houston, Texas, with a place to exercise, pump iron, work out. A group of us used to play handball there during our spare time, day and night. Mike Collins, who played left-handed, was by far the best. Nobody kept any formal score or pulled the rug out from anyone else, but we wanted to beat the other guys, that was for sure. We wanted to do well, just as we all tried to spend as much time flying as we could and flying as well as we could; if anyone was watching, it might just make a difference to getting assigned to a mission.

Throughout 1965 the flight schedule of the Gemini missions—beginning with Gemini 3—proceeded at a rapid pace: five missions in less than eight months. Flight planners were continually revising the flight plans of each mission to include new operations, new hardware and software. But, unless assigned to the crew of a specific mission, all astronauts undertook their training and preparation together, according to which batch they had been recruited in. In April 1966 NASA recruited nineteen more astronauts. So our numbers were growing, and the competition was growing more intense.

Most of our time was spent either at the Cape or in Houston. Each had a very distinct style and atmosphere. At Mission Control in Houston, for instance, there was a pretty well recognized dress code: white short-sleeved shirt and tie. Downtime in Houston was family time. We spent so much time traveling that it was hard to find enough time to be with our wives and kids. Sometimes we were invited as guests of honor at civic events. No civic event in Houston was complete without an astronaut as a guest.

The Cape was more relaxed. In Florida we wore brightly colored Ban-Lon T-shirts—I had a whole wardrobe of them in different colors. We spent a few days at a time there, staying at the Holiday Inn in Cocoa Beach. The motel, once run by the Original Seven Mercury astronauts, and well known for their high-living antics, had been sold by then. But Cocoa Beach still had the air of a boomtown. The program was expanding fast and the number of people employed by NASA was mushrooming, especially once the Apollo program got started. There were a lot of secretaries working for NASA. So there was always a lot of talk. But it wasn't like a scene from a Formula One race. There were never all these groupies hanging around us, as some people, particularly our wives, often thought, although around the time of mission launches some did turn up.

The Mercury guys' reputation for somewhat high living followed us around to some extent. But the complexity of the Gemini missions meant we had less time than they did for extracurricular activities. Most evenings I'd grab a bite to eat in a pizza parlor, reading a program manual or training plan as I ate. Even that once got me into trouble: a NASA official spotted me there with documents marked "Confidential" propped up against the tomato ketchup and reported me for breaching security. But no one made too big a deal of it.

In the long, hot summer of 1965 race riots were raging in Los Angeles, Chicago and Springfield, Massachusetts. The conflict in Vietnam was also growing in intensity; by July that year the US had already dispatched 125,000 troops to Southeast Asia. In those early years of the war many people felt it was just a little brush fire; a conflict which would blow over as quickly as previous ones like the Suez crisis had done. No one appreciated the complexities involved. We had no idea of how the Vietnam war would escalate or of the effect that would ultimately have on funding for the space program. The overthrow of Khrushchev the year before had barely registered, either, though there was some concern that Brezhnev was more of a hard-liner, which might "heat up" the Cold War. Very little of what was going on in the rest of the country—or the rest of the world, for that matter—made much impact on us. We had enough on our hands with the space program. Everyone was trying to perform at his top level. We all wanted to be selected for the next crew.

The selection process was shrouded in mystery. Only once was the astronaut corps ever consulted, and that was over the three-man crew of the first Apollo lunar landing mission. We were told to pick two men and write their names on a piece of paper. There was no need to write three names; it was assumed we'd pick ourselves before anyone else. There was no false modesty or lack of self-confidence in our group. I chose Jim McDivitt and Frank Borman, both from the Air Force and from the second group of nine astronauts. This was called a peer rating. We were never consulted again.

At the end of the day it was Deke Slayton's call. Deke, and, as far as we knew, Deke alone chose who would fly which mission. He was one of the Original Seven astronauts, but had been grounded because of an irregular heartbeat. Since then, though we didn't see him face to face too often, as Director of Flight Crew Operations he had overseen every aspect of the astronaut corps' lives from his office on the ninth floor of the MSC in Houston. Al Shepard—"Big Al"—another of the Original Seven grounded because of health problems, took day-to-day decisions about our lives as chief of the Astronaut Office. Chris Kraft was the lead Flight Director, in charge of the Mission Control Center (MCC), Houston, once a mission was under way. Bob Gilruth was the all-powerful MSC Director, a kindly father figure who would periodically take individual astronauts aside and always ask the same question: "Are you boys satisfied?"

But it was Deke who decided if and when we'd fly. He usually approached his chosen few in a corridor with the casual phrase "I'd like a word . . ."

Alexei Leonov

As 1965 drew to a close, a big party was organized at Korolev's design bureau OKB-1. All big factories and enterprises, at least those run by a sympathetic director, organized such parties, at that time of year. But this one was special.

Around five hundred people from the bureau were invited to the party; fitters, engineers, designers and other scientists. A number of those from the astronaut corps were also invited, including Yuri Gagarin, Pavel Belyayev, Vladimir Komarov and myself. In the enormous hall that normally functioned as the staff canteen, tables were beautifully decorated for a buffet. There was plenty of food and champagne. There were fireworks and balloons. While the men were quite casually dressed, the women wore long gowns. It was a wonderful evening.

A jazz band, made up of engineers and others from the enterprise, played. We danced. Sergei Pavlovich danced with his wife, Nina Ivanovna, and with several other women. He was in great demand that evening and he liked to dance. He was very sociable, not at all the person he was when he was at work. He rarely had the opportunity to attend such parties because of his hectic schedule and heavy workload. But among those he trusted and loved he became almost a different person. He was relaxed. He told jokes. He was the life and soul of the party. We felt very comfortable around him.

That evening I decided to ask Korolev for his autograph. It is something I had never asked anyone before and have never done since. But when I approached him and handed him a photograph of myself, he wrote something which was both warm and wise and which I took very much to heart: "Dear Lyosha, May the Milky Way not be your limit and let the solar wind not pass you by."

Sergei Pavlovich seemed full of vitality that night. He had the physique of an ox, and looked strong and stocky. Yet his health had been failing for some time. He suffered from numerous medical problems, most of which had their origins in the terrible hardship he had suffered during his years of imprisonment as a young man in one of Stalin's remotest gulags.

A few weeks before the party he had been diagnosed as suffering from a bleeding polyp in his intestine. On 5 January 1966 he was admitted for tests to the special clinic in Moscow that treated all top Soviet officials. Several days later he was allowed home so that he could celebrate his fifty-ninth birthday with his family, on the understanding that he would return for an operation to remove the polyp the next day.

On the evening of 10 January 1966 Korolev invited Gagarin and me, together with some members of his family and a small group of academics and scientists, to celebrate his birthday at his home. He lived in a two-story detached house—modest by today's standards—set in a small garden of cherry, apple and fir trees surrounded by flowerbeds full of tulips and carefully tended roses. The house—on a street now named after him in the north Moscow suburb of Ostankinsky—was Korolev's personal sanctuary. He worked such long hours that he had little time to socialize. His wife was an accomplished musician, and both loved the theater, but they rarely had time to go. So Korolev installed a large projector screen in his living room and ordered whatever films and newsreels he wanted to watch to be delivered to his home.

Yuri and I arrived that evening carrying as a birthday present a large bronze statuette entitled *To the Stars*, which everyone in the cosmonaut corps had signed. It was so heavy—it must have weighed over 50 kilos—that we had a real struggle carrying it through the snow to Korolev's house. It kept slipping on my shoulder, and ripped a button from the sleeve of my new overcoat.

That night Korolev proposed many toasts. He was full of praise for his team. "We have great work ahead of us," he said. "With mutual understanding and hard work, we will be able to complete every task that lies ahead." Then, turning to Yuri Gagarin and me, he proposed a more personal toast. "Among us are two young men who, to our great joy, have fulfilled their mission brilliantly," he said.

He then came to thank us both individually. As he shook my hand he said something that made a great impression on me, though I did not understand its full significance. He said he viewed my space walk as the last major work of his life.

As the evening drew to a close and people began to drift away, Korolev drew Yuri and me aside and asked us to stay behind. Servants cleared the long dining table at which we had all been sitting, but left one corner laid afresh. They brought out delicious *pirozhki*, meat and cabbage pies, which Korolev's housemaid loved to cook. A bottle of smooth, three-star Armenian cognac—Winston Churchill's favorite—was set on the table. The servants retired. Nina Ivanovna went to bed.

When the three of us were alone, Korolev began to talk. It was almost as if he were talking to a priest, as if Sergei Pavlovich were in a confessional. He told us the extraordinary story of his life. As Yuri and I sat listening it was hard to believe that the great man who sat before us had endured so much.

Korolev began by telling us about the night he was arrested and sent to one of Stalin's most remote and brutal prison camps. A car arrived late one night in April 1938 at his apartment on Konyushkovska Street. He was bundled into the car so quickly that he was unable to say a proper goodbye to his three-year-old daughter, Natasha, or to her mother, his first wife, Xenia Vincentini. He described the torture and beatings that followed as he was questioned endlessly by a young member of the NKVD secret police.

"Do you want some water?" his interrogator asked, and he said, "Yes." As he was being handed a glass of water, the interrogator smashed the water jug over his head with the taunt: "You scientists are so weak. A water jug can make you faint."

When it came time for Sergei Pavlovich's "trial," he was marched along a maze of corridors before being brought to a halt in front of a pair of double doors. The doors opened and he saw, at the rear of a brightly lit room, a table at which sat three men, all prominent members of the Communist Party. At first, Korolev said, he felt relieved, because he knew who the men were. In the center was People's Commissar Klement Voroshikov. Later this troika, which became known as the Emergency Three, was exposed as having masterminded many of the purges of Stalin's Great Terror, but at the time Korolev believed Voroshikov would give him a fair trial.

It was only when Voroshikov asked Korolev to hand over the document detailing his alleged crimes, that Korolev realized a document had been placed in his hand. It was read out—he was accused of inflating the cost of reconstructing a building where an agricultural institute had been based in order to set up a new institute for rocket engineering and space technology—and he was asked if he was guilty. When he denied the charges, one of the three shouted, "All you bastards say you're not guilty. Give him ten years." The so-called trial had lasted less than a minute.

Many years later, after Korolev had risen to prominence, someone within the KGB showed him the document laying out the grounds on which he had been condemned. Korolev told us he had known those who wrote it. He did not tell us then who they were, although there was speculation later that Valentin Glushko, another rocket engineer who went on to become a bitter but highly respected rival of Korolev's, had a hand in it. While very many scientists and engineers were banished to gulags at this time, Glushko never was. Those who made such denunciations paid little heed to the great damage they did to our country by sending such talented people to the gulags. Most of them were stupid, and simply enjoyed their power.

After his "trial" Korolev was sent to a brutal prison camp on an island about a hundred kilometers from Magadan in the Kolyma region of Far Eastern Siberia, which was later written about by Alexander Solzhenitsyn in *The Gulag Archipelago*. Korolev spent months there in midwinter cutting trees and working in the mines. It was backbreaking work. His health suffered terribly. But, as the Second World War gathered pace, a call went out for rocket scientists to be recalled to Moscow. One of those responsible for gathering them together was Valentin Glushko. Korolev was ordered back to the mainland.

Korolev described in vivid detail the morning he left prison. Other prisoners gave him a warm hat, coat and gloves. He recalled how the huge prison gates had opened and how, as he started walking away in bright sunlight, he looked back and saw all those who had to remain in that hellish place gripping the bars surrounding the dark outline of the camp.

He had no money, but managed to hitch a lift with a passing truck, whose driver insisted that he hand over his boots as payment; in return he was given the driver's ragged shoes. On arriving in Magadan, Korolev found that the last boat of the season had already left for the mainland. The boat later sank in a heavy storm. That was the first in a series of incidents which convinced Korolev he was privileged with some sort of supernatural protection.

It was another six months before the shipping lanes of the Okhotsk Sea would be free of ice and open to navigation once more. Korolev was stranded. He described how he wandered looking for a place to sleep as temperatures sank to −40° C. While searching for shelter, he came across a fresh loaf of bread lying on the path between two army barracks where more prisoners were being held before being sent to the gulag. He devoured the bread and then managed to smuggle himself into one of the barracks. When he woke the next morning he asked the prisoners who had thrown a fresh loaf into the street. They laughed. No one in such harsh circumstances could afford to throw food away and there was no bakery nearby. Sergei Pavlovich was not a religious man, but he took it as a further sign that some divine force was protecting him. All these years later, he told us, what happened that night remains a total mystery.

In order to survive until he could take a ship out of Magadan, Korolev worked as a laborer, helping a carpenter and a cobbler. At the end of May he took a boat to Nakhoda and then boarded a train for Moscow. But by then he was so ill that he was bundled off the train at Khabarovsk. It was feared he was about to die. His body was swollen with scurvy, his gums were bleeding, his teeth falling out. A local took him to the home of an old man, a healer, on the outskirts of the town. The old man put Korolev on the back of a cart and drove him into the hills, which, free of snow by that time, were carpeted with fresh herbs. Hardly able to swallow, Korolev chewed some of the tender stems until his gums bled. The potent vitamin C the herbs contained performed a small miracle: a week later he was well enough to board a train to Moscow again.

When he arrived, however, he was not allowed to return home to his family. He was taken to an isolated building on the banks of the River Yauzer, where other outstanding scientists were being detained. It was comfortable, but it was a prison nonetheless; a *sharashky*, as the detainees used to call it, where scientists, intellectuals and engineers were held on the grounds that they were being protected by the state. It was cruel, but I believe it did guarantee their safety, because it was known that Nazi Germany had hatched many plots to assassinate leading Soviet scientists and military figures.

For the first few years of Korolev's detention, he and the other inmates were not allowed to see their families or friends, even though they could see their homes in the distance, if they climbed to the eleventh floor. They could communicate with their loved ones only by letter. There were no cells in the *sharashky*; the men were accommodated in large dormitories. They were fed well, and in their free time they were allowed to exercise in a fenced-in area they called the "Monkey Place." The group had to dedicate themselves entirely to the war effort. They designed many new rockets, missiles and aircraft. It was not until the end of 1944 that they were free to return to their families.

As the war drew to a close Korolev was commissioned a colonel in the Red Army and was sent to Germany to gather intelligence on the rockets that had been designed and developed there under the guidance of Wernher von Braun. But, by the time he arrived at the secret rocket facility, von Braun had already given himself up to American troops. Vital manufacturing materials and other V-2 rocket engineers were, however, captured by Korolev's team of intelligence agents, including rocket-engineering specialists. After Korolev returned to Moscow, in 1946, he was appointed head of the design bureau OKB-1, which was dedicated to designing new rockets. Many of the German engineers, who had by then been transported to the Soviet Union, were sent to work for him.

Our country was left more devastated by the war than most in Europe. Twenty million of our people had lost their lives. It is hard to imagine how a man who had been so mistreated by the system within which he lived could devote himself to furthering its aims in the way Korolev did. But most Soviet people had an extraordinary feeling of purpose, zeal to rebuild our country, to push technology to the limits to strengthen our defenses so that such a disaster could never be repeated. Rocket science was at the forefront of this powerful push forward.

Many of Korolev's ideas for developing new rockets were revolutionary. Those he worked with were skeptical at first, but he soon gained enormous respect. The priority was to develop the first intercontinental ballistic missiles. Korolev achieved this with his R-7 Semyorka rocket, which propelled a dummy warhead over 6,000 km from Baikonur to the Kamchatka Peninsula in August 1957. The Americans did not manage such a launch until over a year later. It was the beginning of the arms race between our two countries.

But even then Korolev had his sights set on a far more ambitious goal: spaceflight. It was not until 1957, shortly following the record-breaking launch of Sputnik 1 that autumn, that he was at last legally rehabilitated. In the nine years that followed, he proved he was a man of unparalleled vision and determination.

That is the extraordinary story Sergei Pavlovich told us as we sat listening to him talk until around four o'clock in the morning. All that time we were surrounded by reminders of his greatest passions in life. On the wall of his study upstairs was a meticulously annotated map of the surface of the Moon, and he also had a globe of the Moon. In pride of place, in a cabinet on the wall beside them, a small but perfect replica of Sputnik 1 was suspended. Beside that was a table on which was laid the Plasticine he used to make models of whatever aspect of the spacecraft he was working on at the time.

Most striking of all, however, was the painting that hung behind his large mahogany roll-top desk. It showed Lenin walking across the frozen Gulf of Finland, wind beating against his face. The title of the painting was *Walking on Thin Ice*. It was a metaphor for the way Korolev had felt throughout his life.

We had no inkling that night that Korolev wanted to talk because he felt he was close to death. He knew he had to return to hospital the next day, but it was for a routine operation. The minister of health, Boris Petrovsky, who was a friend of Korolev, was to perform the operation, and told him that it would not last long and he would feel no pain.

His wife later said that, after he left for hospital, she found the pockets of his suits and jackets turned inside out. He had apparently been looking for two-kopeck coins to take with him to the hospital for good luck, but had not found any.

When Nina Ivanovna went to visit her husband in hospital the next day a doctor came to see him with the results of some tests.

Pointing at his heart Korolev asked, "How much time have I left?"

"Don't worry," the doctor replied. "You have plenty of life still ahead of you."

But two days after Korolev's birthday party I received a phone call from Yuri early in the morning.

"Quickly, we have to go, Lyosha," he said. "Sergei Pavlovich has died."

I was confused because we had another friend called Sergei Pavlovich, though his last name was Pavlov. "Which Sergei Pavlovich?" I asked.

"What do you mean?" he said. "Korolev, of course."

I was deeply shocked. How was it possible? We had been with him only two nights before. I left immediately and met Yuri at Korolev's home. When we arrived Nina Ivanovna was on the point of fainting and seemed to be slipping in and out of consciousness. There were other friends and relatives there. She was so traumatized she seemed hardly to recognize any of us.

The operation Korolev had undergone was supposed to have lasted for such a short time that he had been given only a small quantity of anesthetic. Complications developed when he suffered excessive internal bleeding. A cancerous tumor was discovered in his intestine. Attempts to administer more anesthetic in order to remove the tumor led to problems with his breathing and heart rhythm. Attempts to intubate him proved impossible; his jaw was malformed—probably as a result of having been broken during his imprisonment in the gulag. A tracheotomy was performed and the operation completed. The surgeon left the operating theater. But he was quickly recalled. Korolev's heart had stopped. Attempts to revive him failed.

There was some controversy about who was to blame for the tragedy. I personally felt it would have been better if the operation had been performed by a specialist surgeon, rather than by the minister of health. But Korolev would almost certainly have died anyway. The years of hardship in Stalin's gulags had taken their toll on his health. His heart had been weakened.

Korolev's body was taken to the great Hall of Columns of the House of Unions, where it lay in state on a high pedestal covered with flowers as crowds of mourners filed past to pay their respects. The hall's white marble columns were draped with red and black ribbons. Symphonies by Tchaikovsky and Beethoven filled the air.

Finally the name of the Chief Designer was acknowledged and there was a military parade in his honor in Red Square. Brezhnev gave a speech in which he said our country had lost its "most outstanding son," who had survived repression and gone on to perform work of the very highest importance.

Yuri Gagarin talked of how the cosmonaut corps had lost its father figure. "The name of Sergei Pavlovich is synonymous with one entire chapter of the history of mankind," he said with great solemnity. "The first flight of an artificial Earth satellite, the first flight by the Moon, the first flight by a human being in space, and the first free walk in space of a human being." That summed up the highlights of Korolev's career for all to hear.

From the columned hall Korolev's body was taken for cremation. Leonid Brezhnev, Chairman Alexei Kosygin, Secretary Mikhail Suslov, Yuri and I then took it in turns to carry his ashes to their place of burial in the Kremlin Wall.

There was a feeling of dread as well as loss throughout the Soviet Union when Korolev died. People understood that we had lost someone of great importance to our country. Although Sergei Pavlovich had never sought public recognition during his lifetime—he had not wanted it to interfere with his work—when he died the great secrecy surrounding his identity was finally abandoned.

Pravda ran an obituary and photograph of the Chief Designer wearing medals which marked him out as twice Hero of Socialist Labor and Lenin Prize–winner. In the many eulogies that followed, his countrymen came to understand that what Korolev had achieved was far greater than anything so far achieved by any single person in America.

Even before the first artificial satellite was launched into Earth orbit, Korolev had been talking seriously of developing rockets to fly probes to the Moon and return them to Earth. He drew up plans for manned flights not only to the Moon, but to Mars and Venus, too. For his fiftieth birthday, in 1957, the year of the Sputnik launch, he was presented by those working for him with a montage showing a photograph of him standing on the Moon with the dogs Strelka and Belka running at his heels. Two years later an

ALEXEI LEONOV

The Leonov family, Siberia (I'm on the far left, first row).

As a cadet at the
Kremenchug School
of Aviation, 1955.

With my wife, Svetlana, 1959.

The original group of cosmonauts.

The first cosmonaut corps at parachute training, April 1960.

Sergei Korolev congratulates Yuri Gagarin on becoming the first man in space, April 1961.

Preparing for the first space walk, March 1965.

The crew of Voskhod 2 (Pavel Belyayev and me) with Leonid Brezhnev and party leaders at the Kremlin Wall, 24 March 1965.

Walking to the launch pad with Vladimir Komarov, 18 March 1965.

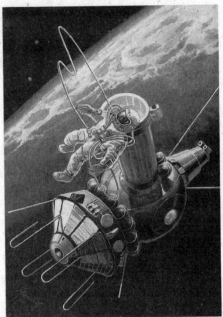

The first spacewalk (EVA) during the flight of Voskhod 2, 18 March 1965 (my painting).

Congratulations from President Brezhnev with members of the Politburo, during the flight of Voskhod 2, 18 March 1965.

Outside Voskhod 2 spacecraft, 18 March 1965 (note balloon inflation of glove).

At the Perm airdrome after
the Voskhod 2 recovery
and our stay in the taiga,
22 March 1965.

With Valery Kubasov at the Biurakan observatory
preparing for the first Salyut space station mission, 1971.

Undergoing testing
for the space station
mission, 1971.

Gathering at Star
City, July 1973 (from
left to right: Gyorgy
Beregovoy, me,
Vladimir Shatalov,
David Scott).

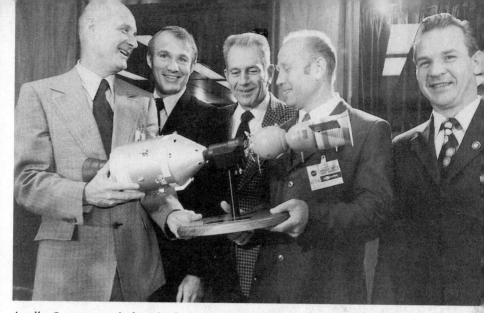

Apollo–Soyuz crews before the flight, May 1975 (from left to right: Tom Stafford, Vance Brand, Deke Slayton, me, Valery Kubasov).

Me and Valery Kubasov before the launch of Soyuz 19 for Apollo–Soyuz mission, 15 July 1975.

Apollo and Soyuz docked in orbit, July 1975 (my painting).

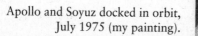

Apollo–Soyuz crews with US President Gerald Ford, July 1975.

Colonel Alexei Leonov, Deputy Commander, Gagarin Cosmonaut Training Centre, 1976.

The artist at work on *Neptune* newspaper.

unmanned Luna spacecraft flew round the Moon and photographed its Far Side for the first time.

By the early 1960s Korolev had begun work in earnest on a manned lunar mission. Although responsibility for various aspects of our space program was, by that time, shared between a number of design bureaus, Korolev's workload was enormous. His primary responsibility had been the development of spacecraft to first circumnavigate and then land on the Moon. It was then that work had begun on developing the Soyuz spacecraft, the basic design of which is still used today.

Responsibility for other aspects of space technology was delegated to designers such as Gyorgy Babakin, Vladimir Chelomei and Valentin Glushko. In contrast to Korolev and Chelomei, who had cooperated, Korolev and Glushko had had very bad relations, partly because of a difference of opinion on technology. Had they cooperated more closely, we would have had fewer problems with the choice of fuel for our rocket launchers and the number of engines the rockets employed. Miscalculations on both fronts cost our program dearly in the years that followed.

The main problem was that Korolev and Glushko had totally different characters. Sergei Pavlovich, for instance, never sank to criticizing his rival personally or professionally—he was far too clever to do that. But Glushko often gossiped about Korolev behind his back, calling him a tyrant and berating him for not being grateful that Glushko had been instrumental in securing his release from prison. Such talk disgusted me.

I always had the deepest respect and a great love for Sergei Pavlovich Korolev. Yuri and I each used to carry a photograph of him in our wallets all the time, alongside photos of our families; I still do.

The death of the Chief Designer was an immense loss to our space program. His deputy at OKB-1, Vasily Mishin, who had worked brilliantly with Korolev, succeeded him. But without Korolev, Mishin was lost. He was a very good engineer, but he had his weak points, one of which was that he drank. He was also hesitant, poor at making decisions and reluctant to take risks. This was to cost us dear.

Compared to Korolev, Mishin was bad at managing relations within the cosmonaut corps, too. When he took over as leader of the lunar program a great rift developed between those who had been recruited at the beginning—all military men, fighter pilots like Yuri and me—and those recruited subsequently, many of whom were civilians. The first civilians recruited, in 1963, included a number of engineers from Korolev's institute, and five women, among them Valentina Tereshkova. A second group of civilians who joined the following year included engineers from OKB-1, such as Valery Kubasov, Konstantin Feoktistov and Oleg Makarov, recruited so that engineers could see at first hand the physical demands on the spacecraft being built.

The harsh physical tests and requirements to which we had been subjected were dropped. Though the engineers were older than we were, we were the real veterans in the corps. But Mishin, who knew many of the engineers well, favored them and did everything he could to promote them, which created great tension. I personally did not get on with him very well at all.

I feared at the time that some of Korolev's great plans for lunar missions would be much more difficult to realize now he was gone. Just how difficult I did not know. But I am convinced now that had Sergei Pavlovich lived just a little longer we would have been the first to circumnavigate the Moon. The events of the next few years, however, dealt the Soviet space program a series of heavy blows.

David Scott

I was standing at the back of the big glassed-in VIP area at the Mission Control Center (MCC) in Houston in late August 1965 when Deke Slayton came over to me and said, "Hey, Dave, I'd like a word . . ."

I had just finished my tour as capcom at MCC for Gordon Cooper and Pete Conrad's seven-day mission aboard Gemini 5. That mission was a turning point. Pete and Gordo had completed 128 Earth orbits, much longer than anything the Russians had achieved. The VIP area was already buzzing with excitement. But those few words from Deke Slayton were enough to quicken my pulse.

"Now Gemini 5 is over, we'd like you to start working with Neil Armstrong on Gemini 8 as prime," Slayton went on. "Pete Conrad and Dick Gordon will be your back-ups. We're looking at about next March for launch."

"Sure, boss. Sounds good to me." I tried to sound as cool as possible.

"OK, get in touch with Neil and he'll brief you on the flight plan, as far as we know it now, as well as the training schedule. And keep this to yourself for now," said Slayton. "Other than the four of you, none of the other guys have been told. We'll probably let all of them know in a couple of weeks. Until then, you guys get going and work everything internal. Any questions?"

"No, sir. Sounds great," I said.

Almost as an afterthought Slayton added, "Oh, by the way, Ed White's up to speed on this too. You'll be doing a big EVA. You might want to spend some time with Ed. He's expecting your call. Good luck. We'll talk to you soon."

As Deke walked off, I just stood there looking through the glass of the big internal window that separated the VIP area from the Mission Operations Control Room. The big board on the wall at the far end of the room still displayed the Gemini 5 orbital track over the Earth. Would it be showing Gemini 8's in a few months? I wondered if this could be for real. It had to be. I knew this was how the selection process worked. I had been chosen so quickly— the first of the group of fourteen to be selected as a prime crew. Some of the first two groups were still waiting for prime crew assignments. But I had to keep it to myself.

I couldn't have been happier and to fly with Neil Armstrong, as my commander, was just perfect. He had almost been one of the two I chose on my peer rating. I hardly knew him, but I admired him. I knew he was a great pilot and a man with cool, calm authority.

I had to find Neil and make sure he was happy with Deke's choice: the commander always has the right to reject a selected pilot. When I caught up with him in a corridor outside the VIP room, he gave me a big grin and held out his hand. That was all I needed.

The next day I went to join Neil in his corner office on the astronaut floor of Building 4. His office mate, Elliot See, was out of town. Just after I arrived, Pete and Dick came and joined us. "Welcome aboard. Glad you're all here. Look forward to working with you," Neil said, all smiles. As we settled into our chairs and listened to him starting to run through the Gemini 8 flight plan I thought, "This is it. We are truly in the inner circle."

Gemini 8 was to be the most complete and comprehensive mission to date. Gemini 6 was to perform the first rendezvous and docking. Gemini 7 was scheduled to complete a maximum duration flight of fourteen days—the longest any Apollo mission was expected to stay in space. But Gemini 8 would combine all the Gemini objectives in one four-day mission: rendezvous, docking, combined vehicle maneuvers, EVA, orbital experiments and a precision guided reentry to the Earth's atmosphere.

I was scheduled to conduct an EVA lasting nearly two hours. This would mean free-floating in space at speeds of up to 18,000 mph for over one full Earth orbit. Forty-five minutes would be in daylight and forty-five minutes at night, during which time I would have to hang on to a tiny platform at the back of the ship until the Sun came up and I could see my way to getting back inside. Such an extensive EVA was designed to test a whole range of new procedures and equipment which would be used on the remaining Gemini missions. "A Walk around the World," the media were calling it.

Space was a big deal for the media then, though it really climaxed later with the Apollo program. The latest mission news from NASA often hit the front pages and Lurton started keeping a scrapbook of newspaper clippings for the kids. I was pretty shielded from the press most of the time, except when I went out touring the country during our "week in the barrel" trips. But as soon as I started intensive mission training those trips were, thankfully, temporarily suspended.

During the time we had launched successive Gemini missions the only Russian spacecraft launched was Leonov's Voshkod 2, just five days before Gemini 3. We weren't too aware of what was happening with the Russian space program. We did not know why it appeared to have stalled. We knew nothing about the death of their Chief Designer—I did not even know there was a Chief Designer. All we knew was what the Soviet authorities allowed us to know. We had no idea what they were planning. But we did not let it concern us on a day-to-day basis. We assumed our bosses were following what the Soviets were up to.

From the moment Neil Armstrong was chosen as commander of our mission it was up to him to decide on our training schedule, using a basic outline defined from previous Gemini missions. For the first few months we made regular trips to St. Louis, where McDonnell-Douglas were building our spacecraft, to verify the hardware and software for rendezvous. On one occasion, we sat in a darkened room on two simple chairs facing a projection screen while a movie of a simulated rendezvous played again and again. This was our first "training" for the complicated maneuver we were expected to perform.

At first many people, including the engineers at McDonnell-Douglas, had difficulty understanding exactly how the rendezvous would work. I was a bit hazy on it, too. There were no textbooks on it; this was breaking new ground. Neil soon put the engineers straight. He had an amazing capacity for assimilating and then explaining the most complex subjects in a very straightforward way, but with a soft, understated touch. I used to call it his "professorial mode." I remember once he stood up in front of a blackboard in St. Louis as we were discussing the rendezvous and said, "Wait a minute, guys. Here's how it's done." He then proceeded to set out the various phases of the profile in the simplest terms possible. It was excellent. It was as if he had been teaching the subject for years and years.

In October that year the Gemini 6 mission suffered its first failure. The Agena rocket with which it was due to rendezvous exploded during launch. This failure and subsequent problems with the Agena cast a temporary pall over our training program. It raised severe doubts about the safety and reliability of the target vehicle, which were to haunt us throughout our Gemini 8 mission. It was a definite blow to morale. Neil and I did not realize, however, how significant the Agena's failure would be when it came to Mission Control advising us on the rocket's unreliability during our mission.

After the initial problems with Gemini 6, its flight program was changed to include a rendezvous with another Gemini spacecraft, Gemini 7. The two were scheduled to come within a foot of each other. But that plan also nearly failed. Just before the launch of Gemini 6, its Titan II booster rocket shut down on the pad, exposing the crew to the risk of a potentially catastrophic explosion.

Like the Agena, the Titan had caused problems from the start. Many of the initial test launches had ended in some serious problem or failure. Vertical oscillations—termed the "pogo effect"—during several tests were so severe that it is unlikely a crew would have survived.

But three days after the aborted launch, Gemini 6 launched successfully and completed the first rendezvous in space, closing to within inches of Gemini 7, launched eleven days earlier. The two spacecraft were not designed to dock, however: only a Gemini spacecraft and Agena vehicle could manage that. This meant it was up to us on Gemini 8 to complete not only the extended EVA but also the first docking of two vehicles in space. It was quite a task. Exciting, exhilarating, but challenging too.

Ed White and I talked a good deal about what I should expect and prepare for during my extended EVA. The risks of conducting an EVA were high. Once outside the spacecraft an astronaut could receive no help from the commander. If he were unable to get back inside, the EVA astronaut would perish. If he became incapacitated or died, the only solution for the commander would be to disconnect him from the spacecraft and let him drift away in space. I knew Ed had experienced severe difficulty closing the hatch after his EVA because his spacesuit had expanded while outside. Introducing a device to allow pressure to be bled out of the suit could have solved the problem, but it was never done. Tight schedules and lack of anticipation meant some risks were overlooked. A slight modification was made to the hatch of Gemini 8 to include a lever enabling it to be shut with less physical force.

On the plus side Ed told me he had experienced no disorientation at all during his space walk. There had been a lot of concern about it, but Ed said he had felt fine, that he knew where he was the whole time. He just hadn't had as much time outside as he had wanted.

"It'll be over before you know it," he said. "It is so exciting, such a gorgeous view. You get to see the whole vista of the Earth below, which you can never appreciate looking through the small portholes of the Gemini spacecraft."

He stressed the importance of me getting fit for my EVA. The spacesuit I would have to wear during this extended period outside the spacecraft was still very stiff and heavy, because of the extra layers needed to protect against solar radiation and the vacuum environment. It had no joints, as later Apollo lunar suits had, which meant it would take a lot of strength and stamina to move around in during the two-hour EVA. Ed had told me I needed to bulk up, get in excellent physical shape, especially arm strength. I embarked on a serious work-out program, jogging every day, playing a lot of handball, spending hours in the gym, lifting weights, training hard.

Neil didn't have to train so hard physically, but he often came to the gym to keep me company. He was a great wit; he cracked me up. As I lay on the floor lifting iron, he'd hop on the exercise bike and switch it to the lowest torque so as not to exert himself.

"You're doing well, Dave. You're doing really well," he'd say, pedaling slowly. "But I have a finite number of heartbeats. I don't want to waste any in the gym."

We had quite a bit of fun training to use a piece of equipment we were to carry aboard Gemini 8, a "low-light-level television camera," which was a very early precursor to the night-vision cameras used extensively today. The plan was for us to activate it at a certain point during the mission so that it could take pictures of the Earth in low-light conditions. To accustom us to using it, it would be mounted inside a Navy plane out at Ellington Air Force Base and the plane would then take us on night flights along the Gulf coast of Texas. The camera was pretty bulky—about the size of a 2-liter bottle of water—and would film predetermined sites on the ground, which we would then compare with map features.

We also used night flights in the T-38 to practice an experiment in astronomy which we were scheduled to perform during the mission. This was to observe what is known as "zodiacal light," a very faint glow on the horizon on Earth seen just after sunset and before sunrise. This subtle light is caused by all the "dust" circling the Sun toward the inner solar system, which forms a dim patch of light, shaped rather like a symmetrical orca fin rising up from the horizon after dusk. The experiment was the brainchild of a Professor Lawrence Dunkelman, or "Dim Light Dunkelman," as Neil and I called him.

"What do you think 'Dim Light Dunkelman' would think about what we can see out there tonight?" I'd say to Neil as we kept our eyes trained on the horizon trying to make out the very faint glow the professor was so absorbed with. Those flights were really enjoyable and not too exacting, unlike other aspects of our training.

There were a great many problems to be dealt with. And, though easy to get along with, Neil was extremely determined; when difficulties arose he dealt with them in a very decisive way. We had a lot of problems with the chest-pack I would be using on my EVA, for instance. It had been redesigned since Ed White's space walk, but just wasn't functioning properly. So Neil and I and our back-up crew flew out to the Air Research Corporation in Los Angeles that made the pack and we spent the day reviewing it from an engineering point of view. Back in Houston Neil and I drew up a list of all its problems—"squawks," as we called them: there were over a hundred. Neil made it very clear that the Crew Systems Division, which was responsible for all such equipment, had to get these "squawks" fixed, and quickly.

Neil was never overbearing, but simply made it clear that when he wanted something done it should be done. He never acted on impulse, was always thoughtful, and you could be sure if he discussed something he had done his homework. He was also so adept technically and had such a strong grasp of engineering that he could explain what needed to be done in clear, concise terms which engineers could understand. The safety of the mission and its crew was paramount. Neil was an "us"–"we" person, never a "you"–"me" person. So as far as he was concerned the chest-pack was a problem not for me but for us. He was very strong on teamwork;

everybody makes a contribution and everybody is equal. Although it was clear that he, as commander, was the boss and he would make decisions, he always consulted others whenever possible and never considered himself to be above or more important than anyone else.

We had a lot in common, of course; we were both passionate about flying. In the rare moments when we weren't poring over plans, training schedules and programs for our mission, we sometimes traded stories about our days as pilots, he with the Navy and me the Air Force. Neil was a decorated veteran of the Korean war, and had had some close shaves. Comparing close shaves is something pilots do. But Neil never talked in a boastful way, as some guys do; he was very matter of fact. He had also gone to Edwards; he was a test pilot with the National Advisory Committee on Aeronautics, precursor to NASA. He loved Edwards just as I did. He had flown the F-100, F-102, F-104 and later the X-15. Neil was never one to strike up idle conversation, but sometimes over a meal at the end of a long day we talked about the thrill of flying high above Edwards in a cloudless desert sky.

Most of our time, however, in the lead-up to Gemini 8 was spent either in St. Louis, where the Gemini spacecraft were manufactured, or at the launch site in Cape Kennedy. At the Cape we went over plans early each morning at Wolfie's, our favorite coffee shop on Cocoa Beach, and at night we'd talk about the mission at a great barbecue place or seafood restaurant we all loved. Developing the smallest details of the flight plan took up a great deal of our time. Neil was more thorough in his approach to training than anyone else I knew. He never skipped any details. We simulated every possible aspect of the mission.

We even once verified the actual docking of the 19-foot-long Gemini 8 spacecraft with its 28-foot Agena target rocket in the middle of an open field near the launch site at Cape Kennedy. The two vehicles, at the time the most sophisticated spacecraft ever built, were put on support scaffolding and slowly rolled through the grass toward each other to check that the electrical and mechanical connections in their docking mechanisms fitted correctly. Neil and I sat in the spacecraft while all these technicians, engineers and scientists ran around verifying connectors and the structural interface between the two vehicles. It was surreal.

On an earlier occasion, in Houston, the spacecraft was placed inside a cavernous 6-meter-tall "thermal vacuum chamber" which not only simulated the vacuum of space but also was heated to the temperature to which our spacecraft would be subjected while exposed to the Sun. Heat lamps were switched on inside the chamber and then temperatures were reduced to below freezing to see which parts of the spacecraft would expand and contract in extremes of temperature.

I also spent hours alone in this chamber, in the suit I would wear for my EVA, going through every aspect of it in minute detail, including preparing to exit and reenter the spacecraft. Sorting out the tangle of my oxygen and electrical umbilical cables while preparing to exit took quite some getting used to in the beginning.

This was just one aspect of training I undertook without Neil, who was to remain inside Gemini during my lengthy space walk. Together with Dick Gordon, my back-up, I also flew up regularly to Wright-Patterson Air Force Base in Dayton, Ohio, to fly what was affectionately known as the "Vomit Comet," the KC-135. Stripped of its seats, the plane would fly in continuous parabolic arcs, shifting from two Gs to zero gravity again and again like a roller coaster, allowing Dick and me, dressed in our pressure suits, to run through maneuvers with a mock-up of Gemini inside the fuselage. That was really hot, hard work. I remember Dick turning to me once, drenched in sweat, and joking, "Isn't this glamorous!" It wasn't glamorous at all. I never felt sick in the plane—I was used to such maneuvers from my test-pilot days—but a lot of the engineers who went up with us were very sick indeed.

Another fun part of training for the EVA was practicing use of the zip gun that would help me move around in open space. It had two small rockets which blew air backward to propel me forward. An ingenious way of testing it was devised, whereby I stood on a thick disk mounted on a large, 20 × 23-foot, air-bearing metal table. Highly compressed air was pumped through holes in the bottom of the disk to remove any friction, allowing me to "fly" across the surface of the table by using my zip gun. It was kind of like being in a fairground bumper car. That was fun, too.

Ed White's brief excursion on Gemini 4 had shown, however, that, although he did not get disoriented in space and his zip gun had worked well, it had been difficult for him to know where he was relative to the spacecraft. He had found it difficult to get back to the spacecraft quickly, if need be, other than by pulling on the umbilical cables and hoping it was possible to grab something on the side of the spacecraft. Since I was to have such an extended EVA, with many procedures to perform on Gemini 8—including moving to the back of the spacecraft and strapping on a backpack—it was felt necessary that I undergo some additional training besides the "Vomit Comet" and the air-bearing table.

So one afternoon, I made my way out to a large water tank at the back of the Manned Spacecraft Center (MSC) in Houston, where the Flight Crew Support Division had prepared a simulated zip-gun exercise not unlike those I had been practicing on the air-bearing table. Whereas movement on the air-bearing table was in two dimensions, this exercise was in three. It involved me lowering myself into a roughly 20 × 20-feet circular water tank with the zip gun and attempting to maneuver myself around 10 feet underwater. I did not have the benefit of underwater equipment—this was a relatively simple experiment and it worked reasonably well—but I did not find the exercise uncomfortable since I had spent so much of my athletic life swimming.

The underwater concept finally emerged as a full-fledged training exercise for later Apollo missions, with a very large water tank being built for the purpose and a full pressure suit worn by astronauts during the training with safety swimmers always nearby. While the concept was being developed—about a year after Gemini 8—we all went through the Navy Underwater Demolition School, in Boca Raton, Florida, to become qualified for underwater training in our pressure suits. For about a week we became Navy trainees in underwater operations, starting out in a swimming pool with two big tanks, and eventually moving out to sea for deeper dives. The course was shortened to exclude such courses as how to attach explosives to ships underwater. And our "final exam" consisted of being dropped off in the ocean some distance from the shore, from which point we had to navigate underwater to a specific point on the shore in about an hour. It was a great experience and for years after that many of us really enjoyed deep-sea scuba diving as a sport.

As launch date approached, during one of our visits to the Cape Neil and I took time one Sunday to drive out to the new area where the Apollo launch complex was under construction. It was amazing. We could hardly believe our eyes. The launch center for our Gemini spacecraft had about twenty consoles. Mission Control in Houston had maybe thirty. But Apollo! Boy, the Apollo Launch Control Center had 240 consoles—*240*!

"No, this can't be," we thought. "They've got to be kidding. Nothing can be this big. Let's get back to Gemini, where we know what's going on."

To get more training and some experience of Neil's job as commander, I also spent one afternoon with one of NASA's private contractors in Dallas simulating the launch aborts he would be forced to perform if a problem with the Titan booster rocket developed during launch, such as a misalignment of the rocket engines, or too much pressure in the propellant tanks, or dangerously high vibrations. Just this one experience gave me some idea of the additional pressure there was on Neil, particularly during launch. It was quite remarkable how little time there was in which to make a decision if anything went wrong.

Throughout the seven months Neil and I prepared for Gemini 8 we spent nearly all our time together. We were so totally focused on what we were doing that there really wasn't time for anything else in life, except, of course, our families. It was a vertically integrated, channeled life, with nothing much else on the outside. Even on Sundays Neil's family and mine met for lunch—my wife, Lurton, his wife, Jan, my two kids, his two kids. But then, in the late afternoon, Neil and I would have to leave our wives and children to enjoy each other's company, drive out to Ellington Air Force Base and fly T-33 trainer jets to wherever we'd got to be first thing Monday morning.

The Christmas before the launch, both our families were invited up to a ranch in a box canyon in Colorado owned by John King, a businessman and great fan of the space program. There were horses for the kids to ride. There was a small ski slope, too. Neither Neil nor I could ski that well, but two and half months before launch there we were out flailing around on the slopes, just asking to break something. It was holiday time, and NASA was not even aware of it.

"God Almighty, what if we had broken a leg?" I now think, looking back. But at the time we were just having fun.

CHAPTER 6

A Violent Tumble

1966

Major David Scott

Arlington National Cemetery, North Virginia

Light snow flurries and fog shrouded Lambert Airfield in St. Louis on the morning of 28 February 1966. Neil and I were in the final weeks of training before the launch of our Gemini 8 mission scheduled for the middle of March. We stayed in Houston or at the Cape. Charlie Bassett and Elliot See, prime crew for the mission following ours, had to fly up to St. Louis that day to check on the progress of their Gemini 9 spacecraft.

Six astronauts had just been assigned to train for the first Apollo missions, and the Gemini program still had to complete its assigned tasks to prove Apollo was possible. We were all pressing hard, staying up late, working every hour we could. No one could afford to get sick or miss a day. The schedule was too tight.

But the bad weather sealed the fate of my close friend Charlie. He was riding in the back of a T-38 trainer which Elliot See was flying. Even though conditions were bad, See tried to bring their plane down visually. The wing of the T-38 clipped the roof of a building. The plane exploded on impact. Charlie and Elliot were killed instantly. It was tragic. We had all come from high-risk environments. I had lost friends before. A year after we joined the space program a good friend from Edwards, Ted Freeman, had died trying to eject after a goose hit the windscreen of his T-38. But the loss of Charlie and Elliot See was particularly sad and demoralizing.

It had a bad effect on the program, too. Gemini was still pretty fragile. We'd had launch vehicles blowing up, and trouble with the Agena target vehicles. It was a crucial time. People started worrying more, fearing we were in a run of bad luck. It was especially tough on the wives. Lurton and I knew Jeannie Bassett and her kids pretty well. They really went through it.

Everything stopped for the funeral. It was held with full military honors at Arlington National Cemetery. The memorial services in Houston, which we attended, were sad, depressing affairs. As a tribute to both men NASA T-38s flew a "missing man" formation over the services, with the lead plane pulling away from the remaining three to symbolize those missed. But we could not go into a deep state of mourning. We simply had to press on with the program.

A week before our launch date Neil and I transferred to the Cape; Lurton and Jan remained in Houston to follow our progress via Mission Control. Once we were at the Cape preparing for the launch, the rest of the world disappeared. We were in our own world, excited. As far as we were concerned this was the best mission so far. We were going to get to perform a whole spectrum of activities, and this was reflected in our mission badge.

The badge—a rainbow of colors refracted from twin white stars through a prism to form the Gemini symbol together with a Roman VIII—had been designed by Neil and me to reflect its many objectives. This tradition of astronauts designing their mission patches had started with the Mercury guys and stemmed from the old squadron patches we were used to wearing as pilots. Military badges and patches are an age-old tradition. Since the time of the ancient Greeks and Romans they have been used as a symbolic way of showing allegiance. But with the space program it was more fun, because we designed them ourselves.

Later on, during the Apollo program, the badges took on such importance that they were left to professional designers. I chose the well-known Italian designer Emilio Pucci to design the badge for Apollo 15. I had been introduced to him some time before that because he had been a fighter pilot during the Second World War and took a keen interest in the American space program. But that lay in the future. If our Gemini 8 mission failed to accomplish its objectives, who knew what would happen to the Apollo program, anyway?

On the morning of the launch weather stations around the world reported that conditions had never been better for manned space flight. While Mission Control in Houston was swabbed with a heavy blanket of ground fog, there were just a few wispy clouds in an otherwise clear blue sky at the Cape. Neil and I were woken at 7 a.m. Some guys had trouble sleeping before a launch, not out of fear, but from anticipation, excitement, but I certainly didn't. Twenty minutes later we were declared physically fit by our doctors and sat down to a breakfast of coffee, filet mignon, eggs and toast with butter and jelly. It was pretty relaxed.

After breakfast we went through a last briefing and were then taken to a launchpad near Complex 19, where Gemini 8 was being prepared. During those last preparations before the launch I could not help remembering the feats of some of the earliest extraordinary aviators. As a tribute to their bravery Neil and I had borrowed two small items from the Air Force Museum at Wright-Patterson Air Force Base to take with us into space. They were small pieces of wood and cloth from an old Douglas World Cruiser called the *New Orleans*, one of two aircraft to make the first round-the-world flight in 1924. We had come a long way since then. But the spirit of those early pioneers was our inspiration.

Neil also carried a wristwatch belonging to Jimmy Mattern, who had attempted the first round-the-world solo flight in 1933. He did not succeed—a frozen fuel line brought his plane down in Siberia—but we wanted something of his to make that trip, albeit at speeds he could never have foreseen. Neil had the watch strapped round the right arm of his spacesuit.

At last we were ready to ride the metal-grilled elevator to the enclosure around the spacecraft hatch, known as the White Room.

My only thoughts as the elevator started hauling us up the side of the launch tower were "I hope there's no hold in the countdown" and "I sure hope there are no problems with the Agena."

Once the elevator reached the top and we stepped into the White Room the atmosphere was very ordered and efficient. This was where suit technicians made all the last-minute adjustments to our equipment and helped us into our seats. The White Room was the domain of Guenter Wendt. He was quite a character, of German origin, very strict; we nicknamed him "the Führer." On every manned mission he was the last guy to shake hands with the crew before the hatch was closed. It felt comforting, a good omen, that Guenter was always there.

He had a great sense of humor, too. To break the tension before astronauts were helped into their spaceship, he and some of his guys sometimes organized a last-minute joke. Two missions after ours, for instance, on Gemini 10, Guenter handed Mike Collins and John Young large Styrofoam mock tools, because pieces of the spacecraft had broken and had to be repaired in the weeks before lift-off.

There was no such light relief in the last moments before we boarded Gemini 8, however. Right at the last minute, there was a minor technical hitch which threatened to delay the whole mission. The connection between my parachute harness and my seat was found to be blocked. A technician, it was thought, must have spilled some material like a plastic sealing compound in there. If the harness could not be connected to the seat we could not launch. It was as simple as that.

Pete Conrad, Neil's back-up, rushed around until he found a dentist's toothpick with which to try and clear the connection out. I remember looking back and seeing Pete sweating like mad digging this stuff out. After a few very tense moments the toothpick did the trick. We were strapped into our seats. The hatch to the spacecraft was closed and the final countdown to the launch continued.

The countdown was one of the most complicated that had, up to that point, been conducted at Cape Kennedy. Exactly ninety minutes before we were due to launch, the Agena rocket, which had given the program so many problems in the past and with which we would rendezvous and dock in space, was to be sent into orbit. Any problem with the launch of the Agena would adversely affect the launch of our Gemini spacecraft. The timing of this simultaneous launch demonstration was crucial.

At 9 a.m. Eastern Standard Time (EST), as we lay strapped into position in our spacecraft with the atmosphere around us slowly being purged with pure oxygen, the Agena blasted away from its launchpad.

"It looks like we have a live one up there for you," Launch Control announced as the Agena climbed rapidly into the clear Florida skies.

"Good show," Neil said, smiling across at me. We were set.

We knew there were a thousand people taking care of every little detail during what was expected to be a four-day flight. You could get spoiled by that sort of attention; people feeding you, dressing you, taking you to your vehicle, strapping you in.

It felt almost as if we were taking a new car out on the road for the first time: a Ferrari, on an open road with nobody around. But everything was familiar. We knew every knob and button with our eyes closed. Everything was within reach. After all the training, simulations—countless flights in the "Vomit Comet," the time we had spent in the thermal vacuum chamber and my own exercises maneuvering around with a zip gun on a floating disk hovering above a giant metal table—it was finally coming together. This was the real thing.

Cradled in my contoured seat it felt almost as if I was being held in someone's arms. It felt comfortable, if confined—the space inside the capsule was little bigger than the front seat of a Volkswagen Beetle. The pure oxygen environment felt clean and fresh; the temperature was carefully controlled, cool. When no transmission from Mission Control broke the silence, there was very little sound.

As the minutes to launch ticked by we were busy starting up systems and checking our controls to make sure everything was working. At 10:41 EST, the last seconds of the countdown to our lift-off commenced.

"T minus twenty and counting," the Launch Director at the Cape began. "Fifteen, ten, nine, eight, seven, six, five, four, three, two, one, zero. We have ignition."

As the powerful engines of the Titan II rocket that would push us beyond the Earth's atmosphere ignited and the giant missile began to clear its support scaffold—at 10:41 a.m. EST—Launch Control transferred to Mission Control in Houston. Chris Kraft had transferred to the Apollo program after Gemini 7, so the Flight Directors throughout much of Gemini 8 were to be John Hodge and Gene Kranz.

"We have lift-off at three seconds," Houston picked up. "Neil Armstrong reports the clock has started. And Dave Scott gives us his first report that everything looks good."

The Titan was smooth. We could feel the thrust building up to over five times the force of gravity. There was some shuddering sideways, vibration, and a slight pogo effect up and down. But it was a solid feeling, a sharp kick in the tail. No real discomfort. In less than a minute we moved beyond the stage at which we could activate our ejection seats in case of an emergency. A further ninety-five seconds and the first stage of the Titan rocket engine shut down after running out of fuel. The strap holding the two stages of the Titan together burst open, allowing the first stage to fall away.

Through our windows we could see bright red and yellow debris as the strap disintegrated, briefly forming a glowing cloud of particles which reflected the sun around our spacecraft.

For a few seconds we were floating against the straps of our harness as the thrust from beneath ceased. Then the second stage of the rocket started to burn and push us, smooth as glass, but harder and harder—up to a force of seven and a half Gs—through black skies, clear of the Earth's atmosphere, before it, too, disengaged and fell away.

Six minutes into the flight and we were 140 miles above the surface of the Earth and traveling at 18,000 mph. We were in orbit.

The first sign of weightlessness was the sight of a small metal washer, dropped inadvertently, hovering in front of our eyes. Then I released my checklist and it started floating around the cabin. Floating in my harness "heads down," I looked away from my computer display through the small window at eye level. It was a fantastic sight.

In this inverted position the blackness of space appeared to be below us. Above it a line of puffy white clouds rimmed a wide expanse of the globe. As Neil rolled the spacecraft "heads up" the scene became more recognisable.

About twenty minutes into the flight, with the Earth below illuminated I could see the Mediterranean, Italy, the clear outline of the Middle East. I could even see the contrails of airplanes and the wake of a boat in the Red Sea. Linear things, man-made things, stood out. Other crews who had made it up here before us had talked about how hard it was to describe the beauty of space. Now I could see what they meant.

My life until then had been the military, West Point, grad school, then test-pilot training. I was used to scientific manuals and technical jargon. This was a totally new domain. Maybe one day, I thought, they would send a poet into space, or an artist. Only they, I felt, could truly express this experience. But I wanted to capture what I could. I pulled out my camera and took my first photograph in space.

Slowly the broad surface of the Sahara Desert became visible, stretching as far as the horizon, where the tops of clouds were beginning to turn pink as we approached sunset. As we passed over the coast of eastern Africa the scene before us was plunged into darkness. In place of our own planet the sky was blanketed with a display of stars like millions of frozen fireworks. Within a short time we passed toward the light once more and could just make out the contours of Australia in the distance with storm clouds gathering on the horizon, illuminated from within by flashes of lightning.

Dragging my gaze reluctantly from the window, I focused again on the complexity of the tasks we had to perform. My main focus on this mission would be my extended space walk the next day. But our primary task on this first day was to rendezvous and dock our spacecraft with the unmanned Agena.

Long before we saw the Agena, on our third orbit of the Earth, we started picking her up on our radar. Then, a little over six hours into our mission, we caught our first sight of her orbiting ahead of us. At first she was just a pinprick of light. Then gradually she became a sleek, silver tube, a spectacular sight, sharp and crystal clear, floating above the Earth, seemingly hanging in space, waiting for us. She looked just perfect.

We began to close the gap between the two vehicles and, as we orbited 1,000 miles west of Hawaii, we began to "station-keep"— maintain the vehicles 150 feet apart. We were checking the Agena for safety and monitoring our instruments for the final alignment.

Neil, as mission commander, would perform most of the docking maneuvers. Just before the sun started to set behind the Earth again, he moved us closer, to a position directly in front of the Agena and we started to move in slowly

"How's it look?" he asked me. "Everything's fine. She's in great shape," I replied. "OK," he called down to ground control. "We're ready to move in."

After Mission Control's go-ahead Neil tweaked and nudged his controls to bring us right in facing the Agena. As the Gemini moved in we watched the three long "whiskers" on the Agena collar carefully to see if there would be any electrical discharge when the two vehicles came into contact. Some thought this might cause great sparks to fly; but there were none. All was quiet as we felt first contact, then a firm clunk and capture as the docking latches joined the two vehicles together. The first part of our mission was accomplished. Docking seemed a piece of cake, as easy as parking a car in an open garage.

"Flight, we are docked," Neil called down to ground control. "She's a real smoothie," I followed.

Neil turned off the Gemini and I started to fly the two vehicles jointly, using the Agena control system.

We still had a few simple maneuvers to complete before we could unstow food for our first rest period. Except for a few quick bites during the rendezvous, it was almost eleven hours since we had sat down to steak and eggs at crew quarters in Cape Kennedy. We had been running on adrenalin. There had been no time for food—no thought of it either. We were hungry now, but it would be many hours before we could satisfy our hunger. We were to come within seconds of losing our lives over the following several hours of high drama, which started with that almost casual comment passed to us from Mission Control via the Tananarieve tracking station in Madagascar.

"If you run into trouble and the Agena goes wild, just send in command 400 to turn it off and take control of the spacecraft with the Gemini . . ."

Over the hours that followed our communications with Mission Control would be limited. We could communicate with Houston for only three five-minute periods every ninety minutes as we passed over secondary tracking stations aboard two ships, the *Rose Knot Victor* in the southern Atlantic, off the southeast coast of South America, and the *Coastal Sentry Quebec* off the coast of China in the western Pacific. All communications between us and Houston had to go via the communications officer on the ships, and this disjointed three-way conversation increased our problems.

It was shortly after the transmission from Tananarieve that I noticed Neil's "8-ball" showing we were in a 30-degree bank, indicating our spacecraft was rolling slowly—even though we could neither feel nor see it, since we had passed into nighttime. Our first instinct was that the Agena, which had suffered problems in the past, was indeed "going wild." At first, after Neil told me to disable it by switching off its control system, the motion stopped. The reason we had been using the Agena engine to fly both vehicles together was to conserve the Gemini's fuel supply. The control systems of both vehicles could not be on at the same time—two sources of motion in a co-joined vehicle might lead to instability.

But even with the Agena control system switched off and control switched to Gemini, after just a minute or so we started to roll once more. Again and again Neil ordered me to switch the Agena on and then off in an attempt to regain control of the joined vehicles. Yet the roll not only continued but began to increase in intensity—to the point that we were starting to rotate on all three axes. As Neil tried to fight the motion with his hand controller, we both knew we were fast running out of options.

When Neil ordered me once more to switch the Agena back on, I looked at the control panel and could see that the fuel in one of the Gemini control systems was down to 13 percent. It was clear we had to disengage from the Agena, and quickly.

"We'd better get off," I told Neil. But we both knew the two wildly spinning vehicles might collide as they separated.

"OK, let me see if we can get the rates of rotation down so we don't re-contact," Neil said. "You ready?"

"Stand by," I replied. I knew from my days as a pilot of high-performance jets how vital it was to record everything as quickly as possible when working through an emergency. I also knew that, once we undocked from the Agena, the rocket would be dead. No one would ever know what the problem had been or how to fix it. So I preset the recording devices on the Agena as quickly as I could: ground control would be able to take control of the Agena and pick up data from it once it passed over the next tracking station. I switched on our window movie camera to record the undocking. There was no time to check settings.

"OK, any time," I said. "We're ready."

"Rog, undock, *now*!" Neil pulled the Gemini back as I pressed the undock switch on my panel. We watched the Agena slide away against the sunlit background of the Earth.

That should have enabled us to stabilize the Gemini, figure out what had gone wrong. But seconds later we began to roll even more violently. It must be a problem with Gemini, not the Agena.

We had spent so many hours training in simulators for every eventuality imaginable, but nobody had envisaged anything like this. Troubles with the Agena were known. We had trained endlessly in how to deal with them. But this sort of problem with Gemini . . . nobody had ever dreamed it would happen. We were stunned.

This spinning roll was going on far too long. The chances of recovering from such a high rate of spin in space were very remote. Both Neil and I were beginning to feel dizzy. It was rather like the feeling you get as a kid when you twist a jungle rope round and round and then hang on it as it spins and unfurls. In space it was not a good feeling. This was definitely not what was supposed to be happening.

Looking at the clock again, I saw it was 7 hours 17 minutes GET, which meant we were about to come within range of the *Coastal Sentry Quebec* (call sign: CSQ). They picked up contact with us first.

"Gemini 8, CSQ capcom. How do you read?"

There was no way they could know what was happening to us. We only had a few minutes to let them know the deep trouble we were in before we moved out of their range again. I knew the worst thing we could do was sound panicked. We didn't want Mission Control getting excited or frightened. We needed them to stay thinking clearly.

"We have a serious problem here," I said, as coolly as I could. "We're tumbling end over end up here. We're disengaged from the Agena."

"OK, we got your spacecraft-free indication here," CSQ replied. "What seems to be the problem?" It was Jim Fucci speaking; I recognized his voice. He was an old NASA hand, very experienced. Boy, was I glad he was the one on the mike. He must have seen straight away how serious our situation was. We were not supposed to be undocked from the Agena so soon. But he sounded quite calm.

"We're rolling up and we can't turn anything off," I reported. "We're in a continually increasing left roll."

"Roger," Fucci replied.

The delay in his three-way conversation between our spacecraft and the CSQ and then between CSQ and Mission Control in Houston meant the Flight Director, "Flight," in Houston was not picking up all of this conversation.

"Did he say he could not turn the Agena off?" Flight asked CSQ.

"No, he said he has separated from the Agena and he's in a roll and can't stop it," Fucci said, then he came back and told us to "Stand by."

We had lost control of Gemini 8 almost completely by this point. Our roll was speeding up. Neil was still working his hand controller to try to slow us down, but he was getting no response. It was like being on a theme-park ride which thrills passengers by spinning at high speed, except theme-park rides don't spin so fast or for so long—if they did, too many of the passengers would black out.

I have often been asked since how I felt during those moments of high tension. Was I afraid? Did I realize I might never see the Earth or my family again? The truth is that we were trained to perform at our highest peak in such situations. Emotion did not come into it. Everything was happening so fast and we had to find a solution. That took every ounce of our energy, every effort of concentration.

Neil's last-ditch option at this stage was to turn to me and tell me to take hold of the controller, see if I could bring the spacecraft under control. My immediate reaction was "Sure, give it to me, I can fix it." But, of course, it made no difference whose hand was on the stick: it didn't respond. If the hand controller had failed, there was little prospect we would ever get home. This was bad, really bad. We were reaching the end of the line.

Then the three-way conversation picked up again, even more confused this time.

I told Fucci, "CSQ. We have a violent left roll here . . . We apparently have a stuck hand controller."

"Did I hear a stuck hand controller?" Flight Control cut in from Houston.

"Say again, Flight?" Fucci asked Mission Control.

It was impossible. Fucci couldn't listen to two conversations at the same time. Houston should have known that.

"Did I hear him say he may have a stuck hand controller?" Houston asked again.

"That's affirmative, Flight . . . He seems to be in a pretty violent tumble rate."

Everything loose in the cabin when the roll had started—the checklist, flight plan, procedures chart—was thrown against the inside walls as a result of the centrifugal force.

The sunlight was flashing through Gemini's two small windows like a strobe light, hitting us in the face. Our roll had increased so that we were spinning at a rate of almost one turn per second. My peripheral vision was beginning to fade. I was getting tunnel vision. G forces were throwing our heads away from the axis of rotation. We were rotating faster than you would in a high-speed spin in a fighter-jet. We were approaching blackout.

Everything was happening so quickly. We had only seconds left before we lost consciousness. We were out of range of any tracking station. We had no time to consult Mission Control, anyway. Neil had a very tough call to make.

"All we have left is the reentry system," he gasped.

I knew he was right. Our only chance was to activate the engines for controlling the spacecraft's reentry into the Earth's atmosphere.

"Do it," I said. We both knew that if this didn't work we were dead.

Activating Gemini's reentry system under normal circumstances was a relatively simple procedure. But in these extreme conditions it was a Herculean task. Neil had to locate the reentry control switch in one of the half-dozen control panels located around the spacecraft. The switch was in the most awkwardly positioned panel of them all—directly above Neil's head. Not only that, but there were over a dozen toggle switches in that plate, and, with our vision beginning to blur, locating the right switch was not simple.

Being fighter pilots, both Neil and I were used to locating every switch in a cockpit with our eyes closed. Before you ever get a fighter plane off the ground you are put through a "blindfold cockpit check," in which you are asked to find a whole series of random switches with your eyes shut. We'd carried this training across to our preparation for the space mission. Neil knew exactly where that switch was without having to see it. Still, with the dizziness caused by the spin of sixty high-speed revolutions for a full minute, reaching above his head to activate the reentry control switch, while at the same time grappling with the hand controller trying to stop the tumble, was an extraordinary feat.

It still amazes me that Neil managed to do what he did. Everything had been happening so quickly. Only our intensive training and Neil's calm and cool demeanor under conditions of extreme danger pulled us through.

Almost as soon as he flicked that switch our spin began to decelerate. It took another thirty seconds or so before the rolling finally stopped. But it did. Books and flight plans were scattered around the cockpit. The worst of this dire problem had lasted about twenty-five minutes in all. But, at last, we seemed once more to be in control.

"You've got it, Neil," I gasped, my vision starting to open up. "You've got it!"

There was no time for jubilation. There was no way of knowing whether the roll would start again. Besides, we had other problems to face.

It would be another fifteen minutes before we passed over the next tracking station in Hawaii. In that time we had to figure out what had gone wrong and what we were going to do next. Neil started testing, one by one, each of Gemini's sixteen rocket thrusters—small bi-propellant rocket engines using both fuel and an oxidizer allowing them to be ignited in a vacuum—which were mounted on specific parts of the spacecraft to move it up–down, left–right, and forward and backward. Rocket number 8 failed.

But it was not a consistent, linear problem. Eventually the flight detectives figured out that sometimes the thruster was on for a brief period when it should have been off, and sometimes it was off when it should have been on. It was really screwed up.

The point about it, though, was that the Agena had been healthy after all, and Gemini was at fault. The shock of what we had been forced to do finally sank in. Mission rules were strict. Once the reentry control system was activated, a spacecraft had to return to Earth as quickly as possible. Neil and I started to discuss whether there was anything of our mission that could still be salvaged. But in our hearts we knew. The mission was over.

Neil turned to me with a kind and gentle face. "Sorry, partner," he said. "Guess we'll have to do it another time." He knew how hard this was for me. We both realized I would not be able to do my space walk, and I might never get another chance.

But the next crisis quickly pushed such thoughts out of my mind.

The Agena vehicle was still on the same orbital path as Gemini. We were in danger of colliding with the rocket as we completed our first orbit since the two space vehicles had separated. That was only an hour away. We had no way of altering Agena's track. We couldn't even change our own direction of orbit, to avoid collision, without calculating precise maneuver instructions. These instructions would have to be fed to us by Mission Control via our next remote-tracking site in Hawaii. We would be over that station for only five minutes, after which we would be out of contact with the ground for another twenty minutes until we passed over the *Rose Knot Victor*. That pass would also last only a few minutes, and then we would have no contact with the ground for more than half an hour until we came within range of *Coastal Sentry Quebec* again.

This meant our entire flight plan would have to be adjusted in a very few disconnected, jerky minutes of contact with Mission Control through those remote sites. At 7 hours 37 minutes GET we entered Hawaii's tracking zone. At first we were told Mission Control was considering bringing us back to Earth on our sixth orbit. That would have given us just ninety minutes to get Gemini ready for reentry. We knew it couldn't be done: our computer had not even been loaded with the reentry program yet.

"That's not enough time," I said to Neil. "We need more time."

Mission Control must have revised their math, too. Minutes later the plan was changed. We were advised of the new instructions just seconds before we lost contact with the ground again. This time we were told we were to reenter the Earth's atmosphere on our seventh orbit. Splashdown would be in a landing zone designated Dash 3. I looked it up in our manuals. Dash 3 was a secondary landing zone in the South China Sea. It was over 6,000 miles away from our primary landing site. There would be no experienced recovery team to pluck us from the sea. No mission had been forced to land in a secondary recovery zone like this before.

I hoped we were not going to be the unlucky guys. I was not superstitious, but I knew some guys were just unlucky; it was a nebulous thing, like a jinx. So far I had got out of all my problems in airplanes, and Neil had got out of all his, but we were new in the space business. Now we had this big problem. Were we going to get out of it? There was no time to dwell on such thoughts. The clock was relentless. The minutes were ticking down to our reentry. I knew that disasters, when test pilots ended up in a pile of ashes somewhere, were rarely caused by one big problem. More often a lot of little things built up. So small problems had to be dealt with as they came up.

We were coming within reach of CSQ again. I had to be ready to receive the parameters to feed into the computer for our reentry. The memory of our on-board digital computer was not large enough to retain the details for our reentry program. Roughly the size of a shoebox and weighing around 57 lb, the computer was hi-tech by the standards of the time. But still it could only store around 160,000 bits, or just over 4,000 computer words, and its memory was already saturated with all the data it had to retain for the complex rendezvous maneuver. The rendezvous data had to be erased before the reentry program details could be fed in from another on-board device, an innovation called an Auxiliary Tape Memory, or ATM. This mini version of the sort of tape then used in mainframe computers was, in effect, an external disk drive. Once I had activated the ATM and connected it to the computer, it loaded in the necessary software ready to receive exact details of our reentry trajectory, which I then fed in by means of a small punch-button keypad.

The details of exactly when we wanted to activate our retro-fire rockets to reenter the Earth's atmosphere and where exactly we wanted to land consisted of nine lines of seven-digit numbers, each preceded by a plus or minus, which we referred to as "up" or "down." I had to feed each number manually into the computer as it was read up to me by CSQ. If one digit were entered incorrectly we would miss our target and might land anywhere—perhaps on top of a mountain in the Himalayas. If we came down off target, over land, we would have to eject. Gemini spacecraft were designed for water landings.

I was glad it was Fucci reading off those lines of numbers. The guys on board *Rose Knot Victor* sounded a little rattled, and so did Mission Control. But Fucci was calm. He read off those numbers as if he was talking about taking a stroll in the park. I entered them quickly so that I could transmit them back to verify with him before we lost contact again.

Twenty minutes later we passed over the Hawaii tracking site and were given a final update on our expected landing point. It was 500 miles east of Okinawa. Two US Air Force C-54s had been scrambled we were told from their bases in Japan—one from

Tachikawa and the other from Naha Air Base at Okinawa—for a search and rescue operation. A destroyer, the USS *Leonard Mason*, which had just returned from a tour of duty off the coast of Vietnam supporting combat missions, had also set sail from Okinawa to pick us up.

We would be landing in full daylight, we were advised, mid-afternoon local time. The forecast was for clear, calm weather, with a light breeze and only three-foot waves. Still we were advised to take motion sickness tablets and drink plenty of water before reentry. Three-foot waves. No problem. We didn't take the tablets.

Retro-fire was programmed to take place at 10 hours 4 minutes GET. It was the most critical phase of the mission. We would get only one shot. We had no chance of aborting the maneuver or ejecting from the craft as we could have done had there been a serious malfunction during the launch. Traveling at 18,000 mph, the timing of the retro-rocket fire was critical, to within fractions of a second.

Under normal circumstances reentry was timed so that the spacecraft was in daylight over a tracking station. This meant that crew and Mission Control could simultaneously count down to the precise second the retro-rockets had to fire so that, if for any reason they failed to do so automatically the crew could fire them manually. Mission Control could then monitor the maneuver and, if one of the rockets still failed, could advise the crew of vital changes needed to correct their alignment on reentry.

But we would be over a remote part of south central Africa at the moment of retro-fire. There would be no way of our knowing if changes were needed.

Nor would there be any possibility of verifying the alignment of Gemini 8 with the horizon prior to retro-fire, another vital safety check to ensure we landed on target. Our retro-fire would be in complete darkness before our trajectory took us into daytime. We were going to have to hope for the best.

Neil and I began counting down the seconds to retro-fire together. "Thirty seconds, twenty-nine, twenty-eight, twenty-seven . . . five, four, three, two, one."

All four retro-rockets fired.

Our speed began to slow immediately and we began to reenter the Earth's atmosphere. We could feel gravity pulling against the spacecraft as soon as we left the vacuum of space. We began to feel pressure on our backs, our heads settling into our headrests. I began to see residuals from the burning heat shield at our backs flashing in front of my porthole, a pink and orange glow. As this haze slowly started to clear, I could see we were coming in over the Himalayas. It was a sensational view.

When our first small drogue chute opened, the nose of the Gemini flipped upward. Our windows would be pointing away from the ground from now on, which meant we couldn't see where we were going to land. In the few seconds between the drogue chute detaching and the main parachute unfolding we were in free fall. I was not prepared for the sickening anxiety this would trigger. But it was brief. The main chute opened, and Gemini jerked to a slow descent.

We still didn't know if our trajectory was taking us away from the Asian landmass. We did not know where we would touch down, on land or water. If it were land we would need to eject. But how would we know? As Gemini rocked back and forth on the chute Neil tried to find out. He swiveled his cockpit mirror to a position in front of his window and peered up. We were heading for water—no idea where.

The pungent smell of fumes filled the cockpit as we drew nearer to landing, a result of the heat shield burning up on reentry. It was warm, too, because the fans used to cool our suits had been switched off. Neil and I had to lift our visors briefly and pop our ears. But the fumes were too strong. We flipped our visors back down and prepared for splashdown.

We slammed into the water with far greater force than either of us expected. But we were down. Back on Earth. We had no idea if we were on target or whether the recovery forces were close by, whether they had tracked our trajectory until touchdown. We had no idea how long it would be before we were rescued. There were no guarantees that they would find us at all. But we were down. And through it all Jimmy Mattern's watch had kept on ticking.

Gemini started bobbing around like a cork. When Mission

Control told us about the three-foot waves they had forgotten to mention the twenty-foot swells. Gemini was a great spacecraft, but it was a lousy boat. Its center of gravity was such that it rolled and pitched with every wave. Both windows were submerged beneath the waves on and off. The seal of the hatch began to drip with condensation.

We wouldn't be able to open our hatches until the rescue forces found us and attached a flotation collar, otherwise the capsule might fill with water and sink. Our spacesuits were uncomfortably hot, mine more than Neil's because it had an extra layer designed to protect me against solar radiation during my space walk. The fumes created by the heat shield ablating steadily seeped into the cabin, making us feel nauseous. We should have taken our motion-sickness tablets. First Neil started to retch, then I did. We had one sick-bag to share.

To conserve our electricity supplies we had to turn off all unnecessary electrical equipment. Meanwhile, we had to try to establish contact with the recovery forces. We extended our high-frequency antenna, activated the receiver and started transmitting the call sign of the aircraft we had been told would be circling near our position when we splashed down.

"Gemini Eight to Naha Rescue One . . . Gemini Eight to Naha Rescue One. Come in Naha Rescue One . . ."

All we picked up was the sound of tinny oriental music from a distant radio station.

About thirty minutes later we heard the sound of an aircraft engine. It seemed close. They had located us after all, perhaps as a result of the green marker dye ejected from our spacecraft on contact with water. But then the sound of the engine faded as the plane moved away. Maybe they had not seen us. Our spirits sank.

We both knew only too well how search and rescue missions at sea operated. A plane would search a designated grid square on a map before moving on to an adjoining square and then the next. The search team rarely returned to the same coordinates, in order to cover as wide an area as possible. If the plane had failed to spot us, they might never find us.

It was not in my nature, or Neil's, to give up hope. We knew we had no option but to stay with the craft. We both fell silent, lost in our own thoughts. We knew we could be in for a long wait.

Then, ten minutes or so later, we heard the aircraft again. This time, as our windows emerged from the water, we could just make out figures being dropped into the heavy seas. At last we could breathe easy. Or at least we could once they got us out of this bobbing, stinking craft.

The swimmers began to attach the flotation collar so that we could open our hatches, but the swell was so heavy and the fumes were so strong that even they became nauseous. It took them nearly an hour to secure the buoyancy of the craft enough for us to open the hatches and draw our first draft of fresh air.

It turned out we had splashed down a little over a mile away from the emergency coordinates—the most precise reentry to date. Our trial was almost over. But not quite. We still had to be pulled aboard a ship before being flown home. Even that proved trickier than expected.

The destroyer USS *Leonard Mason*, dispatched to recover our spacecraft, had been volunteered for this mission by its keen young captain. We later learned that the crew had been less than happy about that. They had been stationed off Vietnam for the previous forty-nine days and had put into Okinawa for a brief spell of liberty, only to be dispatched immediately on a stand-by search and rescue training exercise. It was the last thing they wanted, though their spirits lifted when we were located and the training exercise turned into a real rescue. How many draftees got a chance to rescue a couple of lost astronauts?

But when, three hours later, the destroyer pulled close enough to be able to attach a cable to Gemini, to hoist us from the water, the heavy swell continued to hamper us. Our spacecraft kept crashing against the side of the ship with such force that Gemini's nose was dented. It was a precarious business. Neil and I had no choice but to sit tight in our seats, still in our spacesuits with our regulation issue sunglasses on, clinging to the open hatches for security.

Neil was the first to clamber up on to the deck of the destroyer as I sealed his hatch. When I looked up I could see him waving down at me. But I could not see how he had got on deck. As the waves pulled the craft away from the side of the ship, somebody shouted, "Grab the Jacob's Ladder." But I was in the Air Force. I didn't know what a Jacob's Ladder was. I was beginning to get nervous. Standing in the bobbing, weaving spacecraft crashing against the side of the destroyer was pretty dangerous. And ending up in the water in a spacesuit would definitely not have been good news.

Then a voice boomed down to me, "I got you, sir!" A tall, burly black sailor had reached down to grab me by the arm and haul me on deck. I don't know how he did it—I weighed a lot in my spacesuit. But I was really glad to get out of that situation. I found out some time later that his name was Jacob and he'd once been a professional boxer.

Exhausted, we were taken below to the captain's quarters, which had been converted into an operating room for men wounded in Vietnam. Procedure demanded that we undergo a detailed physical exam. We were confronted by a very young Air Force doctor who, obviously, didn't want to be in the Air Force, let alone sent to Vietnam. And, worse, he hadn't wanted to be out on a destroyer in the middle of the Pacific waiting for a couple of astronauts to turn up.

But when we descended into the operating room, he was all excited. There he was, getting to examine these two guys who they thought were going to die. He had a book out ready on the desk telling him how to give a post-flight physical exam for rescued astronauts. This back-up manual had never had to be used before.

Almost immediately I was summoned back on deck: there was a problem with the spacecraft. One of the swimmers had climbed inside to close my hatch as Gemini was being winched aboard and had got trapped inside. All I could see when I looked down at the spacecraft was the wide, startled eyes of the swimmer, who was trying to wipe the condensation off the fogged inside surface of the tiny window. When I showed them how to lift the bar of the escape hatch from beneath the nose, they pulled him out, limp and not at all happy.

By the time I got below again, Neil had his suit off and was lying on the table with the doctor probing his arm for a vein to take a blood sample. The ship was rolling about and Neil glanced up at me, wide eyed, with a look which said, "Gee, I wish he'd get this over with."

There was nothing wrong with us. There was little the doctor had to do. Eventually we were allowed to tidy up and join the captain at his table for dinner. Boy, were we hungry. The meal was a large oriental supper. I'd have eaten just about anything we were given at that point. We eventually managed to get some sleep as the destroyer powered through the Pacific to get us into Okinawa. There we were met by Frank Borman and Wally Schirra, who were on a Far Eastern tour for NASA, and we were flown directly to a hospital in Hawaii for a more thorough check-up, then transferred to the Cape for a full debriefing. Only after that were we allowed to rejoin our families in Houston.

They had really been put through it. Lurton had been at Houston's Mission Control Center when our troubled transmissions via the *Coastal Sentry Quebec* began. At the first sign that the mission was not going according to plan she had been ushered out of the VIP lounge and driven home to await further news.

Jan Armstrong drove to our house so that both wives could be together. After they had both put our children to bed they had to endure a harrowing four-hour wait for news, while camera crews began to camp out in our front yard. NASA had released very little information. The normal commentary had gone almost silent. People figured we'd died; everyone wanted an exclusive.

Their ordeal ended at last when they got a call from one of the NASA doctors: "They're coming in. They're OK."

They then had to face the crowd of reporters clamoring for news outside.

"I'm a little disappointed, but the primary purpose of docking was accomplished. They'll fly again," Lurton said, mustering more confidence than she probably felt.

"I'm very happy they're home safely." Jan managed a smile. "The stars are very bright tonight."

Even *Life*, which protected our families from the media glare during those frantic hours, wanted to run a dramatic story: " 'Our

Wild Ride in Space,' by Neil and Dave" was what they wanted.

Neil put a stop to it. He called the publisher and said flatly, "You're not going to print that. We're not going to tell a story like that. It ain't going to be."

There was no way we could afford speculation about problems on the program. In no time there would have been congressmen saying, "Gee, those guys almost died. I'm not going to vote more money for NASA, 'cos if I do we might lose our boys and I don't want that on my record."

We couldn't have that. We had to keep things cool. *Life* obliged. It changed the tack of a series of articles it ran in three successive issues. Each was more upbeat than the last. An initial piece headlined "High Tension over the Astronauts" was accompanied by pictures of Jan Armstrong kneeling distraught in front of the television at home at the moment contact with our spacecraft was lost.

Then came a second story "Wild Spin in a Sky gone Berserk"— tempered by pictures of Neil and me smiling broadly as we bobbed in the Pacific shortly before being winched aboard the USS *Leonard Mason*. The last piece showed us both relaxing in armchairs back in Houston. The title of that piece was "A Case of Constructive Alarm."

Everybody was disappointed that we had had to terminate the mission early. But there was immense relief that we had got back safely—everyone knew how close we had come to not making it. The chances of recovering from our high rate of spin in space were very remote.

If we had not recovered, who knows what would have happened to America's space program? Quite likely, it would have been shut down, especially if our spacecraft had been lost so soon after the deaths of Charlie Bassett and Elliot See and the problems with Gemini 6.

Had they lost us they would never have known what happened. They would not have been able to obtain sufficient data to identify the problem as a thruster engine firing randomly on and off—an evaluation which led directly to a rewiring of subsequent spacecraft. All they would have known was that two guys had died in space.

There were plenty of smart guys at NASA, very good sleuths. Maybe they would have figured out that we had been lost tumbling endlessly in space. But it would have been a big-time mystery. The remains of Gemini 8 would have come down maybe nine or ten years later, badly burned through uncontrolled reentry. The press would have gone crazy. There would have been all these senators asking why we had been sent into space in the first place.

I believe it is quite possible the money from Congress for NASA would have dried up. That would have been the death-knell for the space program.

There was no big fanfare, no hoopla or parades, on our return. No dinner at the White House as there had been for the Mercury guys and the crew of Gemini 4—Jim McDivitt and Ed White, the first American to walk in space. We got the standard second-order awards. NASA gave me the Distinguished Service Medal, and the Air Force gave me the Distinguished Flying Cross (not the Distinguished Service Medal, which is a higher honor). I was promoted from major to lieutenant-colonel. As a civilian, Neil didn't even get that.

Not that we cared. Neither Neil, nor I, nor our wives were into celebrity stuff like parades. But there was some talk behind our backs, I later discovered, by those who thought we had screwed up, that we had unnecessarily turned on the reentry control system, which aborted the mission. Monday morning quarterbacks. They were dead wrong. There has never been any doubt in my mind that we did everything right. Otherwise we would not have survived.

As the program progressed and became more complex and demanding, personality clashes were beginning to develop. The competitiveness that had existed right from the start was turning into a more personal rivalry. Men who felt they should have been assigned to missions sooner resented those of us who got to fly before them. Though I personally felt little of this at the time, looking back I guess it was natural. We were all high achievers. Every one of us wanted to be the first man on the Moon. All I wanted to do once our Gemini 8 mission was over was get back into the thick of the space program as soon as possible. That was exactly what happened.

Over the next nine months a further four Gemini missions were flown, with varying degrees of success. Gene Cernan had problems performing a space walk on Gemini 9 when his face-plate fogged up while he hung on to the back of the spacecraft, preventing him from donning his backpack. Mike Collins finally got to perform a longer EVA on Gemini 10. But that mission experienced a problem with the environmental systems control when noxious fumes seeped into the spacecraft. Two months later Dick Gordon had yet more problems with his space walk while Gemini 11 was docked with its Agena; a lack of handholds prevented him maneuvering round the combined vehicles. For a while there were real concerns that an extended EVA was not going to work. On Gemini 12 the EVA requirements were simplified, allowing Buzz Aldrin to complete his space walk without a hitch.

By the time Buzz and Jim Lovell splashed down in the Atlantic on 15 November 1966, Gemini had run its course. All its assigned tasks were accomplished. The program had paved the way for Apollo to secure the ultimate prize: the first Moon landing.

By then I was already in training to grab that prize for myself. Once again Deke Slayton had approached me in a corridor with the words: "Dave, can I have a word?"

CHAPTER 7

Dark Side of the Moon

1966–7

Lt. Col. David Scott

Downey, California

Downey was still a small town in south central Los Angeles County when we arrived there in early summer 1966. Eventually, North American, which was building the Apollo spacecraft there, came to employ over 30,000 people in Downey, but in the early days this was small-time suburbia.

Together with Jim McDivitt and Rusty Schweickart, I had been assigned to the back-up crew of what was at that time referred to as Apollo 204 (it later became known as Apollo 1). This was to be the first manned Apollo flight, an Earth orbit mission which might last as long as sixteen days. It was scheduled for launch in February 1967. Prime crew were Gus Grissom, as commander, Ed White, as senior pilot and navigator, and Roger Chaffee, as pilot and systems engineer. I was Ed's back-up.

From the moment we were assigned, the six of us spent almost every waking hour together out at Downey. Most of the time we stayed at a motel round the corner from the plant. We pretty much lived with the spacecraft as it was being built and tested. There were even bunks at the plant so that we could sleep there; a lot of tests were conducted during the night. It was a twenty-

four-hours-a-day, seven-days-a-week operation. Apollo was behind schedule.

We did not always get back to Houston to see our families at the
weekends, because sometimes it took too long. A couple of times
we went to Las Vegas, which was much closer, and spent the days
sitting round the pool discussing the many problems emerging from
the testing of the first Apollo spacecraft. One evening we even got
to meet Frank Sinatra. He thought the space program was great.

"Hey you guys are doing a great job," he said when we were
taken backstage after one show. "Man, I'd like to be in there doing
it with you."

When we stayed in LA, we often went down to an athletics club
at Long Beach to play handball, eat pizza afterward and drink beer.
Or we'd take up the offer of a friend of the program who had a big
house up on Mulholland Drive, and made it available to us
whenever we wanted. We'd sit by the swimming pool in our sports
shirts and slacks poring over the intricacies of the Apollo program,
trying to figure out how the spacecraft was going to be made to
work.

Apollo was a very, very complex program. No one had ever
designed a vehicle to go to the Moon before. Sometimes there were
engineering paper strip sheets laid out across tables running for
twenty feet. There were many problems, so the budget was hit hard.
No one had anticipated how many problems there would be and
how much it would cost to fix them.

As part of the first Apollo mission, one of our jobs was to re-
write Apollo's procedures, troubleshoot problems when the systems
and subsystems didn't work. They had been written by different
sets of engineers from the various contractors, subcontractors and
NASA organizations responsible for separate sections of the Apollo
program. They all had different cultures, used different terminology
and even had different ways of making technical drawings. If the
procedures were to work, they had to be reconfigured by the men
who were going to fly the spacecraft. They had to operate in the
way they would operate in space—at exactly the right time and in
exactly the right sequence.

It was like taking a hundred-piece orchestra, in which none of the musicians had ever played together before, and trying to get them to play a symphony, straight off, without practice. When we arrived, the Apollo orchestra was seriously out of tune. There were some beautiful instrumental solos, but they were being drowned by too much percussion. Our job was to get those systems playing in harmony.

This harmonization was achieved by placing the spacecraft in a very elaborate test facility in which engineering consoles—test stations—were connected to each sub-system integrated to form a "whole" operational spacecraft. This set-up looked something like Mission Control, but with more consoles and monitors. A test was then conducted electrically by switches to simulate a particular phase of the mission. In theory, all the switches could be activated through the engineering consoles and the input and output of the switches could be monitored on the consoles after the signal had passed through the spacecraft.

But we learned early on that, unless a test was activated by switches inside the spacecraft, and in precise sequence according to written procedures, the signals could bypass a potential fault and the test become invalid. Needless to say, the procedures were very complex and often involved dozens of the 300 switches in the cockpit. Should a switch be activated at the wrong time, the entire system might go down. So our job was to make sure that all procedures worked in harmony and were compatible in the sequence used—both between subsystems and within the whole system of the spacecraft. We also had to sit in the spacecraft and operate the switches manually according to written procedures and following instructions by the test conductor in the control room.

This was a time-consuming and arduous process, especially on the first spacecraft to be manned. Should a switch, circuit breaker, subsystem or system not perform exactly as planned, a hold on the test would be called for the fault to be investigated and corrected. A hold might last for hours or even days, depending on the severity of the problem. The cause of the fault, the way it was investigated and the measures taken to correct it all then had to be carefully documented, reviewed and approved before the test could resume.

Kennedy's deadline of landing a man on the Moon by the end of the decade had imposed such pressure on the Apollo program that there was no time for successive redesigns of the spacecraft's hardware as the systems became more and more refined and integrated. The speed with which Apollo was being developed also meant that a lot of the design lessons learned during Gemini missions were appreciated too late to be incorporated into Apollo— more in terms of hardware than procedure. This was to contribute to a fatal flaw.

In Gemini, for instance, it was discovered it was much better to have boxes containing some hardware such as electrical wiring on the outside of the spacecraft. In Apollo a thick mesh of wiring lined the inside of the capsule. Also in Gemini we learned it was important to be able to open the hatch easily from the inside—this was done by activating latches which released it quickly outward. But Apollo was designed from very early on with a two-part hatch which opened inward and which was locked into place from outside to give it a better seal. This meant it was difficult to open the hatch from inside the capsule. It could be done, but it took over a minute at best and you had to be fit: the hatch was very heavy.

In Apollo the center-couch guy, the senior pilot and navigator, was responsible for removing the hatch. On Apollo 1 that was Ed White, and, as his back-up, me. Ed and I used to do some weightlifting by lying on the center couch and lifting that heavy metal hatch above our heads—the motion that was required to remove just the inner hatch.

There were a lot of problems, too, with the simulators— computer software representations of Apollo's hardware incorporated in a box-like unit designed to look like the inside of the spacecraft. In the early days of the program the main simulators in Houston and at the Cape were hardly working at all. The Command-Module simulator at the Cape was particularly troublesome. Several months before the Apollo 1 mission Gus Grissom had become so frustrated with its unreliability that he had hung a large lemon on it.

We all had concerns about Apollo. North American was less experienced than McDonnell-Douglas, and the Apollo spacecraft was a new and far more complex design than Gemini. Gus wrote a daily memo of concerns to Joe Shea, the Apollo Spacecraft Program Manager. Shea, who had to report up through the management chain to the top in Houston, had to make his decisions based on scheduling and budgetary constraints. Basically we had to accept the overall hardware design of the spacecraft as it was.

But all that changed on 27 January 1967.

Lt. Col. Alexei Leonov

Zvyozdny Gorodok, Moscow

Through the pages of *Life* magazine we followed the progress of the American space program quite closely. It was clear that each successive Gemini mission had taken the Americans closer to the first launch of their latest spacecraft—Apollo—and closer to the goal of a Moon landing.

Even though we had launched no manned missions in the twelve months following Korolev's death in January 1966, a series of missions by unmanned Luna probes had been completed. They had gone into lunar orbit and even landed on the Moon, providing us with vital information for our own planned manned lunar missions.

I was undergoing intensive training for a lunar mission by this time. In order to focus attention and resources our cosmonaut corps had been divided into two groups. One group, which included Yuri Gagarin and Vladimir Komarov, was training to fly our latest spacecraft—Soyuz—in Earth orbit. (Korolev had begun theoretical work on Soyuz as early as the 1950s, and construction of the spacecraft, a modified version of which still flies today, had begun several years before his death.)

The second group, of which I was commander, was training for circumlunar missions in a modified version of Soyuz known as the L-1, or Zond, and also for lunar-landing missions in another modified Soyuz known as the L-3. Vasily Mishin's cautious plan called for three circumlunar missions to be carried out with three different two-man crews, one of which would then be chosen to make the first lunar landing.

The initial plan was for me to command the first circumlunar mission, together with Oleg Makarov, in June or July 1967. We then expected to be able to accomplish the first Moon landing—ahead of the Americans—in September 1968.

Our plans for circumlunar and lunar-landing missions had much in common with the Apollo scheme. The main difference was that we planned for only one cosmonaut to take a landing module down to the surface of the Moon, while another remained in orbit. The reason for this was the limited lifting capacity of the new N-1 rocket booster, which Korolev had been heavily involved in designing to launch the L-3 spacecraft.

To train for the extreme difficulties of a lunar landing we undertook exhaustive practice in modified Mi-4 helicopters. The flight plan of a lunar-landing mission called for the landing module to separate from the main spacecraft at a very precise point in lunar orbit and then descend toward the surface of the Moon until it reached a height of 110 meters from the surface, where it would hover until a safe landing area could be identified. The cosmonaut in the landing module would then assume manual control of its descent. This would involve split-second decisions: he would have no more than three seconds to assess the landing site and enter its coordinates into the on-board computer of the landing module. If no suitable landing site could be identified, the cosmonaut would have to give the command to shift the landing module back into the orbit of the Command Module. For if the landing module touched down on the edge of a crater, for instance, it would become so destabilized that it would never be able to lift away for the journey home.

We practiced for these life-and-death decisions by switching the engines of the Mi-4 helicopters off at a height of 110 meters and bringing the craft down while continually changing its angle to regulate the speed at which its blades rotated. This was a highly risky—and normally strictly prohibited—maneuver.

A great deal of our time was also spent training in Star City with a giant 3,000-ton centrifuge machine. This was designed to allow cosmonauts to experience the immense pull of gravity to which their capsule would be subjected on reentering the Earth's atmosphere. The speed at which a spacecraft would travel in space upon returning from the Moon, for instance, would be 11.2 km per second, which would have to be reduced to 8 km per second in order for it to reenter the Earth's atmosphere for a safe landing. To lower the speed, the spacecraft would have to enter the Earth's atmosphere for a short period, then bounce off, leave it and reenter again. The key to this difficult maneuver was the angle of reentry. If it was wrong the spacecraft would be severely deformed, perhaps even destroyed, by the force of gravity.

Time spent in a simulator of the L-1 capsule, attached to an 18-meter-long arm of the centrifuge machine, allowed us to practice judging the reentry angle and safely experience the enormous forces of gravity to which we would be subjected if the angle was wrong. Several times during training sessions I sustained pressures of 14 Gs, the maximum to which a human being can be subjected on Earth. This put tremendous strains on every cell in my body and caused several hemorrhages at those points where my body was most severely compressed.

Only those of us who had already flown in space could really appreciate the demands and conditions of the missions for which we were training. We continued to make suggestions to the construction team on certain aspects of the design and layout of Soyuz and its lunar mission variants, although most of our suggestions had been made early on, when the craft were still on the drawing board and models of it were being assembled.

Further training involved learning how to fly the spacecraft manually—all our spacecraft had both manual and automatic systems of control. According to the flight plan the automatic system took precedence, but we had to be able to perform every aspect of the flight manually in case the automatic system failed—as it had during our reentry into the Earth's atmosphere aboard Voskhod 2.

After that experience I had argued that, as commander of a spacecraft, what I needed once a flight was in progress was as little communication as possible from the ground—since it served mainly to distract me from what I already knew was necessary—and only manual, not automatic, control.

"Under those conditions," I said, "you can show me any star and I will be able to land on it."

One of the most crucial aspects of manual control for which we had to train was the ability to correct the orientation of the spacecraft at four critical phases of a mission. They would have to be performed according to the position of the stars with the help of a very refined sextant and a stellar-orientation device.

Training to use them accurately involved many hours spent at Moscow's planetarium. But this only allowed us to study the sky over the northern hemisphere. To study the stars over the southern hemisphere we had to fly to Mogadishu, Somalia. There I organized two expeditions out into the desert, where we took many measurements with sextants of the southern hemisphere stars looking south toward the Antarctic during the cool of the night.

David Scott

Several weeks before the scheduled launch of Apollo 1, Gus, Ed and Roger transferred to the Cape, together with their spacecraft, while Jim, Rusty and I stayed in Downey. By then we had been reassigned as prime crew for another more complex mission and another Apollo spacecraft, modified with a lot more computer capability and improved guidance and navigation systems.

On 27 January 1967, Rusty and I were going through some tests in a new hybrid simulator at Downey for the modified Apollo when Jim stuck his head through the curtains. He'd just received an urgent call from the Cape.

"We lost the crew," he said, ashen-faced. "We lost Gus and the crew."

We had no idea what he was talking about. Rusty and I said pretty much the same thing: "What do you mean?"

"There's been a fire at the Cape," said Jim. "They're all dead."

As we clambered out of the simulator, Jim told us all that he knew. There'd been a fire on the pad. The guys couldn't get out. All three had been killed. We were in shock. How could that be? We wouldn't know until we got back to Houston. We were ordered to get back there right away.

Rusty and I got a T-38 trainer jet organized, and flew back to

Houston that night. We had to stop to refuel in El Paso. It seemed a very long flight. Our minds spun with questions. We had both lost friends in airplane disasters before, but this had happened on the ground. What could possibly have gone wrong? How could we have lost these guys on the pad? It was a total mystery. It didn't seem real.

By then it was on the national news. All the air-traffic controllers knew that there had been a fire and the crew had been killed. Normally communications with air-traffic control was very terse and correct, but this time it was different. Since joining the astronaut corps, whenever we flew we used a NASA call sign. So the guys on the ground knew who we were. They wanted to show their sympathy. All along the route we got messages from air-traffic controllers like "Sorry to hear about what happened" and "Good luck to you guys." Their comments seemed to reflect how just about everyone across the country felt that night.

It was after midnight by the time we got to Houston. We went straight home to be with our wives and families. It was too late to go and see any of the families of those we'd lost. Roger Chaffee had lived just down the street, Ed White over in El Lago, Gus in Timber Cove. We knew the Chaffees real well, Ed White's family, too. My children were too young to understand what had happened. But we all went to pay our respects to the families the next day. Then we had to get ready for the funerals.

We were all instructed to wear uniform for the ceremony. We hadn't worn them for a long time, so we had to go scouting around for all the right pieces of clothing and insignia before making our way up to Washington, to Arlington, again, where Roger Chaffee and Gus Grissom were buried. Ed White was buried at West Point, as he would have wished, though unfortunately my work schedule did not allow me to attend his funeral.

Most previous trips by astronauts to Washington had been after successful missions and been light, celebratory, joyous occasions. But on that cold, wet day, 31 January, the mood was very somber, tragic. It was a moving ceremony. President Lyndon Johnson was there. A eulogy was read. The rifle salute was fired. "Taps" was played—that slow, melancholy bugle music written for the

Confederate Army to play every evening during the Civil War. Then the coffins were lowered into the ground. The flags with which they had been draped were folded and handed by the president to the wives.

The ceremony over, we went back to Houston. Everyone began asking themselves the same questions. Now what? What do we do now? Where do we go from here?

Alexei Leonov

The tragedy of Apollo 1 was widely reported in the Soviet Union, and our government sent condolences to the families of the men who had died. The cosmonaut corps also expressed its sympathy by sending letters to the families. Although we did not know any of the men personally, we felt an affinity with them.

The Apollo accident also made us analyze our systems again very carefully. Mishin issued a direct order that every aspect of our spacecraft be reviewed to minimize the risk of fire. Extra insulation was added to exposed wiring, for instance, and television lamps which caused too much heat were replaced.

We discussed the Apollo 1 fire among ourselves a great deal. From a professional point of view, I viewed the deaths of the three America astronauts as a sacrifice which would later save the lives of others. But I was also very angry at how stubborn the American engineers were in continuing to use a pure oxygen atmosphere in their spacecraft. I couldn't understand why they had not switched to the system we adopted after the death of Valentin Bondarenko: regenerating oxygen during a flight.

The Americans must have known of the tragedy that had befallen Bondarenko in a pure oxygen environment. He had been given a big funeral, and the American intelligence services would not have been doing their job properly if they had not informed NASA about what had happened.

We followed the speculation in the American press about the possibility that their space program would be canceled as a result of the Apollo 1 fire. But I did not believe there was any real danger of that happening. A great deal of money had already been spent on their program and the Americans knew how to count their money. Besides, they were very well aware that we were putting all our energy into our own space program. The prospect that a Russian would be the first man on the Moon would have been like a red rag to a bull for the United States.

Then we suffered our own tragedy.

David Scott

President Johnson ordered NASA to conduct an internal investigation into the fire. Again and again investigators listened to a tape of the final seconds of transmissions from the Apollo Command Module, made shortly after 6:30 p.m. on that fateful evening of 27 January.

It was the voice of Roger Chaffee. It had been his job to maintain contact with controllers at the Cape in case of an emergency. His voice was clear and clipped at first: "Fire!"

Seconds later, with more urgency: "We've got a fire in the cockpit!"

His final words were barely intelligible: "We've got a bad fire . . . We're burning up!"

When rescuers were eventually able to prise the hatch from the scorched Command Module, all three men were found to have asphyxiated. Investigators concluded that a spark, probably from worn and damaged wiring in the equipment bay, had ignited flammable material in the capsule. The pure oxygen environment, whose pressure was slightly above that at sea level, had quickly turned the inside of the module into an inferno. In the extreme heat the hatch had become impossible to open, especially with the building internal pressure.

In the wake of the tragedy, NASA's accident investigation board set up various technical committees to look at different aspects of Apollo and to see where things had gone wrong and how the spacecraft could be improved. Each of us in the astronaut corps was taken away from his previous job and assigned to one or more of these committees. I was assigned to the team looking at redesigning the docking system and a new hatch—which opened outward. I also helped to design the entry monitoring system, a visual display to enable the astronaut piloting Apollo as it reentered the Earth's atmosphere to fly a manual reentry.

After the fire, Apollo's board was wiped clean. The music stopped. It was time to get a new orchestra.

There was a change of management, as there was when any serious accident occurred. That was just the way it was. Joe Shea was reassigned, and all senior managers at Downey were reassigned or relieved of their duties. The engineers took it very hard. They felt what happened had been their fault. It was pretty depressing. But the new team who were brought in were good.

One of the key figures in NASA management, who had played a major role before the fire, but whose contribution after it was absolutely crucial in making sure that all parts of the Apollo orchestra worked in closer harmony, was Bill Tindall. Tindall is one of the unsung heroes of the space program, in my view. His major contribution was to introduce a brilliant forum in which information and ideas could be freely and directly exchanged between all those working on different aspects of the program, from the astronauts to engineers, to managers and technicians. "Data priority meetings" were held in a large lecture theater at the MSC almost every Friday. I never missed them if I could help it.

Tindall's unique ability to get people talking and communicating clearly and openly became very important in integrating work on both hardware and software right across the board. His meetings were always lively discussions, where firm decisions and commitments were made on the spot. Bill Tindall's succinct way of summarizing the outcome of the meetings, in what were known as "Tindallgrams," became legendary. The summaries were always clear, informative and succinct, but their tone was often chatty and amusing.

At the end of many Tindallgrams were the phrases "at least that's the way I see it" or "let's vote," which smoothed the way for anyone who wanted to offer a differing opinion or add information which might have been missed. Tindall was also unafraid of lightening the atmosphere a little by poking gentle fun at other people and himself.

He concluded one invitation to a private contractor to visit Houston, to study Apollo's reentry and retro-fire programs: "I hope you get a chance to review it and come down for the picnic. Please bring your own ants." And after a meeting about the problems with Apollo's computer programs, he confessed, "I don't want to minimize the seriousness of this situation. We are in deep serious yoghurt!" Another time, after a meeting on the Lunar Module, he remarked, "I sure will be glad to see that bird fly before it drives us all over the brink."

No matter how light-hearted the tone of these summaries, their underlying message was completely earnest. The fire on the pad meant the whole system was totally reviewed. Without the fire, that would never have happened. Without it, Apollo's many deficiencies would never have been addressed in time to meet Kennedy's deadline. Without it we would very likely have lost a crew during an Apollo mission itself.

Alexei Leonov

By April 1967 we were ready to launch the first manned Soyuz mission. The man chosen for it was Vladimir Komarov, who thus became the first cosmonaut to make a second spaceflight. I knew him very well. He was ten years older than I was, but we had worked together on Voskhod. Unlike me, he was always very serious. He was a first-class test pilot. Everyone understood that this first flight was a high-risk mission. Soyuz was capable of carrying three men. But, because of the risks, Komarov flew alone.

Although there had already been a number of unmanned Soyuz missions, the plan, if everything went well, was for a second Soyuz to be launched the day after Komarov with a three-man crew aboard. The two vehicles would dock—a vital precursor to future lunar missions—and two of the cosmonauts from the second Soyuz would transfer to Soyuz 1 by space walking and return to Earth with Komarov, leaving the second Soyuz to return the same day with just one cosmonaut left on board.

Together with a group of cosmonauts, engineers and designers, I traveled to Baikonur with Komarov for this important flight. Right up to the day before he was due to fly, Komarov was intently reading all the details for his mission. The night before the launch on April 23, we all had dinner together. By this time a new hotel had been built for cosmonauts at Baikonur, and Komarov stayed there that night—not in Gagarin's small house as I had done before the flight of Voskhod 2.

I spent most of that night standing outside the window of his hotel bedroom. I was determined to direct any traffic away from the street outside, so that Komarov could have the best night's sleep possible. The next morning we all took the bus together to the launchpad.

Before any launch a governmental committee comes from Moscow to certify that a vehicle is fit to fly. Since this was Soyuz's maiden manned flight, the committee was extremely high-level. It was led by Leonid Smirnov, then head of the military–industrial complex, and included, of course, Mishin, as chief designer, and Boris Chertok, chief engineer, together with many other engineers and designers. Yuri Gagarin and I were also on the committee.

Once the committee has certified the vehicle is fit to fly, its members remain at Baikonur for the first hours of the mission, after which they return to Moscow to follow the progress of the flight from there. But the senior members of this committee did not return to Moscow as planned.

The launch went perfectly. But within a few hours a major problem developed: one of the spacecraft's two solar power panels failed to deploy, which meant that the capsule was not receiving sufficient power for its guidance computers. This was serious. As soon as it became apparent the mission was running into trouble, an emergency committee was convened, under Chertok's command, to try to work out a solution. I also sat on this committee.

It was soon obvious that the problem could not be solved. Once this was clear we recommended that the flight be terminated early— a little over twenty-four hours into its mission—and that the second Soyuz spacecraft not be launched.

As a cosmonaut and pilot, I appreciated there were advantages in realizing that Soyuz had problems on its maiden flight, because we could fix the problems right at the beginning of the program. It would have been worse had such problems occurred further down the line.

Once the spacecraft began its reentry, we lost contact with Komarov. This is normal. It is only after the landing capsule's main parachute, which contains an antenna, opens that contact can be re-established. But we never did regain contact.

There were serious problems with the capsule's parachutes. The brake chute deployed as planned and so did the drag chute, but the latter failed to pull the main canopy out of its container. While the reserve chute was then triggered it became entangled with the cords of the drag chute and also failed to open.

Before Komarov's flight an unmanned Soyuz test mission with controlled automatic landing had been carried out. The test had passed off smoothly, except for one problem, which would have been avoided anyway during a manned mission: the landing capsule had depressurized because the bottom of the spacecraft burned out. But its parachutes had deployed and the capsule had landed on the frozen surface of the Aral Sea. The cause of the burning was discovered and, since the manned spacecraft had a slightly different design in that aspect, there was no chance of the fault recurring.

As soon as contact with Soyuz 1 was lost, rescue forces were deployed to the area where it was expected to have made an emergency landing, near Orsk, at the tip of the southern Ural Mountains, along Russia's border with Kazakhstan. Shortly afterward I joined a team which left from Baikonur for the site.

When we arrived, we found little more than a lump of crumpled metal. Komarov's 2-meter-high landing capsule had been reduced to a tangled mess little more than 70 cm high. His spacecraft had landed at a speed of 25 meters per second, killing him instantly, and then a fire had started. Lying on the ground near the wreckage were three parachutes. We realized immediately that they had not opened properly.

The cause of the malfunction, it was later discovered, was that the parachute container had opened at a height of 11,000 meters and had become deformed, squeezing the canopy of the main parachute and preventing it from opening. The conclusion of subsequent tests was that the parachute container was not rigid enough. For future missions the container was strengthened and meticulously polished on the inside, so that there would be no question of a similar problem occurring. Every detail was fixed and made reliable. But the life of an outstanding man had already been lost.

Komarov's death had a very profound effect on the morale of the cosmonaut corps. He was our friend. Now he was gone and we had not even been able to recognize his body. Like Korolev's, Komarov's remains were interred in the Kremlin Wall.

Before his death the Soviet press and public had paid little attention to the extreme risks we took. Spaceflight had seemed easy. All the marches, parades, grand music and medals in honor of cosmonauts had made the space program seem like an elaborate exercise. Now people realized that being a cosmonaut did not necessarily lead to fame and public acclaim: it could also lead to death. The public started to appreciate the real dangers involved, as, I believe, the American public did after the Apollo 1 fire. More resources were made available to our space program, although they never matched the huge amount of money available to NASA.

It was quite clear that no future flights could go ahead until all systems on the spacecraft had been completely reviewed. A very high-level governmental decision was taken to carry out a series of tests. This meant that there was no way the original schedule for future spaceflights would be kept.

At the time I thought it would be maybe two years before the next Soyuz spacecraft could be launched. In fact, it was eighteen months.

David Scott

After the Apollo 1 fire there was a real awakening about how tough the goal of a lunar landing was going to be. There were moments when the whole space program was called into question, particularly by politicians. There was more than a small chance that they would close the program down. It was touch and go.

"Why are we involved in this Moon stuff, anyway?" many were asking. "There are a lot of better ways to spend this sort of money. Why are we spending all this money just to get guys killed? Why go to the Moon at all? Just because Kennedy said so? Well, he's gone now."

Hearings into what had gone wrong were held both in the House of Representatives and in the Senate on Capitol Hill. Frank Borman was sent to Washington to represent the astronauts, and did a superb job of explaining how we all felt.

"We are confident in our management and in our engineering and in ourselves," he told the House committee. "I think the question is really: are you confident in us?"

After a long debate the answer was "Yes." Whatever budgetary constraints had previously existed on the space program were loosened. There was more appreciation of the risks involved and the cost of reducing those risks and solving all the problems the program faced. After the fire, a great deal more attention was also paid to suggestions made by the astronaut corps.

But a lot of work was needed in rebuilding morale. For those based mainly at Downey, baseball played a big part in raising our spirits. Pete Conrad and Dick Gordon both loved playing; Dick had once done some semiprofessional pitching, and Pete did some catching. He and Pete were really good friends. One night over beer Gordon decided to challenge North American to a game. He would captain the astronauts' team, and one of NASA's bright new managers, a guy by the name of Joe Cuzzopolli, would captain North American.

A baseball team has nine players, and Cuzzupolli could pick his team from the company's 30,000 or so North American employees at Downey, who included a couple of former professional players. There were only nine astronauts at Downey working on Apollo 9 at the time. The odds were a little against us.

But everyone got into the spirit of it. At lunchtime people at the plant went out and practiced. Shoot, I'd played only a handful of games in my life; I'd played a lot of softball, but not baseball. So I put in a fair amount of practice.

We had a nice, warm spring evening for the game. Some of the secretaries got together as a cheerleading team, and dressed up in little costumes. The North American team killed us. They really killed us. But it was all a lot of fun, and we followed it up with a big party. It lightened the atmosphere at Downey at this very difficult time. Instead of talking about the problems of the program, the deaths and tragedy, everyone talked for a few weeks about little other than The Game. Morale-building exercises like this made a major contribution to the ultimate success of the program. Part of the healing process was about healing attitudes.

These were the first steps out of the valley. They took us away from Depression City, where the sky was gray and the walls were gray, where it seemed there was nothing to look forward to. Downey had been in the deepest part of the valley until then. People there felt personally responsible for the deaths. But the game had been a way of showing that you can't dwell on tragedy for too long. You have to respect it, but you have to put it behind you and move on with life.

Alexei Leonov

In early May 1967, I was invited to Paris by the French government for three weeks. This was shortly after President Charles de Gaulle had taken France out of NATO's integrated command. The previous summer he had visited the Soviet Union. Relations between our two countries were warm and, separated from NATO, France was looking for new allies.

I stayed in a sumptuous room at the Hotel George V, where there always seemed to be someone on hand to light your cigarette. My schedule was hectic. I had to give four or five presentations about our space program each day, before being taken on a tour of the South of France, to Marseilles, and to Bordeaux, Toulouse and Rheims.

The film of Boris Pasternak's wonderful novel *Doctor Zhivago* was being shown in Paris at that time. The book was still banned in the Soviet Union, but I loved the film (particularly the music), which showed the tragedy of the Russian intelligentsia. I was a guest of honor at a party thrown after a showing organized by the French magazine *Paris Match*. It was a warm spring evening when we all entered the theater to watch the film. When we came out it was snowing.

Everyone was shocked. We thought the weather had gone mad. Then I held my hand from underneath an umbrella and realized that it was artificial snow being blown across the entrance. The decorations inside and outside the theater were quite beautiful. A horse-drawn sleigh on small wheels stood at the entrance with actors dressed up in fur coats. It was very atmospheric, a wonderful evening.

There were also big festivities going on to celebrate the work of Jules Verne. There was a huge gathering in one of the city's largest halls, which had been decorated to represent scenes from his many novels. All the guests were supposed to be between the ages of sixteen and twenty-four, but again I was guest of honor. In the center of the hall stood a model of a large cannon, which Jules Verne had imagined could be used to fire a man to the Moon. I was asked to give a speech about Soviet plans for a lunar landing. The young people there asked endless questions. They were full of enthusiasm about space travel. It was all so new and exciting.

Shortly after I returned to Moscow, in early summer 1967, the two new apartment blocks being built for us at Star City were finished. Each of us had had his own projects linked to facilities at Star City, with which we were particularly involved. Yuri Gagarin had overseen the construction of the swimming pool. I had overseen the new training center, the hydro-laboratories and the planetarium.

We had also continued to be involved in the design of the residential blocks. When we were at last able to move in, the apartments were wonderful. Ours was 86 square meters in total. By many standards it was quite modest, but we were thrilled: we had never lived in such luxury.

A huge party was held to celebrate our move. Valentina Tereshkova and I wore our white dress uniforms and stood at the entrance to the two blocks welcoming our guests with bread and salt, as is the Russian custom.

There were four apartments on each building's eleven floors. Each floor had further tables laid with bread, salt and wine. Our wives dressed in their finest clothes to welcome guests. My wife, Svetlana, had prepared a wonderful meal with fine French wine I had brought back from Paris.

Our second daughter, Oksana, was the first baby to be born in Star City, on 15 June that year. It was a happy omen.

David Scott

In late May 1967, Mike Collins and I were asked to go to Paris to represent NASA at the huge, biannual air show. The idea was to show the world that we were back on track after the fire. Both NASA and the Soviets had big pavilions; each was trying to impress the world with its space capabilities. This was the first time, for instance, that the Soviets unveiled their Vostock booster.

It rained the whole week we were there. We weren't used to the cold weather and had to go and buy overcoats. But our wives came with us and we had a great time. We had dinner with the ambassador and also met Charles de Gaulle and his big entourage when they came by the NASA pavilion. We were flown down to the South of France in military transport for an evening of wine tasting in a beautiful château.

Most significant of all, this was the first time I met some Soviet cosmonauts.

We had been told not to meet them—NASA officials said they would embarrass us. But somehow, I guess through contacts between our two embassies, we let it be known that we would like to meet the cosmonauts. We had been fighter pilots, after all, though on opposite sides of the Iron Curtain, and fighter pilots are always greatly curious about each other. They want to know how the other guy does things, even if, or especially if, they are fighting on different sides.

Someone told us that if we stopped by the Russian pavilion at a certain time the cosmonauts would probably be there. This wasn't a formal meeting, we told ourselves. It was just Scott and Collins going by the pavilion at a time the cosmonauts happened to be there. No big deal. It would be OK, we reckoned. So, with our wives, we went by at the appointed time and, by golly, there they were.

Colonel Pavel Belyayev in full military uniform, and Konstantin Feoktistov in civilian clothes. They weren't alone. The world's press seemed to be there, too. They must have been tipped off that we were going to meet. I've never seen so many cameras in my life. They were everywhere.

Belyayev and Feoktistov came out to greet us, all smiles, with their interpreters. Then the press really moved in. There was a lot of jostling, pushing and shoving.

At that point Belyayev took charge. "Everyone out of the way," he said. "We're leaving." He seemed worried about our wives getting hurt.

He was really gracious. He walked us away from the mob to a Russian TU-104 jet parked near the pavilion. Vodka and caviar had been laid on for us on tables set up on either side of the aisle at the front of the plane. I was seated next to Belyayev on one side of the aisle; Mike was with Feoktistov at the table opposite. Our wives sat together with an interpreter by their side.

One of Belyayev's first questions was about how Pete Conrad and Gordon Cooper were doing. His meeting with them in Athens the summer before seemed to have made quite an impression on him. Belyayev was a very positive, thoughtful guy, a real leader; I liked him a lot. Feoktistov was much quieter. He had held back while we were still in the crowd and even in this quieter environment was more evasive, less willing to talk than Belyayev. Mike later nicknamed him "Feo the Fink" because, while the rest of us threw back shots of vodka, he kept watering his down with soda.

We all had quite a few laughs about our mutual dislike of doctors on our respective programs. When asked what he thought of them, Belyayev reached across the aisle and pretended to shoot Mike's hand. We got the impression the Russians were tied up with a lot more medical tests after their flights than we were. We agreed, too, that ground control talked too much during missions.

When we asked Belyayev how much time the cosmonauts had to

spend making speeches, studying or flying their simulators, he joked that he spent most of his time hunting and fishing. It was the only time Feoktistov showed a sense of humor, indicating that Belyayev had a bruise on his arm from the number of times he marked the length of the fish he caught. We invited Belyayev to the Cape to fish. He acknowledged the invitation but was noncommittal.

Then the mood turned more somber. We drank a toast to the hope that there would be no more accidents on either of our programs. Mike and I knew that a few weeks earlier a cosmonaut named Komarov had lost his life. We did not know how, and did not ask. But when Belyayev inquired how the wives of the lost Apollo 1 crew were doing, we asked how Komarov's widow was. Not too well, Belyayev indicated. She visited his grave, he said, every day.

As the meeting drew to a close we invited the Russians to pay a return visit to our pavilion, which they did two days later. Belyayev couldn't wait to clamber inside the Apollo spacecraft on display. He seemed very knowledgeable about how it worked. But why shouldn't he be? We published details of our program quite openly. All our discussions were cordial, though both sides were careful not to seem to be trying to obtain too much technical information.

This was the first time I realized that on their earliest missions the Russians were catapulted out of their spacecraft before they touched down on Earth. We knew their craft were designed to touch down on land, rather than splash down in the ocean as ours were, but I had not realized that before that they returned to Earth on their own personal parachutes. The Russians seemed very interested in how much control we exercised over our craft during a mission, particularly during reentry. Belyayev mentioned that their craft were designed for automatic reentry. He then muttered something about being the first cosmonaut to bring his mission back to Earth on manual control in some sort of emergency landing. I gave this revelation little thought at the time.

It was rather like when you go to visit the other guys' practice field for the first time, and neither side talks about the plays their opponents might use that Saturday. We were all more intent on establishing a good rapport. We ended with a toast to more cooperation between the United States and the Soviet Union in space.

For us to be able to sit down in a casual, private environment like this was a very big deal. I had never expected to meet a cosmonaut, yet we had had the chance to sit down and really talk turkey. I didn't feel any animosity. Any feelings of rivalry were subsumed by our mutual interest in what the other guys were up to.

It was as if we were all members of an elite club. Being a member of that club dominated all other considerations. It subsumed politics. It rose above the bitter fray of the Cold War.

CHAPTER 8

Did You See God?

1968–9

Lt. Col. David Scott

Manned Spacecraft Center, Houston, Texas

Nineteen sixty-eight was a pivotal year in the space programs of both the United States and the Soviet Union. The two most powerful countries on Earth were pressing to demonstrate to the human race that their ideology was best, by being the first to reach the Moon. Who would win? That was the burning issue at the start of the year.

It was a lousy year in a lot of respects for the United States. The Vietnam war was really taking its toll. It was going badly, and the Vietcong had scored a major victory with the Tet offensive in January. The assassinations, later that year, of Martin Luther King and Bobby Kennedy came as a tremendous shock too. These were tragic events. Behind the Iron Curtain the situation seemed to be deteriorating, too. A crackdown on attempts to introduce liberal reforms in Czechoslovakia had led to the restoration of hard-line communism. But, bad as the news was both at home and abroad, most of us in the astronaut corps remained largely isolated from such events. We were in a tunnel. Totally focused.

Both American and Soviet space programs had suffered severe setbacks. Ours had been delayed at least a year by the Apollo 1 fire. All our energy was devoted to trying to solve the problems we were facing—and there were many. We were really going full tilt, working fourteen or fifteen hours a day. Everything had been torn apart and rebuilt. Even the names of the Apollo flights were changed because of the fire. Each flight in sequence would be assigned a mission type or purpose.

Before the fire, each flight had simply had the number of the Saturn launch vehicle or the Command Module the crew was working on. Afterward, it was decided that the lost crew had been the first Apollo flight. The next three flights were due to be unmanned tests of the new Saturn V booster rocket with unmanned Command and Service Modules and the smaller Saturn-IB booster rockets with unmanned Lunar Modules. The Saturns, the brainchild of Wernher von Braun, would be numbered Apollos 4 to 6. The first manned flight would be Apollo 7.

Furthermore, the new Apollo plan laid out a sequence of escalating "missions," designated from A to J to mark its type or purpose. A and B were unmanned missions. C marked the first manned mission of the Command Module in Earth orbit. The D and E missions were to fly the Lunar Module in Earth orbit, while F would mark the first mission to the Moon and G would attempt the first lunar landing. The H and J missions were to extend the length of time a crew spent on the Moon's surface and greatly expand the scientific work they carried out while there. The I missions were to have been orbital "science missions," but were canceled.

The whole program was incremental. The objectives of each type of lettered "mission" had to be accomplished, no matter how many "flights" that took, before the program moved up to the next letter or level of complexity. This meant that it would not be clear until very shortly before the first attempted lunar landing who would be the first man to set foot on the Moon.

Amid all the turmoil that followed the fire, word went out one afternoon in April 1967 that Deke Slayton wanted everyone currently active in the astronaut corps to come to a meeting the next morning. Such meetings were usually called by Al Shepard, chief of the Astronaut Office. The fact that it was Deke who had scheduled this one meant something out of the ordinary was happening, though none of us had any idea what it was.

As the eighteen of us who were not off traveling elsewhere filed into the small conference room in Building 4 at MSC the next morning, there was a real buzz of anticipation in the air. Most of us had on the formal shirts and ties we usually wore in Houston, though a few of the guys involved in simulator training later that day were looking a little more casual in sports shirts and Ban-Lons. But there was little else casual about our demeanor. We joked and chatted as we took seats round a long table or stood leaning against the walls, where a few mission charts left over from a previous meeting were pinned. But each one of us stiffened perceptibly and the room became hushed pretty much as soon as Deke walked in.

After a short preamble he came straight to the point. "The guys who are going to fly the first lunar missions are right here in this room," he said. "And the first lunar landing crew is looking at me right now."

The significance of that was clear right away: no astronauts recruited after the first three groups would get ahead of us in the line of those scheduled to make lunar landings. The selection of nineteen new astronauts had been announced earlier that month, but they were not to be included in these early plans (many of them, including Al Worden and Jim Irwin, did eventually fly lunar missions).

From that moment on we knew we were firmly in the loop. That was one of the most exciting, if briefest, meetings I'd ever been at. After Deke left the room we spilled out into the corridor with broad smiles on our faces. I felt I was in pretty good shape to be among those who would one day step out on to the surface of the Moon.

"It's going to be interesting," I said, turning to Rusty Schweickart, "to see how many missions we'll need before the first lunar landing."

It wasn't clear then which of us would attempt the first G—lunar-landing mission—but each of us was assigned to either the prime or back-up crews for the C, D, E and F missions. Even before the Apollo 1 fire, Jim McDivitt, Rusty and I had been switched from back-up on that mission to prime crew for the first D mission due to perform a rendezvous and docking between Apollo's Lunar Module and Command Module in Earth orbit. Originally this was to be the Apollo 8 mission. It later became Apollo 9—a highly significant switch.

The original schedule for each mission, around the time of the meeting with Deke, called for Apollo 7 to launch in October 1968, with Apollo 8 flying in December 1968 and Apollo 9 the following spring. All would be Earth-orbit missions. The first circumlunar flight was not scheduled until late spring 1969. If successful, this would still allow a lunar landing—in time to beat Kennedy's deadline—by the end of that year.

But by the summer of 1968, even before Apollo 7 launched, beating Kennedy's deadline had become secondary. The real deadline was to beat the Russians; not just to a lunar landing but to the first manned flight round the Moon.

Both countries were entering the last lap in the race to the Moon. The stakes could hardly have been higher. At this stage the outcome was still very uncertain. Only a select few in senior management at NASA—those with access to CIA files—knew just how close the Russians were to the finishing tape.

George Low, the Apollo spacecraft program manager at NASA in Houston, apparently had access to the intelligence files. He very likely knew the Russians were preparing for a series of unmanned Zond missions round the Moon as a precursor to a manned circumlunar flight in late 1968—six months before an Apollo crew was due to fly round the Moon.

The political implications of the Russians making this first circumlunar flight were unthinkable. Little was known about the difficulties the Soviet space program was facing. All that was clear was that we had to speed things up. Low did just that. A major change to our program was about to be made and it directly affected me.

Colonel Alexei Leonov

Zvyozdny Gorodok, Moscow

Shortly after Yuri Gagarin returned from his first spaceflight a reception was held in his honor, and the head of the Russian

Orthodox Church, Alexis I, was present. "When you were in space," he asked Yuri, "did you see God?" Yuri said he had not. "Please, my son," Alexis replied, "keep that to yourself." A little later Nikita Khrushchev posed the same question. Out of respect for Alexis I, this time Yuri said that he had. "Dear Yuri," Khrushchev entreated, "please don't say a word about that to anyone."

Yuri told me this story with a big grin, his large blue eyes lighting up his broad face. It is just one of my many happy memories of the man who became my dearest friend. We took many holidays together. One was in the midst of my intensive training for the Voskhod 2 mission. Yuri knew how hard I was working and suggested we take a little time off with our wives so that I could relax. He was very thoughtful like that.

It was August 1964 and we took a ten-day sailing trip along the River Moskva on a boat named *Friendship*, which had been given to Yuri by the Danish government after his historic flight. We agreed that the only provisions we would take with us were tea, salt and sugar. We would find our own food along the way. So we spent a lot of time fishing, duck shooting and hunting for mushrooms.

The Moskva and Volga rivers are connected by a network of canals and at one point, while we were negotiating a lock, we sailed right up behind a much larger boat, whose captain immediately recognized Yuri. Tossing a ladder over the side, he invited Yuri to climb on board to meet his passengers. This left me in charge of navigating our boat through the lock, something I'd never done before. We soon ran into trouble.

As Yuri stood on the deck of the other boat, receiving the congratulations of its passengers—he was like a god to many people at that time—I took the helm of *Friendship*. This left Svetlana and Yuri's wife, Valentina, struggling to secure the boat to the sides of the lock with mooring ropes. But they had no experience of doing it, and our boat began knocking violently against the side of the lock, slightly damaging her hull.

Earlier that day we had bought some honey from villagers along the banks of the river. With nowhere to store it, we'd poured it into a frying pan, which we had left on deck. As the boat hit the lock wall, the pan of honey slid to one side, spilling its sticky contents.

I saw Svetlana and Valya slipping about, so I let go of the wheel and tried to help them.

With no one at the helm, the boat started rocking even more violently. I lost my balance and fell overboard. Wondering what the commotion was, Yuri and the passengers on the neighboring boat peered over the railings and saw me flailing in the water. They started waving and roaring with laughter. The captain of one of the boats behind us then started shouting for us to get a move on. We were creating a logjam in the lock, he said. This made Yuri and the passengers in front laugh even louder.

When Yuri eventually climbed back down to join us, I apologized for damaging his boat. But he was so happy with the reception he had received, and so amused by our predicament, that he just slapped me on the back and carried on laughing.

These were the most innocent and happy of times during the early years of the space program. Although we had mourned the death of Bondarenko, we had not suffered the tragedy of losing one of our cosmonauts during a spaceflight. We were riding high. We were ahead of the Americans in the space race. We had accomplished so many firsts that nothing could shake our confidence.

Then Korolev had died. Then Komarov had crashed to Earth in Soyuz 1 and been killed. Under Mishin, the program was slipping behind schedule.

The Americans, we knew, were also suffering. The deaths of the Apollo 1 crew had dealt a heavy blow to the astronaut corps and the US space program. Then came the assassinations of Martin Luther King and Bobby Kennedy. We could not understand how Bobby Kennedy could have been killed so soon after the assassination of his brother. Our country seemed very calm and stable by comparison. Stalin's Terror was over. We felt safe, secure.

David Scott

The first time I got a good look at the Saturn V and Saturn-IB rockets that would propel Apollo into space was during their assembly at the Kennedy Space Center. Early on in the program I had been to the Marshall Space Flight Center in Huntsville, Alabama, where Wernher von Braun and his rocket team were working on the Saturn V, and witnessed the test firing of one or two of its engines. Even before that, while still at test-pilot school in the early 1960s, I had felt the powerful vibrations of those engines, built by the Rocketdyne Company of Los Angeles, being tested at a test site miles away from Edwards Air Force Base.

But now they were fully assembled I could hardly believe their size. The Saturn V stood over 350 feet tall. Its weight once fueled, plus Apollo, was 6 million pounds—the thrust of the rocket was 7.5 million pounds. My initial reaction was almost "Something this huge can't possibly work," even though I had every faith it would.

The first test of an unmanned Apollo atop a Saturn V, in November 1967, had shaken the world—literally. Vibrations from the ignition of 135 tons of propellant, as the rocket and Apollo module cleared the launch tower and roared into the Florida sky, shook tiles from the recording studio ceiling of veteran news anchor Walter Cronkite over three miles from the launchpad.

From a physical point of view that sort of power shakes a person's insides. It was awesome. But from an emotional point of view it was comforting to know there would be that much force propelling your spacecraft into orbit.

There were some initial problems with the Saturn V. Vibrations had been so severe in the second unmanned test that two of its engines had shut down prematurely. But Wernher von Braun was confident the problems could be ironed out. The first two manned missions—Apollo 7 and 8—were due to use the smaller Saturn-IB. But even before Apollo 7 launched in October 1968 George Low proposed a major change to our space program.

The change was visionary. It was the turning point. But, at first, others in senior management at NASA were aghast at its audacity.

The original plan, if all went well with Apollo 7, was for Apollo 8 to fly the Lunar Module together with the Command and Service Module in Earth orbit, performing complex rendezvous maneuvers between the two vehicles. Not only would it test fly the Lunar Module alone in Earth orbit for the first time, but it would also perform Apollo's first rendezvous maneuvers between the Lunar Module and Command Module and involve quite risky EVAs. This was the D mission to which I had been assigned. It was a really juicy mission, a test pilot's dream.

After that would come the E mission, the first to use the Saturn V, with the Lunar Module in high Earth orbit to test the reentry system of the Command Module from a near lunar distance. Given all the problems with the Lunar Module, however, the prospects for the D and E missions did not look good in summer 1968.

Prompted by fears that the Soviets would beat us to a circumlunar flight if we waited for all the problems with the Lunar Module to be ironed out, Low proposed bringing forward plans to send a spacecraft around the Moon. The Command and Service Module (CSM) would go alone. And not only would it loop round the Moon, but it would enter lunar orbit, circling the Moon ten times before returning to Earth. This meant the CSM's engine would have to work, very precisely, twice in order to succeed. If it did not, the Apollo crew would find itself permanently orbiting the Moon.

It was a bold and risky plan, but it seemed the only way of winning the race. After it was approved, Deke Slayton called Jim McDivitt into his office to offer him the chance to fly with Rusty and me on this first circumlunar mission. Jim declined. As far as I was concerned he made the right decision. The circumlunar flight would be an exciting mission, if successful. But it was not going to offer the same challenges of EVA, rendezvous and docking that we had been training so hard for.

This meant Frank Borman—originally slated as commander of the mission after ours—became the commander of Apollo 8 with Jim Lovell and Bill Anders as his crew, becoming the new C mission. We would now fly the mission after this, Apollo 9, still the D mission. The shake-up also meant that the crew, originally slated as back-up to Apollo 9, would back up Apollo 8 and then very likely fly the Apollo 11 mission. If all went according to plan, Apollo 11 would be the first mission to attempt a lunar landing. Commander of that crew was Neil Armstrong.

However, it was an incremental program and no one knew

whether such risk-taking would pay off—to do so, each mission in succession had to succeed. When it did eventually launch, in October 1968, Apollo 7 was not without its problems.

The crew of Apollo 7—Commander Wally Schirra, Walt Cunningham and Donn Eisele—all came down with colds early on in their eleven-day flight. They felt miserable. Worse still, they were irritable, which led to some confrontation between the crew and Mission Control.

Much of the conflict concerned a television program based on a camera mounted inside the Command Module. Wally was due to do a live transmission from space around twenty-four hours into the mission. But when he got the order from Houston "OK, it's time for your TV show, Wally," he decided it would interfere with his flight plan. He didn't want to do it. And he told Mission Control so in no uncertain terms.

"I'm commander of this mission and I'm going to do it my way," he said in the end. In principle, he was right. He was in command. The role of Mission Control is often misunderstood. Houston did not "control" the missions. They advised, but had no way to control the spacecraft—that had to be done on board. The commander of the ship was in ultimate control. On the other hand, you had to have the proper respect for Mission Control. They had far more data and information as well as an appreciation of certain aspects of how a mission was going that the commander of a spacecraft might not have.

I have to defend Wally's decision, though not his manner. He was under a lot of pressure. This was the first manned mission after the Apollo 1 fire, and he knew he had to get it right. It was a big responsibility. But he was too abrasive; he didn't handle it right. An open confrontation like this did not go down well. It really hurt the controllers' morale and those of us who relied on Mission Control during flight were quite embarrassed. Chris Kraft, then Director of Flight Operations, didn't appreciate it at all. He vowed, privately, that none of the crew would ever fly again. And they never did. That was the end of their careers as astronauts, although Wally had announced his retirement before the mission.

In November 1968 Richard Nixon was elected president, and the administration changed from Democrat to Republican. Initially, this was good news for NASA—the Republicans were known for spending more money on projects like space and the military. Even if Nixon had not been elected the space program would have continued, but NASA's budget had to be approved by Congress every year and if the budget were cut the program would slip. We were close to Kennedy's end-of-the-decade deadline. We could not afford to face monetary problems now.

Shortly after Apollo 7 was launched, the Soviets launched two Soyuz spacecraft into Earth orbit. Though the first, Soyuz 2, was unmanned, these were the first such missions since the death of Komarov. *Time* magazine had a big story, "Race for the Moon," on its cover. It was known the Soviets were developing a rocket even bigger than our own Saturn V, with enough thrust to get their spacecraft into lunar orbit. Then in November 1968 they flew an unmanned Zond—Zond 6, a lunar version of the Soyuz—round the Moon; it took dramatic pictures of a half-Earth over the lunar horizon.

The race was closing. Even if Apollo 8 beat the Russians to make the first manned circumlunar flight, there were real fears that they might still beat us to the first Moon landing. Tension was so high that large sheets with the handwritten countdown dates to the next possible lunar launch window—the dates when it was feared the Russians might attempt a lunar landing—were taped up in strategic points around Mission Control in Houston. You can fly to the Moon only every twenty-eight days, during a certain phase when the lighting is right. The angle of sunlight on the surface of the Moon has to be around 12 degrees—much lower and the shadows are too long, much higher and the bright light causes a "wash-out" of features on the lunar surface.

Yet we couldn't go any faster than we were. We were working at peak.

The shake-up in scheduling and redesignation of Apollo 9 as a much more complex and demanding mission meant our training program had to be stepped up considerably to include many more simulations of the situations we would or could face during the spaceflight. During part of the mission I would be flying the CSM alone, while Jim and Rusty took the Lunar Module on its first test flight. This meant that not only did we have to train for the wide spectrum of individual tasks we would be called on to perform, we also had to simulate the many emergencies we might have to face while the two vehicles were apart. For me, as Command Module pilot, this meant learning to fly two major profiles—rendezvous and reentry—on my own. For the most dire of the emergencies we might face would be a major failure of the Lunar Module so that Jim and Rusty could not get back to the CSM and I had to rescue them by performing the rendezvous myself. If for some reason the Lunar Module were uncontrollable I might even have to pilot the Command Module back to Earth alone.

Bringing Apollo home as a one-man show involved my mastering many aspects of all three jobs performed by the crew, Jim's as commander, Rusty's as systems engineer, my own as navigator. The sheer logistics of operating in all three positions, let alone learning the complex procedures this would require, was challenging, to say the least. Simulations of the emergency run both at the Cape and Houston required me to physically switch between the three couches—the left-hand couch of the commander, the center couch of the navigator and the right-hand couch of the systems engineer.

In order to cope with both eventualities I set the procedures up so that I started the rendezvous or reentry sequence in the right-hand couch, making sure all the communications, electrical and environmental systems were working correctly. I then slid across to the center couch, setting the computer to get the rendezvous or reentry program running, before settling into the commander's couch to perform the rendezvous, or in the case of reentry the de-orbit maneuver with the big engine.

Once strapped into the left-hand couch I had to make sure I had access to all necessary systems. I also had to simulate flying the Command Module manually with a joystick for reentry by aligning the spacecraft very accurately and lighting the big engine on the Service Module at exactly the right time to slow us down for de-orbit and reentry into the Earth's atmosphere. Then I would have to operate switches to separate the Command Module from the Service Module and orient the former so that the heat shield was pointing in the right direction to prevent the spacecraft from burning up on reentry. I must have practiced this emergency simulation half a dozen times. It was pretty exacting. Just shifting between the couches as I performed the complex procedures took quite some physical effort. In space, of course, mobility would be aided by zero gravity.

In addition to this I had to run through complicated simulations of what would happen if the Lunar Module (LM), rather than being destroyed, developed a serious fault and was unable to perform the scheduled rendezvous with the Command Module at the end of its test flight. I had to run through the different trajectories I could fly to go and rescue the LM. The rule was that, if the LM were unable to initiate the correct maneuver in the rendezvous sequence within a minute of when it was scheduled, Jim and Rusty would stop trying to perform the maneuver and simply wait for me to come and rescue them. That exercise saw me sitting in the simulator running through the different trajectories I would fly, while monitoring an image of the LM—transmitted via a television screen in front of me—as if I were looking out of one of the Command Module's windows toward the stricken vehicle in space; the image showed alternately what I would be able to see of the LM during daytime and then at night, when all there would be to indicate its position was a flashing light.

Two other eventualities for which we had to plan and practice were a failure of the Command Module and Lunar Module to dock and a failure of the tunnel linking the two vehicles to open correctly after docking. If either happened, Jim and Rusty would have to perform an EVA to reenter the Command Module. One day, after Jim, Rusty and I had already performed a big rendezvous simulation at the Cape early in the morning, we had to drive down to Patrick Air Force Base, get in our T-38s, fly them to Houston and then run through a long and complex EVA simulation.

For this we had to unstow our pressure suits inside mock-ups of the Command and Lunar Modules: life-size models of the two sections of Apollo without any electrical or computer connections. We then had to practice the lengthy procedure of putting our suits and helmets on and hooking up our hoses, after which Jim and

Rusty simply had to walk, with their full kit on, from their mock-up of the Lunar Module to mine of the Command Module. In space such a transfer would carry the considerable risk of Jim and Rusty having only limited oxygen supplies. The risk was especially high for Jim, who would have to rely on the emergency oxygen supply from Rusty's backpack; if he didn't make the EVA transfer within forty-five minutes or so he would die. That simulation ended well beyond midnight after a training day lasting nearly twenty-four hours. It was totally exhausting.

None of the simulations ran smoothly. It was rarely a question of going through a process step by step. To add to the challenge, engineers and technicians monitoring each test would add failures and faults to make the situations as complicated as possible and see how we dealt with them. Very often the simulators malfunctioned, too, and the exercise had to be put on hold until things were fixed. There were many problems with these complex simulators before Apollo 9, especially linking them with Mission Control for "integrated sims"; our launch was almost postponed because we could not get enough training, especially for the rendezvous profile. After every sim was finished we were debriefed and often had to explain why we had failed to deal with a particular situation. It was all pretty intense.

Alexei Leonov

At the same time as training intensively for our space missions, most of us in the cosmonaut corps continued with our studies at the Zhukovsky Academy. In addition, Yuri Gagarin had been elected a member of the Supreme Soviet. He attended its sessions when he could, but he had his duties as deputy director of the Cosmonaut Training Center, and his international standing also meant he was called on to attend functions and give speeches representing the corps both at home and abroad. I don't know how he managed such a heavy workload, except that he was young, fit and, I suppose, a workaholic.

His hectic schedule meant that he was not always able to put in as many hours flying as he wished. Sometimes months passed without his being able to pilot a plane. Whenever any of us went for longer than a month without flying, we were required to take a plane up accompanied by a military commander who could check that we were still competent to fly in difficult conditions.

In spring 1968, as a result of a number of trips, a short holiday and an exam session at the academy, Yuri had not flown for about three months. This meant he was subject to a routine test to verify his flying skills. On 27 March 1968, he took off from Chkalovskoye air base in a MiG-15 with Vladimir Seregin, a veteran commander.

That morning, I was commanding a training session of cosmonauts preparing for lunar missions. We had taken a large helicopter out from Kirzach Airfield, about 80 km from Chkalovskoye, and I was putting my group through their paces with a series of parachute jumps. The weather was very bad that day: the cloud cover was low and it was raining hard. My team had performed just one jump when the weather deteriorated even further. The rain turned to sleet, and conditions were so bad that I canceled the session and requested permission to return to base.

As we stood around waiting for permission to come through, I heard two loud booms in the distance. We immediately started discussing what the noise had been, whether it had been an explosion or the sound of a jet plane exceeding the sound barrier. My impression was that it had been both.

Twenty minutes later we received permission to return to base and I piloted the helicopter back toward Chkalovskoye. Midway through the flight, I heard radio controllers urgently repeating Yuri Gagarin's call sign: "741, 741, come in 741."

There was no reply.

At first, I thought they might be trying to reach me—my call sign, 841, was similar to Yuri's—so I radioed back. "841. I can hear you. Are you calling me?" No, it was not me they were trying to reach, I was told. Nothing more.

As soon as we touched down at Chkalovskoye, an officer rushed over to tell me that the fuel in Yuri and Seregin's plane would have run out forty minutes before, but their plane had not returned to base. I made straight for the command post, and said to General Kamanin, "I know it is terrible even to suggest such a thing, but I heard an explosion coming from that direction." I drew a rough sketch of the relevant area. "I think we should send a helicopter to check."

A plane was dispatched immediately. Twenty minutes later, its pilot reported that he could see smoke coming from a patch of forest close to the area I had indicated. He was ordered to land his plane in the nearest clearing and proceed by foot to where he had seen the smoke. The ground was covered in thick snow, and it took him almost two hours. When he reached the site he found the wreckage of a plane.

A search party was dispatched. A torn fragment of Seregin's flying jacket was found, but, apart from the wreckage of the plane, very little else. There was no sign of Yuri. At first, it was thought he might have ejected. It was getting dark, but throughout the night four hundred or so soldiers searched the area with torches, calling his name.

As dawn broke, the crash site was reexamined. Shreds of Yuri's flying jacket were recovered, together with small fragments of flesh. They were placed in a surgical bowl and shown to me.

A few days before, I had accompanied Yuri to the barber to have his hair cut. I had stood behind Yuri talking while the barber worked. When he came to trim the hairs at the base of Yuri's neck he noticed a large, dark brown mole.

"Be careful not to nick that," I told him.

"I'm always careful," he replied.

Now, looking down at the fragments of flesh lying in that metal bowl I saw that one bore the mole.

"You can stop searching," I told the rescue workers. "It's him."

I cannot describe my feelings. All that remained of that extraordinary man, whom I had loved as a brother, lay in a metal dish. Death had never seemed closer, nor more terrible.

In the years since, I have made many drawings and paintings of that awful crash site; the torn trees, smoke, wreckage.

Within days of the crash, a government committee of investigation was established, under the leadership of the minister of defense.

Gherman Titov and I were assigned to it as representatives of the cosmonaut corps. Different possible causes of the accident were closely examined. One theory was that Yuri's plane had gone into an uncontrolled spin after maneuvering to avoid a flock of birds. Another was that it had collided with a hot air balloon, the remnants of which had been found not far from the crash site. The latter was the conclusion of the accident investigation committee. But neither theory was ever proven.

After a while rumors started to circulate. One claimed that Yuri had been drinking before he flew. Another speculated that he and Seregin had been taking potshots at wild deer from their plane, causing it to spiral out of control. Yet another claimed that Yuri was not dead at all, but had been thrown into prison after tossing a cognac in Brezhnev's face. Another had it that he was languishing in a mental asylum. Such rumors drove me crazy.

At the height of this ridiculous speculation, I received a call from Sergei Mikhailovich Belotserkovsky, our beloved tutor at the Zhukovsky Academy.

"We have to do something to defend Yuri's honor and reputation," he said. "If we don't, who will?"

We tried very hard to have the investigation into the crash reopened. I wanted to conduct my own independent inquiries. We spoke at numerous scientific symposiums stating that we did not believe the reasons for the crash had been looked into thoroughly enough. In the beginning Titov stood by us, but with time even he distanced himself from the controversy.

Yuri's wife, Valya, became quite angry at our attempts to have the investigation reopened. She said we should let the matter rest. She did not appreciate what we were doing because we had shielded her from all the ugly rumors concerning the flight. Eventually we had to tell her of the ridiculous claims being made about her husband. "Would you prefer it to be said that Yuri was drunk?" I asked. "Or that he is still alive, but mad?"

It was twenty-five years before all the documents concerning the crash were declassified and I could gain access to them. When I studied them carefully I found a document I had written at the time, describing the one-and-a-half- to two-second interval between the two booms I had heard. The document had been altered: in handwriting that was not my own the interval between the two booms had been changed to twenty seconds.

The documents also revealed that, in their last transmission the crew had reported they were flying at a height of 4,200 meters. Yuri said they had finished their maneuvers and were returning to base. The crash site was at almost exactly the position from which that last transmission was made, indicating that they had been forced into a flat spin immediately.

At the time of the accident, it was known that a new supersonic Sukhoi SU-15 jet was in the same area as Yuri's MiG. Three people who lived near to the crash site confirmed seeing such a plane shortly before the accident. According to the flight schedule of that day, the Sukhoi was prohibited from flying lower than 10,000 meters. I believe now, and believed at the time, that the accident happened when the pilot jet violated the rules and dipped below the cloud cover for orientation. I believe that, without realizing it because of the terrible weather conditions, he passed within 10 or 20 meters of Yuri and Seregin's plane while breaking the sound barrier. The air turbulence created overturned their jet and sent it into the fatal flat spin.

To complicate matters, Yuri and Seregin's MiG had been fitted with external, expendable, 260-liter fuel tanks, the purpose of which was to allow a plane to fly much further in combat. The tanks were designed to be dropped before entering a combat zone, where complicated maneuvering was called for, because they severely compromised the plane's aerodynamic performance. Yuri and Seregin were not expected to perform such maneuvers that day, but it was clear from the way their plane had chopped through the treetops that they had tried to recover from the spin and it seemed they were short of doing so by a matter of just one and a half to two seconds.

The investigating committee would never have admitted at the time that that is what had happened because it would have meant admitting that flight controllers were not adequately monitoring the airspace close to sensitive military installations. I believe ordinary people were unable to accept the real explanation because the technical details of Yuri's plane being intercepted by an SU-15 jet were too complicated for most to understand.

But now nobody repeats any nonsense about Yuri being drunk, irresponsible or mad. After the many years I have spent talking about what I believe to be the truth of what happened that terrible day in March, the explanation that it was the result of an approach by a supersonic jet, is, at last, widely accepted.

David Scott

On Christmas Eve 1968 Apollo 8 passed round to the Far Side of the Moon. Lurton and I, together with several other astronauts and their wives, spent that evening with Frank Borman's wife, Susan, and their two sons. It was a time when everyone was supposed to be joyous and happy, but the atmosphere that Christmas Eve was very tense.

We were concerned that Frank had apparently been struck down with a flu virus in the first day of the flight. After suffering diarrhea and vomiting for twenty-four hours—a challenge for the spacecraft's waste-management system—he recovered, though it did cause a great deal of concern about the crew.

When Apollo 8 had come round to the Near Side of the Moon for the first time the previous day, Mission Control had just one question for the crew: "So what does the ol' Moon look like from sixty miles?"

"The Moon is essentially gray," Jim Lovell replied. "No color. Looks like plaster of Paris."

Then, as Apollo 8 continued into lunar orbit, the crew took it in turns to read the first verses of Genesis. "In the beginning God created the Heaven and the Earth . . ."

Now, nearly four days into the flight, at the most critical moment of the mission, when Apollo 8 was ready to leave lunar orbit and return to Earth, we waited for news with Susan Borman in her kitchen. Over a billion people worldwide were estimated to have turned on their radio and television sets, waiting, like us, for news of the mission. After twenty hours of orbiting the Moon, Apollo 8 prepared to light its big engine to depart lunar orbit and head for home, over a quarter of a million miles away.

Because Apollo 8 was at the Far Side of the Moon there would be no contact with the crew for thirty-seven minutes. If the transmission blackout lasted longer than that, everyone knew it meant the single engine on the Command Service Module had failed and they weren't coming back.

The engineers were confident, but the wives were far removed from their technical world. Susan was very worried. Several of us were sitting with her at her kitchen table. Almost to the second, thirty-seven minutes after we had last heard from Apollo 8—at twenty-five minutes past midnight on Christmas Day—a transmission from the spacecraft was received in Mission Control.

"Please be informed," Jim Lovell's voice crackled across the airwaves, with an upbeat festive message, "there is a Santa Claus."

They took some amazing photographs during their mission, including our classic first shot of the Earth rising from behind the Moon—Earthrise. It was the first time human beings had glimpsed another celestial body up close with the naked eye—quite extraordinary! But somehow, from the way they talked later of their first impressions of the Moon, I don't think those guys had time really to appreciate a lot of what they saw.

Apollo 8 was a big step in the Cold War space race. It meant we were no longer behind the Soviets. We had scored a "first"—we had sent men into orbit round the Moon. This was a huge accomplishment. In a sense we had won the race. But in retrospect we know now that the race was very, very close, especially during the autumn of 1968. Some of our engineers who worked on lunar trajectories tacked some informal notes on their boards that indicated a Soviet launch window to the Moon in early December, three weeks before the scheduled launch of Apollo 8. Few others paid much attention or were aware (except certainly NASA senior management). Others were aware that in September, just a month before Apollo 7, a Soviet Zond 5 spacecraft had flown around the Moon but was only partly successful upon its return to Earth. The Zond was a Soyuz manned spacecraft modified for a lunar mission. Then on November 10, Zond 6 flew around the Moon, unmanned, and returned to Earth capturing a dramatic view of Earthrise from the Moon. And a few in NASA probably knew that the next Zond spacecraft was on the pad ready for launch during that early December window. But for several nontechnical reasons, it was not launched. If a Soviet Zond crew had returned high-quality photos of the Earth from the Moon three weeks before Apollo 8, the balance of the race could very well have changed dramatically. But the United States was not satisfied to rest there with a manned circumlunar mission. Now we needed to move on to the greater technical challenge of actually landing on the Moon. Our Apollo 9 mission was up next, and we really had our work cut out for us.

Jim, Rusty and I transferred to the Cape three weeks or so before the launch of Apollo 9. Living conditions had improved since I had been there for Gemini 8. We each had our own bedroom. We shared a bathroom and a big, comfortable living room—nicely lit, soft chairs, homey. A small court had been built for us to play handball at night. There seemed to be an unlimited budget, too, for Lou, the chef who prepared our food. He was great, a former cook on a tugboat, who served us marvelous meals every time we sat down.

Apollo 9 was scheduled to lift off from the Cape at the end of February 1969, but they slipped the launch three days because, during physicals, they detected McDivitt had a reduced white-cell count, indicating he might be about to get a cold. We were furious. We were ready to go. We weren't sick. Right after they told us they were slipping the launch, Rusty and I put on our jogging clothes and ran a circuit round the launch complex. We'd show them we were fit. Nobody got sick.

After an early supper on 2 March Rusty and I decided to drive out to the launchpad and take a look at the launch vehicle. It looked spectacular at night. All the searchlights were on it. It looked gorgeous against the black sky. The atmosphere on the pad was electric. Everybody was busy preparing for the following day. Fueling, checking, elevators going up and down, technicians busy.

"Man, look at that," I said to Rusty. "We're gonna ride that tomorrow."

Alexei Leonov

In the months after Yuri's death, morale in the cosmonaut corps was at a new low. We continued training intensively for circumlunar missions, but Mishin's extreme caution continued. Following the disaster with Soyuz 1, exhaustive tests had been carried out on a redesigned parachute system, and the engineers were fully confident the new system was flawless. Still Mishin hedged his bets.

Compared to Korolev he was an uninspiring leader, bad at making decisions, hesitant. However, he was a highly skilled engineer, even when he had been drinking, as he proved on one crucial occasion.

Mishin's plan was for several unmanned launches of modified Soyuz capsules—known as Zonds—in the first half of 1968, followed by the first manned launch of a circumlunar mission later that year. Meanwhile the manned Earth-orbit Soyuz program would continue to test a series of complex rendezvous and EVA maneuvers.

In March 1968 the first unmanned Zond spacecraft—Zond 4— had successfully flown into deep space to simulate the return from a circumlunar mission. In September, another Zond was sent for the first time to loop round the Moon with an experimental cargo of two steppe tortoises. After a successful launch, Mishin left Baikonur to fly to Yevpatoriya in the Crimea, where the main flight control center was located at that time. During the flight, he had a good meal and toasted the success of the launch with quite a few glasses of wine.

Shortly after the flight got under way the spacecraft started to develop problems. It was not at all clear what the problem was at first (later it emerged that the capsule's stellar orientation device was malfunctioning). I was already in Yevpatoriya monitoring the flight when Mishin arrived and was told there was a malfunction. He immediately asked me for a technical drawing of the failing device, which I had prepared. Despite having drunk rather a lot on the flight, it took him only a few minutes to identify the part of the unit that was causing the problem. He circled it with a pencil saying, "Look, here it is," and was later proved to be exactly right. He had very good intuition.

Once the problem was rectified, the mission was a spectacular success. After one Earth orbit the Zond-5 capsule began its trajectory toward the Moon. On 17 September it looped round the Far Side, taking pictures of the Earth from above the lunar surface. Although the capsule experienced problems on reentering the Earth's atmosphere, when it splashed down in the Indian Ocean the tortoises were recovered alive and well.

The following month, the first Soyuz flights since Komarov's death were successfully accomplished. First an unmanned Soyuz capsule was launched, followed the next day by the launch of the Soyuz 3 spacecraft with Gyorgy Beregovoy. The two craft performed a rendezvous, but no docking, and Beregovoy returned to Earth after sixty-four orbits.

After the successes of Zond 5 and Soyuz 3, many of us cosmonauts were pushing for the next flight to be a manned circumlunar mission. Media reports from the United States confirmed that NASA was planning its first circumlunar mission at the end of the year. But instead Mishin launched another unmanned mission—Zond 6—in November. Again the spacecraft passed round the Far Side of the Moon, taking unique pictures of previously hidden portions of the lunar surface. But again the capsule experienced a problem on reentry. Plans for a manned circumlunar mission were once again postponed.

I was in Moscow, busy working on our L-1 circumlunar program, when the news came through that the Americans had sent a manned spacecraft into orbit round the Moon. News of the Apollo 8 mission was everywhere: on television, on the radio, in the newspapers.

I suddenly had the feeling that everything was slipping through my fingers. I could see my dreams going up in smoke.

It felt as if everything I had been preparing for so hard over the past few years had been a waste. I was filled with a deep apprehension that they might cancel our circumlunar program.

Still, I admired the way Apollo 8 had accomplished its mission. It was brilliant. There was no denying that. This was officially recognized by our political leaders when Frank Borman became the first astronaut to be formally invited to visit our country.

A big party was thrown in his honor at Moscow's Metropole restaurant. The place was so crowded you could hardly push your way through the throng. Some men were there in military uniform, some had come straight from work. Borman, by contrast, wore a dark jacket and white shirt. Round his neck, instead of a tie he wore a thin string, like a cowboy's, tied with a bright blue stone in the middle. He cut a very striking figure. Everyone wanted to stand near to him. To touch him.

When I eventually met him I congratulated him on his mission. I of all people, I said, knew how hard it must have been. I did not tell him that I, too, had been training for a circumlunar mission. But I felt as if he knew. He was very gracious. He congratulated me on my space walk with Voskhod 2. We began to talk about how the Moon had looked when he was in orbit. We discussed good locations for lunar landings.

Then Borman was called upon to make a speech. He proposed several toasts to future cooperation in space between our two countries, a desire I shared. He was very polite and very tactful. He tried to answer all the questions that were put to him, after which he was presented with a Tula gun. His wife, Susan, dressed all in blue, then stood up and said she wanted to offer a personal gift of her own. She removed from her finger a ring with a big turquoise stone at its center and handed it to Gyorgy Beregovoy, who had been promoted to a senior position in the cosmonaut corps after his Soyuz 3 mission. Everyone stood up and applauded this gesture of goodwill.

We had been looking at each other as enemies for so long. We had hardly ever met American astronauts face to face. But when such meetings did take place we had the feeling that we were just the same, that, like us, the American astronauts had their joys and sorrows.

When Richard Nixon was elected president in the United States it seemed he had a greater interest than his predecessor did in developing closer relations with the Soviet Union. After a sharp escalation in the arms race, there were signs that some limitations on this military proliferation were going to be imposed. First, a treaty was signed banning nuclear weapons from outer space. We took this as a good omen that the animosity between our two countries might ease.

Two weeks after the Apollo 8 crew had returned to Earth, however, any interest I harbored of becoming involved in coöperation between our two countries almost became irrelevant. Once again, I almost lost my life.

It happened at the height of celebrations in Moscow after the return of the crew of our latest space missions, Soyuz 4 and 5. In the middle of January 1969 a reception was organized at the Kremlin to honor Vladimir Shatalov, Yevgeny Khrunov and Alexei Yeliseyev. A motorcade was to bring them to the Kremlin through cheering crowds gathered around Red Square. While the cosmonauts rode at the head of the motorcade in an open limousine, behind them, in a Zil-117, rode Leonid Brezhnev and Alexei Kosygin.

Valentina Tereshkova, Beregovoy and I rode in the car behind Brezhnev. Behind us came a long line of cars with diplomats and other officials.

After crossing the River Moskva, the limousine carrying the crews of Soyuz 4 and 5, together with the car in which Brezhnev and Kosygin were riding, abruptly veered off to the side to approach the Kremlin through a different gate. The rest of the motorcade, with our car now in front, proceeded toward Borovitskaya Gate.

As we drew close to the Kremlin, a man suddenly stepped out from the crowd toward our car. He was dressed in police uniform but was brandishing pistols in both hands. Without warning, he started shooting in our direction. Our driver sustained severe head injuries, and the car veered out of control.

I had been sitting directly behind the driver, Beregovoy to my right. The second I saw blood spurting from a hole in the driver's neck I turned my head sharply to one side. A shot shattered the window immediately to my left, and shards of glass flew into my face. Another bullet grazed the breast of my military coat, and a third was fired at the level of my stomach, again just missing me and embedding itself in the ashtray of the opposite door. Yet another passed just behind my back. Even with my head bowed I could see the gunman firing wildly until he ran out of ammunition and was overpowered by the crowd.

Our car had fourteen bullet holes—it is incredible that I was not killed. Had the car not veered away from the gunman, one of his shots would certainly have killed me. But I escaped without serious injury, Tereshkova and Beregovoy too.

The gunman, it was later revealed, was a young lieutenant from Leningrad who had sought refuge in Moscow with his brother—a policeman, whose uniform he was wearing—after stealing two military pistols from the barracks where he had been stationed. He said he had wanted to kill Brezhnev. He gave no particular reason.

The authorities in Moscow already knew, when we set out for the Kremlin, that two pistols had been stolen in Leningrad. All trains had been stopped and the main roads in and out of the capital had been closed for fear that there might be a terrorist attack. That was why Brezhnev's car had taken a different route. We were told this after the event.

After the shooting all motorcades for cosmonauts returning from space missions were suspended.

David Scott

We didn't hear about Yuri Gagarin's death right away. Then, when the media carried reports of his funeral, there were so many rumors flying around about what had happened to him that we were left wondering what the truth was. There were stories that he had been drinking and had then died in a hunting accident. But the reports that he had died in a plane crash were the most consistent. How sad that the first man to have flown in space should die in such a way, we all thought. We had lost astronauts in plane crashes, too: Ted Freeman, Charlie Bassett, Elliot See. Thoughts of such losses had to be pushed to the back of our minds. In spring 1969 I was about to embark on my second space mission.

As the sun came up on 3 March—launch day for Apollo 9—the scene could not have been more different from the intense activity we had witnessed the night before. The launchpad was deserted. It was like a ghost town. The rocket was fueled, ready to go. Apart from the guys in the White Room at the top of the launch vehicle, waiting to help us into the Command Module, there was no one there. It would have been too dangerous to have a lot of people around. But it was really striking to see the site so quiet.

While Jim and Rusty went ahead into the White Room I stood alone for a few minutes on the gantry outside; the structure could not take the weight of all three of us together. As Command-Module pilot, I was in the center couch, so I was the last to board. As I waited for Rusty to get into the spacecraft and slide into the right couch, Jim to the left, the Sun was coming up. Standing up there, 360 feet above the complex below, out in the open with a clear view of sky, I could see it was a pretty morning, a beautiful scene.

"Here we go," I thought. "This is it."

The launch of Apollo felt totally different from that of Gemini, much more dynamic. The force and vibrations of the Saturn V rocket beneath us were incredible. Two and a half minutes after lift-off, shut-down of the first stage was violent! It was like being compressed at the top of a giant spring. We were hurled forward into our straps toward the control panel. It seemed as if we were in the middle of a train crash, everything rattled and shook. There was a deafening roar.

Through our portholes we could see all sorts of debris sweeping forward to engulf the spacecraft as the first engines shut down. Once the second stage lit, we were thrust back into our couches again and it was relatively smooth. The third stage ignited eleven minutes after take-off and fired us into Earth orbit.

Our mission was scheduled to last ten days. The first five were crammed with a complex series of maneuvers involving both Lunar and Command Modules. Less than three hours into the flight we had to retrieve the Lunar Module from its storage pod in the launch adapter below us. This meant we had to separate the Command Module from the top of the rocket, to which we were still attached, turn the Command Module round and point it back toward the tip of the rocket to dock with the Lunar Module.

This was my job as Command-Module pilot. But, almost

immediately, I ran into problems. The rocket thrusters that should have enabled me to move us sideways didn't work. We started drifting.

"Uh oh, not again," I thought. After the problems Neil and I had had with faulty thrusters on Gemini 8, this was the last thing I needed.

If we couldn't pull the Lunar Module out from its storage pod, we didn't have a mission. We had this short moment when it seemed it wasn't going to work. While I maintained our position near the Saturn IV-B, Jim and Rusty tried to figure out what was causing the problem. As they ran through checks of all switches and indicators on the panels to the front and side of them, Jim noticed that several of the "attitude" engine indicators—"talk-backs"— were showing they were closed instead of open. Where the talk-backs should have been clear they were crosshatched like the sign outside a barbershop. This meant that the propellant valves to those engines were shut.

After all the preflight testing of the valves, this was certainly not supposed to happen. At one point ground control thought one of us must have bumped several of the switches closed as we were jostled around during launch, but we were strapped into our seats so tightly this was impossible. Later analysis concluded that the valves had flicked shut as a result of the shock caused by staging. Another in-flight surprise! Once Jim had recycled the switches to open the valves I was able to maneuver the Command Module for docking and pulled the Lunar Module out. We were all set.

A little further into the mission we ran into a further problem, however. Rusty had difficulty adjusting to zero gravity and on day three suffered a severe bout of motion sickness, or "space adaptation syndrome." Frank Borman had apparently suffered from it, too, on Apollo 8, but he had not undergone postflight testing. So when Rusty was sick, twice, very little was known about the problem.

After our mission was over Rusty volunteered to undergo research into it at the US military's special center on airborne sickness at Pensacola Naval Air Station in Florida. It was a brave undertaking. It involved him being subjected at least once a week for several months to a regular torture chamber of tests involving rotating chairs, tilting rooms, balance beams and sickness-stimulating movies. Eventually it was discovered that perhaps as many as 40–50 percent of those who flew in space experienced the problem—the build-up to, and effect of, which is identical to sea sickness—to the extent that it disrupted activity. No reliable predictors of the problem were found; some people were sensitive to it in tests and were fine in flight, others were resistant on the ground and sensitive in flight.

What emerged from the research was that adaptation to a weightless environment—or any motion—was achieved most rapidly by repeatedly moving the head to stimulate the symptoms to a certain extent, but not to the extent of actually being sick. A hierarchy of "malaise" levels was mapped out, with barfing being, say, level 4. The adaptation strategy involved repeatedly moving the head to, say, level 2a and then stopping. Rusty's strategy during our mission—the natural reaction of most people stricken with such symptoms—was to keep his head as still as possible for as long as possible in order not to trigger nausea. What he did not know was that this simply delayed adaptation until he was really needed on the third day of our mission.

The plan had been for Rusty to perform a two-hour space walk on the third day, after he and Jim had transferred into the Lunar Module and the two spacecraft had been depressurized. That was to demonstrate that the crew could transfer from the Lunar Module to the Command Service Module via an EVA in case the transfer tunnel between the two became blocked with the docking probe. But after Rusty was sick—though he then showed every sign of being over the problem—the plan was modified. The EVA was shortened to allow him to stand on the outside "porch" of the Lunar Module, while I opened the hatch of the Command Module and filmed him, without actually making the transfer.

For some time there had been people in the astronaut corps pushing for us to be allowed to give our spacecraft names. "No names, only mission number designations," NASA had said. They gave no particular reason. At times, I think, they just lost sight of the human dimension to the program. But when it came to Apollo 9 we had two spacecraft, the Lunar Module and the Command Module. To communicate between the two, we had to have separate call signs.

During training we had nicknamed the Lunar Module "Spider" because of its spindly legs and Spider became its call sign. The Command Module had arrived at the Cape on the back of a truck, wrapped up like candy in light blue Cellophane, earning it the call sign Gumdrop. When the media got wind of these call signs they adopted them as names for the two spacecraft.

Bowing to the inevitable, NASA allowed all Apollo crews after us to name their spacecraft. The crew of Apollo 10 used *Snoopy* and *Charlie Brown*. After that NASA got a little more esoteric, insisting that the names bear more relation to the missions, hence the use of *Columbia* and *Eagle* for Apollo 11's historic mission. But there we were: *Gumdrop* and *Spider*. Five days into the mission and the time had come to fly the two vehicles separately.

This was the climax of the mission. We had to prove that the Lunar Module and Command Module could fly independently for several hours and then relocate one another—rendezvous—and dock once more to fly as a single craft. It was a vital precursor to a lunar landing.

As Command-Module pilot I would fly *Gumdrop* while Jim and Rusty took *Spider* off on its maiden flight. *Gumdrop* and *Spider* would then fly separately for five hours and, at the height of their different trajectories, the two craft would be a hundred miles apart. This was pretty risky for all of us. Jim and Rusty faced a much bigger potential threat. *Spider* had no heat shield, so it couldn't reenter the Earth's atmosphere and land safely. If anything went wrong with it, it would be up to me to rescue them. Throughout training I had had to learn to perform every function of the Command Service Module spaceflight single-handedly, including rendezvous.

After the two craft separated I was alone. I had been used to flying alone as a fighter pilot. I had spent over a thousand hours in F-100 Super Saber jets during my time in Europe. I had enjoyed that. You got to control everything. But there was a big difference between flying 40,000 feet up in a Super Saber and orbiting 300 miles above the surface of the Earth in space. I had the comfort of knowing, however, that Mission Control could do a lot more to help me with any problems I faced in space than air-traffic control could ever have done during my time flying fighter jets. And there were problems.

Right in the middle of our rendezvous maneuver I lost sight of *Spider*. During each forty-five minute period of daylight I had, until then, through my sextant been able to see *Spider*'s shiny covering reflecting in the sunlight. During the night periods I had been able to track its flashing light. But now, during one of these dark periods, it had disappeared.

Did they have a problem? Did I have a problem? Were they in the wrong trajectory? Was I in the wrong trajectory? Was there a problem with our instruments? Maybe their light was faulty. Should I assume they were in the right place? Or should I go find them? Mission Control had no problem tracking both of us and said we were both where we were supposed to be. So, all I could do was sit, wait and hope *Spider* came back into view once we reentered daylight.

As the Sun came round once again, there she was. Peering through my sextant I could see her awkward silhouette against the bright sky.

"Hey, amigo," I radioed Jim. "I've got you in my big eye."

"Oh, boy," Jim said with obvious relief. "Am I ever glad to hear that."

But as *Spider* and *Gumdrop* moved closer together, ready to dock, it looked like we were running into another serious problem. Jim was supposed to perform the docking from the Lunar Module. But as he approached the Command Module the sun was in his eyes and he couldn't align the vehicles. Using my alignment device, I talked him through the maneuvers and he was right on target, though there were more than a few tense moments. As the probe at the top of a short tunnel on the Command Module was inserted into the drogue mechanism at the top of a short tunnel on the Lunar Module, three small capture latches engaged and the two vehicles performed a "soft docking."

"I have capture," I radioed Houston. The probe was then

retracted mechanically to clamp the two vehicles together firmly by activating the series of hook latches and I notified Houston that we were "hard-docked."

Once a firm seal between the two spacecraft was confirmed, I activated the switch to pressurize the tunnel connecting the two vehicles. I was then able to open *Gumdrop*'s hatch to the tunnel, collapse and remove the probe manually, before opening a valve in *Spider*'s hatch to equalize the pressure between the two vehicles. Once this was complete, Rusty opened *Spider*'s hatch to clear the tunnel and Jim and Rusty were able to transfer back into the Command Module. If any of the steps had not been completed successfully, Jim and Rusty would have had to don their pressure suits, as would I, to transfer by EVA to *Gumdrop*'s hatch, which I would have opened and prepared for them; a very time-consuming and difficult procedure.

Our method of transfer marked one of the major differences between the Soviet and American concepts for lunar landing. The Soviets had planned for their cosmonauts to transfer between the lunar orbiting spacecraft and the lunar lander by means of an EVA. But our approach enabled the crew to transfer between the two vehicles through this connecting tunnel, in shirtsleeves.

After the relief and elation of completing the rendezvous, we felt we could relax a little. The difficult part was behind us. During the second half of the mission we had more time to reflect. Perhaps more than during my short and extremely stressful mission with Neil on Gemini 8, I think I really appreciated during this second time in space the astonishing beauty of the stars and the planet Earth.

At one point we dimmed the lights in the spacecraft so that we could get the best possible view of the amazing vista that lay before and below us. Most striking of all was the clarity with which it was possible to see streaks of lightning piercing the clouds during thunderstorms in the Earth's atmosphere, both along weather fronts and in swirling tropical air masses. Between flashes we could also see subtle streaks in the sky below us. They were so subtle that we weren't sure at first what they were, or if the others had noticed them too. It was Rusty, I think, who first mentioned them.

"Yeah, I've been seeing them, too," both Jim and I said and quite quickly we realized what they were: shooting stars, meteors burning up as they entered the Earth's atmosphere. It only took us a few moments to realize that if we were seeing them "down there" they must have passed by us "up here" first.

We took hundreds of photographs during the last five days of the mission, and I spent quite a time talking with Rusty about astronomy, just as I had during desert survival training when we first joined NASA. He was so knowledgeable on the subject. He was a really cultured man. He had brought quotations from Elizabeth Barrett Browning and Thornton Wilder along on the flight. I had brought along a selection of flags, patches, tie-pins and a Snoopy comic strip. Rusty had also wanted to listen to a cassette of classical music during the mission, Vaughan Williams's cantata *Hodie*. I wasn't really a fan of classical music at the time and hid his cassette in one of my pockets as a joke until the ninth or tenth day. He never forgave me for that.

Jim McDivitt was quite different from Rusty. Much more conservative, a real military man. We all got along well together. But Jim was tired by the end of the mission. On our last night, while Rusty was asleep, Jim and I were going through the final preparations for reentry when he turned to me and said, "Davy, this is my last mission. I'm really, really tired." And he was. He had commanded a very difficult mission, carrying tremendous responsibility, and he did it brilliantly.

He went into management at NASA almost as soon as we got back. Frank Borman left the program after Apollo 8 as well. It was easy to burn out on missions, get permanently tired, and not want to fly again. The great NASA team made them look too easy. They were really, really hard.

We nailed the touchdown on Apollo 9: we came down right on target and splashed down next to the carrier, where there was a chopper ready to winch us from the water and a great meal waiting for us. We stopped over in the Bahamas on the way back to Houston, where our families were waiting for us on the tarmac. Almost as soon as I stepped off the plane my daughter Tracy wanted to show me a short essay she had written for an English class while we were away. It was all about us going to the Moon as a family one night. I don't know where she got all her material for the essay—she was only nine at the time—but there she was, talking about us all bouncing around on the Moon and breathing oxygen from tanks.

"There was lots of dust," she wrote in conclusion. "We decided that, next Saturday, we would go to the zoo."

Alexei Leonov

Over the following few months we continued to monitor the American space program closely. Nothing could match the frustration I had felt after hearing news of Apollo 8's circumlunar mission, so I observed the next mission with a little more detachment. It was clear that the Apollo 9 mission was a very complex one, involving considerable risks. But I knew very little about its crew.

All I knew was that Jim McDivitt was commander, with David Scott and Rusty Schweickart as his crew. As the mission progressed we followed it on our television and in the press. I had great admiration for the good job they were doing. When the capsule landed back on Earth it was quite clear the Americans had taken yet another major step toward landing a man on the Moon.

By this point I knew we were not going to beat them. Although no firm dates had been announced, we believed there would be one more Apollo mission before the ultimate goal of lunar landing was attempted. We would never be able to accomplish the lunar missions we had planned on schedule.

In addition to our problems with the Zond spacecraft, the new N-1 rocket designed as a launch vehicle for the L-3 lunar landing missions was badly behind schedule. I had had personal experience of the problems of its precursor, the UR-500K Proton rocket, used to boost the lighter Zond capsules toward the Moon, when observing one of its launches at Baikonur in the spring of 1968.

Just before the lift-off, I had been observing the last-minute preparations near the launchpad. As the countdown started I loaded the team I was with into a bus and drove some distance away to watch. We stayed quite a distance from the launchpad for safety reasons. Shortly after lift-off the giant rocket keeled back toward the ground and its propellant exploded into a dense cloud of yellow gas. The cloud spread rapidly in our direction. I immediately herded my men back into the bus and tried to speed away from the dense fumes, but the road was a dead end. We had to make a U-turn and drive back toward the explosion. Fortunately, the cloud of gas swept past our vehicle, but it was an unpleasant experience. Nothing, however, compared to the misfortunes that subsequently befell the N-1 rocket on which our hopes of a lunar landing mission rested.

The first test launch of the N-1 had taken place in February 1969. It had been aborted after eighty seconds when the rocket's thirty first-stage engines shut down prematurely. This brought the rocket crashing back to Earth over a hundred kilometers from the launch site. Again I was at Baikonur and witnessed the failed launch, but all I saw was a flash in the distance and a fire on the horizon. Strict safety precautions meant no one was hurt. Although this set our program back, it was not totally unsurprising on a maiden launch. We knew the Americans had had problems with their giant Saturn V rocket. To encounter teething problems with such complex and ambitious feats of engineering was to be expected.

The N-1, originally designed by Sergei Korolev, was the most powerful rocket ever built by man. It weighed 1,800 tons, had a thrust of 4,500 tons and stood 110 meters tall. But it had a fatal flaw. Its thirty engines were arranged in two concentric circles. When they were all fired up at the same time, a damaging and destabilizing vacuum was created between the two circles. This had not been discovered prior to launch because we had no facilities to test all thirty engines together. What happened at the second launch of the N-1 five months later was a much greater blow.

It was shortly before midnight in Baikonur on 3 July 1969 when the thirty engines of the second N-1 ignited. Just seconds after the elongated cone of the giant booster lifted away from its support structure, I once again saw a fireball engulf the rocket. The N-1 tilted over and collapsed back toward the launchpad with such a deafening explosion that windows in buildings tens of kilometers away were shattered. Again, strict safety precautions meant there were no human fatalities, but the launchpad, one of Baikonur's main facilities, was completely destroyed.

Although Korolev had contributed to the design flaws of the doomed N-1, had he lived he could, I believe, have rectified its faults, and we would have kept on schedule. Even before the N-1 rocket was needed for a lunar landing, I firmly believe, we would have beaten the Americans to complete the first manned circumlunar mission months before it was accomplished by Apollo 8.

David Scott

When we got back from our mission, not a lot of fuss was made. I don't think the public really understood the significance of what we had done. In an effort to conjure up more interest, public relations at NASA commissioned a documentary to be made about Apollo 9. It was quite light-hearted, set to the music of The Beatles' *Yellow Submarine*.

Jim, Rusty and I loved it, and thought they'd done a good job of explaining some of the complexities of the mission, but NASA hated it. They thought it trivialized what we had accomplished. They never allowed anything like that to be made again.

President Nixon invited us to the White House with our wives for a private dinner shortly after we returned. "Ten days that thrilled the world" was how he'd described our mission.

At first, it was a pretty relaxed evening. A lovely meal had been laid out for us in a small room of the White House. There was a lot of discussion about our kids, families, and life in Houston. When the main courses were finished and it was time for coffee, however, the conversation took a more serious turn.

Besides Vice President Agnew, Nixon had invited along a pretty outspoken Democratic senator, who started challenging us about all the money being spent on the space program. It turned into a heavy-duty debate, polite but very demanding. As we talked, the vice president threw in the occasional comment. But Nixon said nothing. He just sat there, listening.

The senator was very bright, articulate and uncompromising. "You'll never make a lunar landing by the end of the decade," he insisted. "We've wasted all this money. What do you think you're really going to get out of it? Why spend money on exploration when you can spend money on the homeless?"

It didn't seem like it at the time, but looking back this was clearly Nixon's way of drawing out of us the astronauts' point of view about why we thought what we were doing was such a great idea. We talked about the benefits for science and technology, exploration, spin-offs for the American economy. We argued then, as people continue to argue now, that investing in space exploration has to be a conscious decision by society to invest in the future. If you don't spend money on space, it doesn't automatically follow that money will be spent on the needy and homeless.

At the end of the evening Nixon summarized very accurately both sides of the debate. He did not take a position. But after the ladies had left us to cognac and cigars, and we stood around chatting, he was clearly pretty enthusiastic about the program. After all, it gave him a high, positive profile at a difficult time in the history of our country.

The unspoken political undercurrent to our discussion was the importance of the space program in winning the Cold War. I did not say it directly to the senator grilling us, but underlying my thinking were very fundamental questions: "Do you want us to win this race? Do you want to live in a free society? Or do you want to live under communism?"

The bottom line was our profound belief that we had to demonstrate democracy was a better system under which to live.

CHAPTER 9

The Eagle and the Bear

1969–71

Colonel David Scott

Mission Operations Control Room, Manned Spacecraft Center, Houston

We were on a roll. We had accomplished so much on Apollo 9 that the scheduling of missions was right on track. Next up on Apollo 10, Tom Stafford together with Gene Cernan, succeeded in taking their Lunar Module to within 50,000 feet of the lunar surface, and John Young in the Command and Service Module made sure they got back. When the crew got back we sat watching the film they had made of their flight down toward the Moon; the lunar craters and mountains—some over three miles high—were truly beautiful and impressive to observe so close up.

Almost from the moment Apollo 10 returned in May 1969 it was clear that the next mission was aiming for the ultimate goal, a lunar landing. All the preliminaries were over. Now we were going for the big one. We had to beat Kennedy's deadline.

There was a lot of optimism. But there was apprehension, too. Apollo 9 and Apollo 10 had gone so well. There were real concerns that things might not continue to go so smoothly. "Can we really be that fortunate?" some asked themselves. "Can Apollo 11 really get down to the surface of the Moon OK?"

I felt elated that Neil Armstrong had been chosen as commander of Apollo 11, with Mike Collins and Buzz Aldrin as his crew. From my experience of flying with Neil on Gemini 8, I knew what an excellent commander he was—cool and calm under the most extreme circumstances. I guess all of us in the astronaut corps thought for a few moments, "Too bad it's not me." But any tension that may have existed between Buzz and Neil, about who was going to be the first to step down on to the surface of the Moon has, I believe, been blown up out of proportion.

There was no doubt it would be Neil. The way the door of the Lunar Module opened would have made it very difficult for Buzz to get out first. Besides, Neil was commander. Neil would step out first. That was it. End of discussion.

I had never thought I would be in the first crew to attempt a lunar landing: there were too many people further up the ladder. After Apollo 9, I had been assigned as back-up to Pete Conrad, commander of Apollo 12. I used to joke with Pete that if Apollo 11 didn't make it I'd break his leg and get to command the next crew to try to land on the Moon.

Neil did have a few close shaves in the run-up to Apollo 11. Part of the training for a lunar landing involved piloting an unwieldy jet and rocket engine machine called the lunar landing training vehicle (LLTV), which simulated the extreme difficulties of bringing the Lunar Module down to the surface of the Moon. The awkward-looking craft was nicknamed the "Flying Bedstead." It was absolutely essential training, but it was dangerous. Neil once bailed out of the LLTV just seconds before it crashed and burst into flames at Ellington Air Force Base.

A few days before the launch, I rang Neil at the Cape to wish him luck. One of his favorite lines, before our Gemini 8 mission, was "If you can't be good, be colorful." Well, we'd certainly been colorful on Gemini 8. So this time I repeated one of Lurton's comments to me before Apollo 9.

"OK, Neil, last time was colorful. This time be good."

After Gemini 8, Neil and I had spent little time together. All of us in the corps tended to spend most time with each successive crew to which we were assigned. It worked that way for our families, too.

But Jan Armstrong and my wife had remained firm friends. A few days before the launch of Apollo 11, Jan called Lurton. She wanted to know if we could help make arrangements for her to watch the launch away from the crowd of over a million people who were expected to descend on the Cape for the big event.

Jan had been invited to watch the Apollo 11 lift-off from the deck of a big motor cruiser, owned by North American, which would be moored off the Florida coast, but she needed some help in making the arrangements to get there. I knew the North American people pretty well, so I made a few calls and set things up for us to fly together to Florida on one of their corporate jets.

On the night of 15 July 1969, Jan, Lurton and I flew down to Patrick Air Force Base in Florida, where we were picked up and driven to a friend of the Armstrongs' to grab a few hours' sleep. Early the next morning we made our way out to the boat.

It was a beautiful day. The launch was spectacular. It went perfectly. Jan was pretty cool, as I recall. We stayed on the boat for quite a few hours afterward, listening to transmissions from the crew on a NASA "squawk box" and waiting for the huge crowd at the Cape to disperse. There was great relief that the launch had been smooth. But no bottles of champagne or boxes of cigars were opened; they were reserved for the end of the mission.

When we got back to Houston, I followed some of the mission from inside the Mission Operations Control Room, (MOCR), the center of activity behind the glass window of Mission Control's VIP room, where the Flight Director and his team coordinated every second of the flight. As back-up commander of the next mission, I had open admission to this area of highly restricted access.

I didn't go in to MOCR too often, because the priority was always to limit those in that nerve center to personnel who strictly needed to be there during the mission. The room was full of rows of consoles at which dozens of controllers, working in shifts, sat facing a giant screen showing the orbital track of the flight in progress. In the center toward the back of the room, with a clear view of the entire operation, was the Flight Director, the conductor of the symphony. In front and to the left side of "Flight"—during early missions this was Chris Kraft—was the astronaut acting as capcom. Close to the capcom sat the NASA doctor on duty.

All doors leading into MOCR were manned by a NASA guard who checked the identity badge of anyone wanting to enter. Spanning out to the sides of corridors surrounding this nerve center were many smaller control rooms dedicated to individual aspects of the mission. Each back room "mini-MOCR" was also equipped with rows of consoles, lining the walls or down the center of the room, linked by computer and telephone lines to those within MOCR. And each mini-MOCR was in turn a hub for a network of regional control centers spread out across the country at the sites of the private contractors or other government facilities which had built or were supporting different aspects of the Apollo program. It was a vast, complicated web and worked superbly. But the heart of it all was MOCR.

When I did go into MOCR I sat over on a step to the side of the room, quietly observing and listening with a headset to everything that was going on; chairs for those with official duties were always full during these periods. I was on that step on the afternoon of 20 July 1969, watching and listening as Neil and Buzz brought their Lunar Module, *Eagle*, down toward the surface of the Moon. The atmosphere was electric.

Things got very tense when the crew started having computer problems in the minutes before they were due to land. Their computer flashed a "Program Alarm" light indicating that it was overloaded and not able to keep up with the calculations necessary for landing. The computer did "re-cycle" in the event and managed to keep up, but not before it had set off the alarm light four more times. That was a real heart-stopper.

Mission Control did not realize at the time how hard Neil was struggling to find a level, unobstructed spot to put the Lunar Module down. It all happened so fast. It wasn't until all the data were analyzed afterward that the real difficulties were appreciated. The Lunar Module's early guidance software was not too precise nor was the knowledge of the Moon's gravity characteristic. It was guiding the craft toward landing in a large crater a couple of hundred feet wide. Neil almost ran out of fuel trying to get *Eagle* to touch down safely beyond a boulder field on the rim of that crater.

The Sea of Tranquillity, where *Eagle* eventually landed, had looked smooth and level in all the photographs taken before the mission. What these images had not shown were the steep slopes and giant boulders of the crater, where *Eagle* could easily have come to grief.

It took several hours, during which Neil and Buzz got suited up and checked that all their systems were working, before they were ready to open the hatch of the Lunar Module and go outside. When Neil did take his first step down to the surface of the Moon, I didn't give too much thought to what he would say. No announcement was scheduled. I thought his first priority would be to get a soil sample, in case there were any problems and he had to get back into *Eagle* straight away.

But it was a historic moment and Neil had obviously given some thought to the first words he would utter on the surface of the Moon. I found out later that people had sent him passages from the Bible and Shakespeare, suggesting quotations he could use. But the eleven words Neil chose were perfect.

"That's one small step for man . . . One giant leap for mankind."

It was right on, typical Neil. A lot said in a few words.

Kennedy's deadline had been met. We'd beaten the Russians. As Kennedy had said, it was the Soviet Union that had chosen space as the arena in which to demonstrate the superiority of its system. But we had shown them our system was better. In the race to the Moon we had started out behind, but now we had passed them. We had won the race, hands down.

Whereas to the outside world this first Moon landing was a single point in history—and it obviously was an incredibly momentous event—for us on the inside at NASA it was a very significant stop along a very long and winding road. It was not a snapshot, but rather the freeze-frame highlight of a feature film. Whoever could have predicted that when man took his first steps on the Moon much of mankind would watch it on television? Faint though the image was, the world could watch it in real-time.

As a close friend of Neil's I felt very happy for him. I knew how hard he had worked to make Apollo 11 a success. But my feelings were also marked with relief and elation—"Fantastic. It works. Now I wonder when it will be my turn to go."

Colonel Alexei Leonov

Space Transmission Corps, Komsomolsky Avenue, Moscow

Everything was set up at the army engineering research center known as the Space Transmission Corps on Komsomolsky Avenue in Moscow on the morning of 21 July 1969. The facility was fully equipped with the latest intelligence-gathering and surveillance equipment. During the early hours of that morning—evening in Houston—all television monitors and radio receivers were tuned in to what was happening a quarter of a million miles away from our own planet.

There were a few cosmonauts present—maybe ten of us from the first group—and many senior military officers, together with intelligence experts specializing in the lunar program. The room was packed. We were all transfixed by the crackling transmissions from Apollo 11 commander Neil Armstrong as he guided his lunar Landing Module, *Eagle*, down toward the surface of the Moon.

Two months before I had watched the commander of Apollo 10, Tom Stafford, bring his lunar lander to within a few kilometers of the surface of the Moon. It was clear then that the next Apollo mission would attempt a landing. I thought the Americans might time it to coincide with the Fourth of July, but when Apollo 11 launched less than twelve days later, on 16 July, we knew this was it. Their goal would be to land on the Moon.

By this point, it was clear that we were in no position to carry out such a mission. The problems we had had with the N-1 rocket meant we would not be able to attempt a lunar landing that year, or even the following one. So it was with mixed emotions that I stood watching events unfold on our television monitors that July morning.

When Apollo 11 had soared away from Cape Kennedy I had kept my fingers crossed. I wanted man to succeed in making it to the Moon. If it couldn't be me, let it be this crew, I thought, with what we in Russia call "white envy"—envy mixed with admiration. I was envious that America had asked a great deal of the crew of Apollo 11 and that here they were, accomplishing what was expected of them. But I was full of admiration, too, for what they were doing.

As I watched the grainy black-and-white images of Neil Armstrong taking his first tentative steps down the ladder of the lunar lander, it was the most amazing feeling. I held my breath as he touched the lunar soil very lightly with the tip of one foot before lifting his other foot away from the limb foot pad and letting go his grip of the flimsy metal ladder to stand full square on the surface of the Moon.

Only two live television broadcasts have ever made such an impression on me: Apollo 11's lunar landing and, more recently, the terrible events of 11 September 2001, when we saw so many people dying before our eyes in the twin towers of the World Trade Center in New York.

The first showed the best of mankind. It was a celebration of the strength of human intelligence and courage. The second showed the depth of evil to which man can sink. I feel, very deeply, that had the men responsible for the second act witnessed the first—an American citizen taking the first steps on the surface of the Moon—they would never have dared to carry out such evil.

On the morning of 21 July 1969 everyone forgot, for a few moments, that we were all citizens of different countries on Earth. That moment really united the human race. Even in the military center where I stood, where military men were observing the achievements of our rival superpower, there was loud applause.

Very soon this atmosphere of celebration was overtaken by professional talk. We cosmonauts began discussing how easy it appeared to walk on the surface of the Moon, how easy it was to jump. We would have to take this into account, we agreed, when we went there ourselves.

One of the officers, who knew I had been training intensively for a lunar mission, came up to me and patted me on the back.

"That's how it's done," he said. "That's the task that lies ahead of you."

People started looking at the Moon in a different way from that point on. I had spent many hours myself staring at its silvery globe through a telescope. I had drawn it many times. I knew its many craters, its oceans and mountains. But that night, when I looked through my telescope at home, all I could think of was that man had actually walked on its surface that day. That achievement filled me with pride for all humanity.

At a gathering of cosmonauts a few days later, we drank a toast to the safe return of the crew of Apollo 11.

Some in our country were not so pleased, however, with what Apollo 11 had achieved. Mishin was very upset, and began looking for excuses for our failure to achieve what the Americans had done. He attributed their success to the fact that so much more money had been made available to NASA than to our own space program. Some estimates have since concluded that the Apollo program cost $24 billion in 1969 terms, compared with an equivalent in rubles of $10 billion spent on our own lunar missions.

This complaint ignored what I firmly believe to be true: that we could have been the first to complete a circumlunar mission if Mishin had not been so obsessed with completing unmanned missions round the Moon before entrusting cosmonauts to our lunar spaceships.

Our party leaders made no public comment on Apollo 11, although they did send a letter of congratulations to the United States. Following the party line, our newspapers gave very little coverage of the mission. *Pravda* barely noted the launch of Apollo 11, and the Moon landing was reported only on an inside page.

More was made of the fact that the Americans had taken flags from every nation to the surface of the Moon, which was acknowledged with thanks. It was only after news of Armstrong's Moon walk was in the world's press the next day that a short clip of the mission was shown on Soviet television.

Not showing live coverage of the Apollo 11 Moon landing was a most stupid and short-sighted political decision, stemming from both pride and envy. The Soviet Union had been working for this moment for so long that there were those who felt they could not show another achieving our goal. But our country robbed its own citizens by allowing political considerations to prevail over genuine human happiness at such events.

A great deal more attention was given in our media to the announcement in the following months that America's new president, Richard Nixon, was committed to withdrawing US troops from Vietnam.

The war in Vietnam had been one of the biggest events of the decade. So many people were dying there, both soldiers and civilians. It was clear that only political will could bring an end to the slaughter. To admit that such great human sacrifice had been in vain was potentially disastrous, but the American people had had enough. Neither Russia nor America should ever have interfered with other countries as they did Vietnam.

David Scott

There was great exhilaration when the crew of Apollo 11 returned from their mission. They'd brought back moon rocks. They were healthy and well. After a period of quarantine they went off on a world tour.

In one sense, their success took some pressure off Apollo 12, which was due four months later, but in another way it increased the pressure. Sometimes a second mission is more difficult to perform than the first. When a first flight goes so successfully, people can tend to let up a little on their work. But it was vital to keep everyone on the ball, keep the program on track.

As back-up commander of Apollo 12, with Al Worden and Jim Irwin as my crew, part of my job was to keep some levity in the game, keep things light and loose, relieve the tension when I could. In the last days before Pete Conrad and the crew of Apollo 12 were due to launch we got a cartoonist to draw some sketches to stick round their flight plans. We stuck some pictures of *Playboy* Bunnies in there, too, which brought a few laughs when they opened the documents in space. Pete Conrad and his crew were a fun-loving bunch. They were very good friends.

It also fell to the back-up crew to organize a party on the return of the prime crew. We really threw ourselves into that task. We had a 30-minute movie made, a spoof of the serious documentaries that NASA favored. Fighter squadrons sometimes did that. None of the bosses knew it was going on. I even went out to an old Hollywood film archive to get some footage of comedy sketches on space by the Three Stooges.

We also asked NASA's photo lab to help us out, on the quiet. There were always cameramen around, shooting film footage of the crew's training sessions. We asked them to pass on to us any material which showed the crew doing something funny. The cameramen really liked being on the inside loop of something secret and fun like this.

After the mission we took excerpts from Pete Conrad and Al Bean's transmissions on the lunar surface and included them in our documentary as a voice-over. It wasn't hard. They had been quite funny, humming to themselves as they went about collecting rock samples. We filmed some scenes which were supposed to show them on the surface of the Moon. One showed a figure dressed up as Pete sitting on the ground, cuddling a rock in his lap like a teddy bear, with a real voice-over of his from the lunar surface saying, "Hey, Al, I gotta a rock . . . Look it's a rock, Al, a rock."

We also decided we had to have a scene of Dick Gordon circling the Moon alone in the Command Module, while Pete and Al were down on the lunar surface. With the sort of dramatic music NASA always used playing in the background, someone solemnly announced, "Dick Gordon: the lonely vigil."

To mock up this scene we had the simulator building opened up one night and went in there with a cameraman, a secretary dressed in a green bikini and Jim Irwin—Al Bean's back-up—dressed up as Dick and wearing a monkey mask. As the announcer asked "Is Dick really alone?" the secretary stood behind Jim, stroking his back.

We made only six copies of the movie, one for each member of the prime and back-up crews, then we destroyed the master tape. We were afraid NASA would be angry that we had used too many official resources. But they saw the funny side of it, too. When we screened it at the party for the Apollo 12 crew after their mission it really brought the house down.

We also observed another nice tradition at this "pin party" (as it was called). Just as the military awards "Astronaut Wings" to those who have flown in space, the NASA Astronaut Office awards a gold "Astronaut Pin" to those in the astronaut corps who have flown in space. This is a small lapel pin that depicts three branches of a shooting star passing upward through an orbit. When an individual is officially selected by NASA to become an astronaut, he or she is presented with a silver astronaut pin. The silver changes to gold after his or her first flight, quite an achievement. The astronaut pin was designed by the Air Force to honor Gus Grissom after his Mercury MA-2 mission. It seemed to be such a good idea that the Astronaut Office adopted it for all subsequent spaceflights, and the Air Force selected it as a symbol to be superimposed on the shield of its pilot wings, and the Navy did as well. Alan Bean was proudly awarded his gold astronaut pin at this Apollo 12 pin party (Pete Conrad and Dick Gordon had already received their gold pins after the Gemini 10 mission).

Then came Apollo 13 and the entire mood of the American space program changed.

Alexei Leonov

Even after America's astronauts had first successfully circled the Moon, aboard Apollo 8, we had wanted to continue, of course, with our own circumlunar program. We had been training for it hard, and there was no reason to abandon it just because the Americans had achieved it first. Lunar missions were a vital part of our program, and those of us in training for these flights were firmly convinced we should go ahead and carry them out.

I had appealed directly to the Central Committee of the Communist Party and the head of the military–industrial complex that we be allowed to continue with our program. At first they agreed, deciding to launch the first of two remaining Zond spacecraft on an unmanned mission. If all went well, it was decided, the second mission would be manned and I would be in command.

But shortly after the crew of Apollo 11 splashed down back on Earth those circumlunar plans were abandoned.

The unmanned circumlunar mission, Zond 7, launched on 7 August 1969, carrying turtles and white mice, had gone brilliantly at first, sending back many beautiful pictures of the Moon, but on reentry there was a problem with the heat shield. The animals died. Such a problem would not have occurred with a manned mission, because of a slight difference in design of the Zond craft for manned and unmanned missions.

But the decision was taken not to continue with the circumlunar program. Then our manned lunar-landing program was canceled, too.

It was a political decision, of course. The Politburo were afraid to take the responsibility of losing another cosmonaut in space. The risks were too high.

A number of spectacular unmanned lunar missions still went ahead. The most successful was Luna 17, which delivered the first automated lunar rover, Lunokhod, to the surface of the Moon in autumn 1970. The vehicle roamed the Moon for nearly eleven months, transmitting back to us fascinating information.

The cancelation of our manned lunar program was a devastating personal blow. I was very upset and angry. I felt I had wasted the best years of my life on a project I was not being permitted by our political leaders to realize.

Even though, ultimately, it was not Mishin who had made the decision, a lot had depended on the way he presented circumstances to the Politburo. If he had argued firmly that everything was ready, perhaps their decision would have been different. But because the chief constructor did not seem adamant, it was easier for them to cancel the program.

I am quite sure that if Mishin had been in charge at the time of my Voskhod 2 mission he would have found a reason to delay it. He would probably have used our spacecraft for an unmanned mission. He would have delayed the whole program.

But if Korolev had still been alive I am certain our lunar programs would have continued. Under his leadership I firmly believe we would have been the first to circumnavigate the Moon. There would have been no guarantees that we could then have landed a man on the Moon, but Sergei Pavlovich really knew how to learn from past mistakes: he would have learned all the lessons possible from the early failures of the N-1 rocket, for instance, and corrected them properly. Had America's first attempt to land on the Moon failed, he would have taken the risk of sending his own crew to attempt a lunar landing.

I argued very hard that we should continue with our work, but the higher powers were adamant. The lunar groups which I had commanded and trained with for three long, hard years were disbanded.

The order came that we were to switch our focus to an orbiting space station which had been under development for some time. I was appointed commander of this new program, which was given the name "Salyut," as a salute to Yuri Gagarin.

There was no use, as the Russian saying goes, rubbing ashes on my forehead—crying over spilt milk. I knew I had to accept the decision. If I did not, there was a risk that I would never fly again.

But the switch to Salyut meant I had to get to grips with the extreme demands of an entirely new program. Besides the intensive physical training involved, I had to take technical drawings and detailed plans for the program home to study at night. The strain was so great that my hands sometimes used to shake.

David Scott

Just a few days before Apollo 13 was due to launch in April 1970, Ken Mattingly, the mission's Command-Module pilot, was exposed to the German measles. That night Deke Slayton came by my room at the Cape.

"Hey, Dave. We've got a problem with 13," he said. "I wonder what you think about it."

"OK," I said, "what is it?"

After explaining that he was thinking of replacing Mattingly with Jack Swigert, or possibly replacing the whole crew, or delaying the launch, Deke asked what I thought about the Mattingly–Swigert switch.

"This close to launch I don't think that's a very good idea," I said. "I don't really know what you should do, but I don't think switching Mattingly for Swigert is a good idea at all."

I knew Mattingly had a great reputation, and I didn't know Swigert very well—hardly at all, in fact. But I knew the back-up crew had had little access to the simulators over the previous few weeks. In the immediate lead-up to a launch, the prime crew get priority with everything. Teamwork was important, too. The business of being able to make a "no-look pass," which Neil and I had developed on Gemini 8, was vital. You couldn't achieve that with a last-minute replacement.

"I don't think Jack Swigert's ready—not from a personal performance point of view, but from a training point of view," I said.

"OK," Deke said. "Thank you," and he left.

The next day I found out that Mattingly had been replaced by Swigert.

"Oh, boy," I thought. "That's going to be difficult."

It was Apollo 13 commander Jim Lovell's call, of course. From what I understand, he was given a choice between flying with Jack Swigert replacing Mattingly and not flying at all. If they had tried to slip the launch by a few days they would have missed the lunar launch window—the phase during which the lighting on the Moon's surface makes for optimum landing conditions—and would have had to wait a whole month.

But I didn't see what the hurry was. There was no pressure on the system for Apollo 13 to go right then. We had met Kennedy's goal. We had landed on the Moon. We could have waited a month to see if Mattingly got German measles. If he had, the extra time would have given Swigert the chance for more intensive training. The last-minute crew switch increased the risk of the mission significantly, in my view.

After Apollo 13 launched, I was sitting in our living room at home listening to the NASA squawk box transmissions, which I had piped into our stereo system, when I heard the first indication of trouble.

"OK, Houston." Swigert's voice had a note of urgency. "We've had a problem."

There was no indication of how significant the problem was, but it seemed something had happened when the crew carried out a routine stir of one of the Command Module's oxygen tanks.

I had nothing particular to do that evening, so I decided, as part of the learning process, to make my way over to Mission Control Center (MCC), and see how the problem with the oxygen tank was diagnosed and resolved.

That was the only way I could watch what was going on. There had been little television coverage of the mission because after the climax of Apollo 11's lunar landing, the networks had started to lose interest in the space program.

The whole world had tuned in to Apollo 11, and a lot of people had watched Apollo 12, too. By the time it got to 13, I guess it was thought the mission was going to be just more of the same thing. It was no longer news. So the networks had decided not to transmit much live footage. The Apollo 13 crew were not aware of it, but the transmissions they had been making for the audience back home were, on one occasion, shelved in favor of *The Doris Day Show*.

When I got to the VIP room of MCC, there was talk of an explosion aboard Apollo 13. It was clear the situation was very serious. It was a time of great confusion. It was also time for a shift change of the team working in MCC. Those who had been on duty when the problem occurred were tired, but no one wanted to leave. Everyone was totally focused on finding out what the problem was and fixing it.

It was discovered, much later, that one of the oxygen tanks in the Service Module had been dropped during the manufacturing and assembly of the spacecraft. This had not been followed up well and the accident had left some micro-fractures in the tank. Then, during pre-launch testing, as was also discovered later, a faulty heater switch had overheated and fused closed. Subsequent current melted the wiring insulation. When the fans were turned on inside the damaged oxygen tank during the mission, to cool the oxygen, the tank blew up, damaging the electrical supply of the Command Module, where the crew lived.

DAVID SCOTT

Left: Growing up in Dad's footsteps, March Field, California, 1947 (Dad, Mom, Tom Jr., and me, top right).

Below: The good times in Holland—playing poker (and flying)—Gunnery Team, 32nd Fighter Squadron, Soesterberg Air Base, Netherlands, 1959 (From left to right: me, Frederick C. Blesse, Ed Schmidt, Jack Cebe-Habersky).

Above: Cadets and future officers (starched!)—1st Regimental Staff, West Point, 1954 (James Serber, me, Harry Sullivan, Fred Bowman).

Right: Ready for a "zoomy"—as a space pilot with Starfighter (F-104), Aerospace Research Pilot School, Edwards AFB, 1963.

Preparing for landing anywhere on Earth—Jungle Survival School, Panama, 1964 (part of our astronaut group of fourteen).

Amazing what you can do with a parachute!—Desert Survival School, Stead AFB, Nevada,1964.

Topsy-turvy (a weightless turnover)—practice for manoeuvering in space with the "zip gun"— zero gravity in the "Vomit Comet," Gemini 8 training, 1966.

Wrapped, packed and wired for orbit—Gemini 8 cockpit, 1966.

Dressed for the occasion—Gemini 8 crew leaving for the launch pad, 16 March 1966 (Neil Armstrong and me).

On our way!—GT-8 Titan II ICBM launch, 16 March 1966.

Apollo docking in our hands, Command Module to Lunar Module—press conference, Apollo 9, 1969 (Me, Jim McDivitt, Rusty Schweickart).

Saturn V in all her splendor—the night before the Apollo 9 launch, 2 March 1969.

Traveling light (not with vacuum cleaners!)—departing for the pad, Apollo 9, 3 March 1969 (Jim McDivitt, me, Rusty Schweickart and safety).

Photo opportunity with an outside view—Apollo 9 EVA, 5 March 1969.

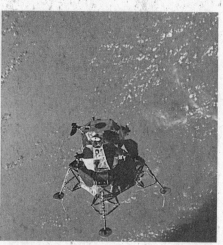

The connoisseur's test flight, ten days in orbit, ten hours solo—*Gumdrop* and *Spider* after undocking, 8 March 1969.

Back to my world—returning to Houston after Apollo 9 (Doug is in my arms, Tracy is approaching me).

Two weeks before launch, always worth a final check—me and my spacesuit, Apollo 15 Lunar Module and Lunar Roving Vehicle, July 1971.

Drilling for NASA— full dress rehearsal, geology "rock pile" behind crew quarters, Cape, February 1971. Note the reflection of the observers in my helmet visor.

"Flying Bedstead"— Lunar Landing Training Vehicle, Ellington Field, Houston, February 1971.

Parking not a problem! Lunar Rover with view up Hadley Rille, Moon, August 1971.

The dusty lunar explorer—collecting souvenirs on the steep slope of Hadley Delta Mountain, August 1971.

Just like home—Hadley Base: Jim Irwin, our flag, Lunar Module *Falcon* and Lunar Rover with Mount Hadley Delta in the background, August 1971.

Splashdown, back to Earth and a warm handshake—recovery by Lt JG Fred Schmidt, leader of Swim Team 2.

Paying our respects to the team—Apollo 15 crew arrival on USS Okinawa, 7 August 1971 (me, Al Worden and Jim Irwin, all looking thinner!).

We found what we came for!—the "genesis rock" in the Lunar Receiving Laboratory, Houston, August 1971.

This meant the crew had to transfer to the Lunar Module—designed for only two people—and the Lunar Module had to be used as the lifeboat to bring them back to Earth.

Interestingly enough, during our Apollo 9 mission we had carried out maneuvers to demonstrate that the Lunar Module could be used to bring the Command Module home. So, as soon as it was realized the Apollo 13 mission would have to be aborted and the crew brought home in the Lunar Module, I was involved in a lot of the intensive simulations to verify exactly how it could be done.

There was such a succession of problems with the spacecraft—not only with its electrical supply due to lack of oxygen, but with the supplies of oxygen, and water, not to mention a surfeit of carbon dioxide—that we were lucky to get the Apollo 13 crew back alive. It was a close call, right on the edge.

After all the celebrations to mark the crew's safe return had died down, there were renewed calls in some quarters for the lunar program to be canceled. There was a real fear that, next time, we might not be so lucky, that lives would be lost.

There was also continuing pressure from Congress over how much the space program was costing. Even before Apollo 13 flew, the budget for NASA's manned spaceflights was being questioned. The program was beginning to lose momentum and it was getting increasingly difficult to persuade Congress to continue voting for funding.

Proposals for Apollo to be followed by a manned mission to Mars, for instance, were considered too expensive and were axed. Even the intermediate aims of a reusable space shuttle and manned space station were openly debated. But, to a great number of people, it was unthinkable that the Apollo program be canceled, especially on such a low note.

The conservative approach after Apollo 13 would have been for NASA to fly—as planned—two more H missions, identical to 13, with their goal of limited lunar exploration, and then quit while they were ahead.

Instead, NASA, with the support of Congress, took a very bold decision. In summer 1970 it was announced that the last two scheduled missions in the program—Apollos 18 and 19—would be

258 TWO SIDES OF THE MOON

canceled. But the scope of the three missions before them—Apollos 15, 16 and 17—would be greatly expanded, and their scientific activity significantly increased. This directly affected me.

Very soon after Apollo 12 splashed down in November 1969, Al Shepard had called me into his office. Al had been appointed head of the Astronaut Office in 1964 when a problem with his ear disqualified him from flying. Eventually, in 1969, the problem was rectified through surgery and Al was appointed commander of Apollo 14. But before that he was heavily involved with Deke Slayton in crew selection.

"You're going to be commanding Apollo 15, with Al Worden and Jim Irwin as your crew," said Al. "Dick Gordon, Vance Brand, and Jack Schmitt will be your back-ups. Is that OK with you?"

OK? It was more than OK. It was great. This would be my mission. Of course, with the position of authority came all the responsibilities of coordinating every aspect of the mission for the prime and back-up crews. The commander had the option of rejecting any individual as a member of his crew, but I knew, from working with them on back-up to Apollo 12, that both Al and Jim were exceptional. Dick, Vance and Jack would make another great team. There was no need for a long discussion.

"That's fine," I said. "Let's go."

Alexei Leonov

For some time I had been director of training for the entire cosmonaut corps, which consisted of instructors, trainers, engineers and technical staff. Head of the unit was my former commander on Voskhod 2, Pavel Belyayev. But on 20 December 1969 Pasha was diagnosed as suffering from a duodenal ulcer. He had not felt any symptoms until it perforated and by the time surgery was performed it was too late. On 10 January 1970 Pavel Belyayev died. After Yuri Gagarin he had been my closest friend. I never again formed such close friendships as I had enjoyed with those two men.

Shortly after Pasha's death, I was appointed head of the training unit in his place. It was extremely difficult to combine running the Salyut program with the many duties of overseeing overall cosmonaut training and heading the entire corps.

Twenty cosmonauts were assigned to me to begin training for the Salyut program. Twelve of them were appointed to four three-man crews. We undertook even more intensive physical training than before, because the physical toll of spending prolonged periods in space was beginning to become apparent. We had to be prepared.

In the beginning, it was thought that weightlessness was beneficial, because it imposed less strain on the heart and other bodily functions. There had even been some speculation that space hospitals might be developed, because it was thought some particularly tricky operations might benefit from being performed in zero gravity. But gradually it became apparent that long periods of weightlessness caused considerable weakening of the muscles and immune system.

In autumn 1969 the Soyuz program, which had flown only four manned missions up to that point, had been pushed into high gear in preparation for Salyut. In October 1969 alone three Soyuz missions were launched. A great deal was learned about long-duration flight and complex maneuvering from these missions. In June 1970, Soyuz 9 set a record for a long-duration mission: eighteen days. One of the crew members was so weak on his return that he could not even carry his own helmet when he stepped out of his landing capsule.

From then on it was obvious that exercise equipment would have to be incorporated into a space station so that cosmonauts could keep up their muscle strength.

The long periods cosmonauts were expected to spend aboard Salyut also meant that the practical aspects of the program's medical and biological training were greatly extended. We had to learn how to perform emergency medical procedures in case anyone fell sick in space. We learned, for instance, how to extract teeth and how to administer and interpret electrocardiograms and encephalograms. We learned to take blood samples from both the finger and the veins.

The program also included a wide range of astrophysical

research, including the study of the solar corona and the constellation of Orion, so we studied not only astronomy but also the physical processes within the celestial bodies. We cooperated at astrophysical observatories with leading academics in the field such as Andrei Severny in the Crimea, Viktor Ambartsunian and Grigor Gurzadyan in Byurakan, Armenia, and Yevgeny Kirillovich Kharadze in Abastumani, Georgia. After completing each course we had to pass a series of exams.

After Apollo 11, we heard very little about the American space program until the Apollo 13 disaster hit the headlines. That whole drama lasted for seven days. Again we had no television coverage, but this time there was a lot press coverage of the crew's problems.

Our training program was so intensive by this time, however, that I paid little attention to outside events. By April 1971 we were ready to launch the 18-ton Salyut 1, the world's first orbital space station.

The plan was for a Soyuz spacecraft—Soyuz 10—to be launched with a three-man crew aboard four days after an unmanned Salyut had been sent into orbit. The Soyuz spacecraft would then dock with the space station and its crew would make an internal transfer into Salyut. I was commander of the back-up crew for Soyuz 10 and, once again, made the journey to Baikonur to remain on standby.

Salyut 1 launched perfectly and was followed into space four days later by the prime crew aboard Soyuz 10. But when the two vehicles rendezvoused and tried to dock the Soyuz capsule developed a problem with its docking mechanism, and the Soyuz crew had to perform an emergency undocking and return to Earth. After the cause of the problem was identified, another Soyuz mission was scheduled for launch on 6 June 1971, with the same flight program as the previous mission but with a new docking mechanism. As commander of the back-up crew on the previous mission, I was designated commander of Soyuz 11. We had just a month to prepare for this mission. My crew were Valery Kubasov and Pyotr Kolodin.

All my possessions—my underwear, pajamas, sketchpad and colored pencils, strung together with a sturdy thread attached to a bracelet I could wear when drawing—had already been stored aboard Salyut 1, in case I was required to launch in Soyuz 10. They caused some consternation when a problem developed with Salyut's ventilator system.

Mishin rang me up and demanded, "How come your coloring crayons are up there causing problems?"

"They can't be," I assured him. "They're in a box that's taped shut."

For some time people thought my belongings had caused the system to malfunction; that the crayons and the thread that held them together had become entangled with some operational part of the spacecraft. Apart from this technical hiccup, I thought the omens were good for Soyuz 11: my designated number in the astronaut corps was 11. But fate was to play a very cruel trick.

Shortly before the mission Kubasov developed a problem with his lungs. It turned out later that he was allergic to a chemical insecticide used to spray trees near the cosmodrome in Baikonur.

At first, a member of the back-up crew, Vladimir Volkov, was transferred to my crew in Kubasov's place. Then, eleven hours before the launch, the entire crew was changed. It was feared that Kubasov might have a lung infection, and might have transmitted it to Kolodin and myself. I thought it unlikely, since everyone around us had had to wear masks for two weeks before our mission, precisely to prevent such problems. If Kolodin and I had not already come down with an infection, I didn't see any reason why we should fall ill.

But the decision was final. I would not be commanding the mission after all. I would be replaced by Gyorgy Dobrovolsky and Kolodin by Viktor Patsayev. Kubasov's replacement, Volkov, would remain with the mission.

I was both angry and astounded at the decision. The back-up crew were even more aghast. They had not had as long to train and prepare for this mission as we had, and were not expecting to fly for another two months. If you look at photographs taken of the replacement crew just before the launch, they even look a little scared.

That night I made a sketch of Patsayev which I called *Patsayev's*

Eyes; they looked troubled. Patsayev was a very gentle and considerate man. He even came up to me to apologize for what was happening.

Our cancelation as crew for the mission left me with a bitter sense of *déjà vu*. It felt like a repeat of the terrible disappointment when our circumlunar and lunar landing programs had been canceled. For Kolodin the disappointment was even worse. After all, I had flown in space before, whereas it would have been his first space mission. He cried with frustration.

I protested to Mishin that this was our project. I stressed that I had been involved in the design of the spacecraft. I had been training for so long. Everything was ready. Even my drawing materials were waiting for me on the space station. But Mishin would not listen, nor would the military committee who had come from Moscow to give final approval for the mission. The doctors had not yet determined the cause of Kubasov's lung problem.

"Don't forget," Mishin argued, "you have been sharing a room with Kubasov. You might even have drunk from the same cup that he used. We cannot take the risk that you will develop lung problems, too."

He was right, I now realize, but at the time the disappointment was almost unbearable. Mishin tried to soften the blow by sending me on vacation to the Black Sea, but I stayed at Baikonur long enough to observe the first part of the Soyuz 11 mission. I witnessed it launch and dock, successfully this time, with Salyut 1. Everyone stood and applauded when the crew at last transferred into the space station.

I was eager, once the crew entered the space station, to hear their analysis of what was causing the problem with the Salyut's ventilation system: I wanted confirmation that it was not my crayons. With Mishin standing right beside me, I switched on the transmitter in the space station so that I could talk to the crew.

"Could you please confirm that my crayons are still in their box?" I asked.

"Yes, they're still there," Dobrovolsky radioed back.

"Well, maybe my underpants are causing problems, then?" I inquired, turning to look at Mishin.

"No, they're stowed away, too."

Vindicated, I left Baikonur for the Black Sea. Close examination of Salyut's systems later revealed that some small fibers had been sucked into a ventilator filter, causing it automatically to switch off.

For twenty-three days the mission went remarkably well, despite the fact that the crew had had so little time to prepare for their packed schedule of biomedical and scientific experiments. When we returned from our two-week vacation to the Black Sea, the mission was nearing completion.

We had to start training immediately for the next Soyuz mission, which was due to launch a month after the return of the Soyuz 11 crew. But as Soyuz 11 prepared for reentry to the Earth's atmosphere I was following events closely in the bunker of a new Mission Control center in Kaliningrad, near Moscow. It was customary for the commander of the back-up crew to monitor events at this time, together with the head of the mission and the chief constructor, to maintain radio contact with the crew and to pass on any advice and necessary instructions. So I was there monitoring the control of all on-board systems in a logbook.

As the crew went through the control of positioning the air vents located between the landing and orbital modules, I advised them to close the vents and not to forget to reopen them once the parachute had deployed.

"Make a note of it in your logbook," I instructed them.

Although this deviated from the flight regulations, I had trained for a long time for the mission they were flying, and in my opinion this was the safest procedure. According to the flight program the vents were supposed to close and then open automatically once the parachute had deployed after reentry. But I believed there was a danger, if this automatic procedure was followed, that the vents might open prematurely at too high an altitude and the spacecraft depressurize.

It seems the crew did not follow my advice. Unfortunately my intuition proved right. The pressure equalization vents did open prematurely—before the capsule reentered the Earth's atmosphere—and the capsule depressurized. Soyuz 11 landed where expected, but when the recovery team opened its hatch they found the crew dead. Their bodies were still warm, and the recovery team tried to resuscitate them, but it was too late. According to their cardiogram records, Dobrovolsky had died two minutes after the landing capsule had depressurized, Patsayev after a hundred seconds and Volkov after only eighty seconds.

On every mission following Soyuz 11, the flight program was changed to incorporate my advice. The vents were also redesigned.

When the rescue forces reported that the crew were dead, I was instructed to fly to the landing site immediately with fellow cosmonaut Alexei Yeliseyev. We were appointed members of the government committee dealing with the aftermath of the disaster, and our main task was to secure the spacecraft and take photographs of the scene. It took us about three hours to reach the site, by which time the bodies of the crew had already been removed. Their blood-soaked seats, and signs that attempts had been made to resuscitate them, were the only evidence of the tragedy.

The committee concluded that had the crew been wearing spacesuits they would have survived the depressurization. After that it was decided that during launch and reentry cosmonauts would always wear spacesuits. This meant that, in future, Soyuz spacecraft would be able to accommodate only two people, not three.

The loss of the Soyuz 11 cosmonauts was a terrible blow to the morale of the whole corps. Everyone understood that we were in the business of testing spacecraft, and the deaths of these three men undoubtedly saved the lives of subsequent crews, because of the substantial modifications made, but their loss was a tragedy. Not only was I deeply saddened by what had happened, but I was frustrated, too. Had I been allowed to fly in their place I am sure my crew would have survived.

I never told anyone that the crew had failed to follow my instructions and that this had led directly to their deaths. But many years later Victor Patsayev's wife, Vera, who was one of the leading experts at the design center, and who had access to radio exchanges prior to the launch, recognized the crew's tragic mistake of not following my advice and made that fact public.

Dobrovolsky had two daughters; the younger was a friend of my own daughter, Oksana. Volkov had a son, and Patsayev two boys. I tried to avoid seeing them after the accident. I could not bear to look into their eyes. Even though it was not my fault, I blamed myself for what had happened. It was not until much later that the children learned how desperately I had tried to avert the tragedy.

David Scott

The significant shuffling of missions following the Apollo 13 failure in the spring of 1970 had enormous implications for Jim, Al and me on Apollo 15, which was due to launch in July 1971.

When I was assigned as commander of this mission, at the end of 1969, the flight had been scheduled as an H mission—essentially the same as the previous lunar landings. Eight months later it was upgraded to the first J mission. That meant we would have to accomplish a great deal more during our twelve-day mission. Instead of spending one and a half days on the Moon, during which two walking EVAs would be performed—as on an H mission—our stay was extended to three days. During that time Jim and I would conduct three much longer EVAs—totalling about twenty hours—with the use, for the first time, of the Lunar Roving Vehicle.

In addition to developing new procedures and testing new equipment, Apollo 15 was to be a highly scientific mission. Most of the time we spent outside the Lunar Module would be dedicated to carrying out a detailed geological investigation of the lunar surface. Because of the length of our EVAs we would have to carry more sophisticated equipment, our backpacks would be of longer duration, and our spacesuits better. So much new equipment was to be put into use for the first time on our mission—not least the lightweight, collapsible, but still cumbersome Lunar Rover—that the Lunar Module had to be upgraded and its performance enhanced to accommodate everything.

We would be the first crew who would truly have to adapt to living on the Moon. We decided to plan our schedule in rough accordance with the circadian rhythms of a working day in Houston. We knew we would have to perform at our peak, and so would have to get some sleep. Previous lunar stay missions had lasted thirty-six hours, so the crew could just about get away with not sleeping. But for three days that would be impossible.

Sleeping on the Moon was not as easy a proposition as it might sound. In order to rest properly, we decided, we would have to take our cumbersome spacesuits off. We would then set an alarm clock, turn the lights off and draw the blinds.

It would be permanent daylight on the Moon during our stay. For fourteen Earth days our landing site on the Near Side of the Moon is bathed in sunlight, forming a lunar day. For fourteen days it is dark. We would be landing on the Moon during the early morning of the lunar day to avoid the intense heat—around 230° F—of high noon. Landing then also meant that the sun would be behind the Lunar Module as we approached touchdown from our east–west orbit. The low sun angle would produce long shadows on the surface terrain, which would make it easier for us to identify the landmarks near our landing site.

We would be sleeping in hammocks strung from the walls of the Lunar Module. One night we tried simulating this at the Cape and it didn't work at all; we were too heavy. In one-sixth gravity, on the surface of the Moon, it would be much easier—like sleeping on a feather comforter—but on Earth it was pretty uncomfortable.

Attempts were also made to simulate driving the Lunar Rover in reduced gravity. At first the spindly vehicle was strung up on supports from the side of a big building in Houston, but this proved a poor imitation of what we might encounter on the Moon. To develop and practice our geology tools and procedures, tons of volcanic cinders, rocks and boulders were unloaded at a large area behind the training building at the Cape, and we attempted to make it roughly resemble the lunar surface. We spent many hours learning to master these tasks in that area, which came to be known as the "Rover Racetrack." It turned out to be pretty easy to drive the rover on Earth, but on the uneven lunar surface it would be another matter.

In case there was a problem with the Lunar Rover, we had to design and prepare an entirely separate program of geological expeditions, or traverses, which could be carried out on foot. The risks associated with such long distance EVAs on foot, should the rover not function correctly, were not the only dangers we might face. The very real risks of astronauts being exposed to dangerous levels of radiation caused by solar flares—periods of intense solar activity—have been staple fodder for science fiction writers for years. Such activity can be fairly reliably predicted, and our mission was expected to avoid it. Still, the timing of Apollo 15 did coincide with a period of high solar activity. There were no guarantees we would not be subjected to solar flares.

We also had to develop an effective system of orientation to use on the Moon, in case we got lost, which might prove fatal. Our supplies of oxygen would be limited, the simple rover navigation system was unproven, and we would lose sight of the Lunar Module over the horizon after traveling just over a kilometer. With no magnetic field on the Moon, a compass would be useless. Instead we designed a sun compass resembling a sundial, which was mounted on the back of one of our checklists.

With all the new objectives of this mission it was as if the Apollo program had expanded from a 100- to a 200-piece orchestra. There were all these new instruments, and each one had to be harmonized.

As Command-Module pilot, Al spent most of his time out at Downey, preparing for the complex part of the mission he would perform alone in lunar orbit while Jim and I carried out our three days of exploration on the surface. No American had ever before spent so long alone in space in such a complex spacecraft. Al would, in effect, be performing alone all the maneuvers carried out by some of the earlier three-man Apollo missions.

While Al was in California, Jim and I spent most of our time on Long Island—where the Grumman Corporation was assembling and testing our upgraded Lunar Module—at the Cape or in Houston. We ran through ever more complex and demanding simulations of what we could expect during our landing and lengthy lunar expeditions. Part of the strategy was to put us through every kind of conceivable emergency. As it turned out, we were to face at least one emergency even NASA had not anticipated.

Once again I got to enjoy the delights of the KC-135 "Vomit Comet," in which I had trained intensively in the run-up to Gemini

8, with Dick Gordon who had been my back-up on that mission and was my back-up again on Apollo 15. Over the course of our training Dick and I—with Jim and his back-up Jack Schmitt— regularly went to Ellington Air Force Base, in Houston, or Patrick Air Force Base at the Cape for trips in the KC-135, which was specially equipped to allow us to simulate some of the tasks we would be performing in the zero gravity of space or the one-sixth gravity of the Moon. Sometimes Al Worden joined in these training sessions, in preparation for an EVA he would be performing during our return from the Moon.

As the plane, roughly the size of a Boeing 707, flew continuous parabolic arcs, shifting from two Gs to zero gravity again and again like a roller coaster, we practiced some of the more complex activities we would be performing. To prepare Jim and me for our time on the lunar surface, a mock-up of the forward hatch and front porch of the Lunar Module was bolted inside the stripped-out fuselage so that we could practice entering and exiting in our pressure suits. Another mock-up, of the seat section of the Lunar Rover, was bolted to another part of the fuselage so that we could practice our movements in the seat of the vehicle, again in our pressure suits.

The logistics of simulating these maneuvers in the very brief periods during the flight when the plane went through spells of zero gravity and one-sixth gravity was complex. A loud buzzer would echo through the belly of the plane to signal the beginning and end of these periods. Zero gravity lasted only 15–20 seconds as the plane went over the top of a parabola. One-sixth gravity lasted for about the same time but was flown as a separate profile by the KC-135. As soon as the buzzer sounded, we started performing the small section of whichever maneuver it was we could practice in those 15–20 seconds, such as climbing on to the rover and putting our seatbelts on. The buzzer sounded again to signal that the period of one-sixth gravity was over and we had to wait for the next period to carry on with the activity. If the plane were in a dive, the buzzer would sound again as the pilot began to pull out of the dive to warn us we were about to experience a force of two Gs. This was to make sure we were not in a compromising position in which we might hurt ourselves as the plane bottomed out of the dive before climbing again.

In addition to the handful of engineers from our Flight Crew Support Team, who accompanied us on these flights, there were always a couple of safety personnel monitoring our position and helping to get us in a safer position if it looked as if we might get hurt. I never was hurt and never got sick, but others did go distinctly green around the gills, especially the members of the media who were occasionally allowed to join the flight for observation. Those who became ill could sit strapped at the back of the plane to observe our training. The flights could last for a couple of hours and there was no way they could be cut short to alleviate the suffering of a member of the press. They were simply given enough sick-bags to last them through the flight.

The last zero-gravity flight was the longest. We must have performed some 130 parabolas. We didn't know it beforehand, but it turned out Al had arranged this so that we would set a new record. When we landed, there was a photographer there, waiting to take a picture of the whole crew to mark the event.

I also spent hours learning to pilot the lunar landing training vehicle—the "Flying Bedstead," which Neil had had to bail out of prior to Apollo 11—at Ellington Air Force Base. At the start of the Apollo program there had been four of these Flying Bedsteads for the use of successive crews, but they proved so unstable and subject to technical faults and accidents that by Apollo 15 there was only one left.

Each time an LLTV crashed after developing a fault and forcing its pilot to eject, there was a great hue and cry among the management at NASA. Many felt the awkward-looking craft were too dangerous, exposing astronauts to too high a risk during training. But I, and all the other commanders of Apollo lunar landing missions, considered we would be running a much greater risk by piloting the Lunar Module to the surface of the Moon without having had the benefit of the experience of flying the LLTV. It was always a little hair-raising. No other flying experience could compare—a helicopter would be about the closest thing.

We did spend a short period of intensive training flying

helicopters; Dick Gordon and I spent two weeks together at the Navy's helicopter school in Pensacola, Florida, and had a great time concentrating just on flying. To maintain our proficiency after that, we flew NASA helicopters at Ellington and I often spent Saturday mornings practicing landings and flying in the chopper in the areas around Clear Lake.

But flying the LLTV was the best way of learning to fly in the coordinated mode, using both hands, that was needed to control the forward–aft, up–down and left–right motion of the Lunar Module. The pilot of an airplane has to contend with only four degrees of freedom: an aircraft only flies forward, though it can roll, pitch and yaw. A helicopter has, like the LLTV, six degrees of freedom, forward–aft, up–down, left–right; but it is still easier to fly than the LLTV, since it does not have the variable-thrust rocket engine of the "Flying Bedstead," which we had to bring in to land safely from a height of about 500 feet. One feature of the LLTV which caused more than a few problems was the hydrogen peroxide in the engines used for attitude control. On humid days in Houston the vapor from the engines created quite dense clouds, which meant flying virtually blind for short periods. But it was invaluable training, and I flew the LLTV two or three times a month as we drew closer to launch.

Another vital focus of our training was the intensive preparation for the geological investigations we would be carrying out. This meant field trips. I loved them. I loved to be outdoors. It was a chance to get away from the simulators and other hardware for three days at a time. A chance to have a few beers in the evening after a hard day in the field.

In the evenings after many of these trips we discussed the names we wanted to give to some of the craters and mountains of the Moon that we spent so much time studying. Technically, only the International Astronomical Union can approve names for landmarks on the Moon, and, as it turned out, they didn't like some of the names we chose. They wrote us a long, very serious letter stating their objections to, for instance, "No-name Crater": "You can't call a crater that. It's just not logical."

But the names we chose stuck, and were used by all the geologists and scientists working with us at NASA. In the end they were recorded in all the official documents and that essentially carried more weight than the objections of the IAU.

There was little appreciation at first of how complex our geology missions were. I'd had an interest in geology for some time: the art of interpreting millions of years of history in rock samples and structures seemed a natural progression from the interest I had long harbored in history and archaeology. But it proved pretty hard to get some of the NASA guys interested in observing our performance during this aspect of the program. I never could get Deke Slayton to go on geology trips, for instance, or even in the simulators for the mission. It was really difficult to communicate with him at that time. He seemed rather detached.

I remember one meeting, in particular, with Deke late one evening in his office. I had to argue my corner very hard to get him to agree to us taking just one extra piece of geological equipment, a light aluminium rake. Deke just didn't get it. He chewed his cigar and slumped back in his chair digging his heels in as I explained it would enable us to pick up small fragments of rock. He was equally resistant to my idea of taking a telephoto lens for high-resolution distant shots of geological features we would not have time to study at close quarters.

I understood his concerns about the extra weight the items would add to the LM, which was already heavier than that of any previous mission because of the rover and additional equipment we needed for our longer stay on the lunar surface. Weight was a crucial issue on every spacecraft and was closely monitored by NASA's Configuration Control Board. Early on in the Apollo program technicians had been asked to review every single item carried aboard, and, in an effort to rid the spacecraft of even the smallest unnecessary weight, they reduced the number of Band-Aids in the first-aid kit. But I was confident that the two extra items I wanted to take could be accommodated. For certain aspects of the mission we required less propellant in the LM than previous missions, for instance, because we would be making a direct rendezvous with the Command Module after our lunar stay, rather than orbiting the Moon once before the rendezvous.

Deke eventually came around to my way of thinking, but I still had to spend time arguing my case at higher levels. There had long been resistance among senior managers and engineers at NASA to including more science on the Apollo missions. Their position was, I think, "Seen one rock, you've seen 'em all. Just get the geology out of the way. Get the rocks and come home." But there was a gradual shift in opinion once the major hurdle of landing men on the Moon and safely returning them to Earth had been overcome. That was due in large part, I firmly believe, to the brilliance and enthusiasm of Professor Lee Silver, our main guide and mentor in the study of geology.

Lee Silver was an inspiring teacher and a really nice guy. He made learning entertaining. He knew how to get across the complexities of what he was talking about. He knew how to fill us with enthusiasm for all that could be achieved in the geological study of the Moon. Lee worked closely with our geology team leader, Gordon Swann. They had been close friends for many years and were both great to work with. But it was Lee who usually led the geologists who took us out in the field, preparing trips to areas resembling some of the geology we might expect to find on the Moon. At least once a month throughout most of our twenty months of training, Lee and his team took us out to ever more complex and demanding locations.

We went to Hawaii, the Rio Grande, the Mohave Desert, and the Orocopia Mountains, the Coso Hills and the San Gabriel Mountains in California. We were taught to observe carefully and analyze what we saw. We were shown how to pick out key rock samples, which explained how that particular piece of the Earth's crust had been formed. On the Moon, our mission was to find rocks which would explain the mysterious genesis of our nearest neighbor in space.

In the beginning we had little idea of what was expected of us. When asked to describe what I saw on one of the first trips to Orocopia, for instance, I got not much further than saying, "Boy, there's a lot of stuff on the other side of the hill." Lee Silver helped us tune in to the language of geology, and soon we were describing the composition of that "stuff" as granite, basalt, sandstone or conglomerate, and its shape as angular, sub-angular or rounded. On the long drive out to a site he would suddenly have us pull over by the side of a road cutting through the mountains and get us to describe the rock formations we saw.

One of Lee's favorite tasks was to get us to collect a "suite" of rocks, a selection which represented the geological setting and diversity of the area. There was always a fair amount of friendly competition between the prime and back-up crews, so Jim and I would compete against Dick Gordon and Jack Schmitt to present the most impressive suite to the professor. Jack was the first trained geologist recruited to the astronaut corps and his expertise spurred us to learn. It also meant he and Dick were a tough team to beat. On a trip to Hawaii, Jim and I got caught in a thunderstorm on a pretty remote mountainside and approached our geology task with a little less enthusiasm than usual; consequently we missed something important. Jack and Dick hadn't missed it, of course, and said mockingly that we "needn't think we could get out of a task because of a little freezing rain." It was all good-natured, but kept us on our toes none the less.

Once our knowledge was sufficiently advanced, we started conducting geological traverses as if we were on the surface of the Moon. We carried backpacks of a similar shape, with radios attached so that we could communicate with a mock-up Mission Control—another member of the geological team sitting in a tent on the other side of the mountain we were exploring. These exercises were valuable in getting us to describe the landscape in detail to people who could not see it.

During the months with our geology mentors a new concept of field geology emerged, which might be called "planetary field geology." In some ways it was more of an art than a science, since it had to be conducted under very different and much more demanding conditions than terrestrial field geology: in a hostile and unforgiving environment and under extreme time limitations offering no opportunity to return to the same location—at least not in our lifetime. A terrestrial field geologist can often spend weeks or months in one location and return later to complete and reevaluate findings, we had to make instant analyzes, and decisions on the scientific value, of the objects we found, without time to relish their meaning. In effect, we would have to rely on instinct and training in picking a sample and would have only about five seconds to look at it, and maybe ten seconds to describe it, before bagging it and moving on.

Key to fulfilling such an ambitious mission was deciding where on the Moon we would land and explore. There were a number of good landing sites, each of which had its keen advocates. Finally, a big meeting was convened in Washington to decide our landing site. The choice had narrowed to two possible sites; one called Marius Hills and another known as Hadley Rille.

Hadley Rille was a 1000-feet deep, 1.4-km-wide channel on the shore of one of the lunar seas, the Mare Imbrium or Sea of Rains. It was near one of the Moon's most magnificent mountain ranges, the Apennines, which was thought to contain parts of the Moon's primordial crust.

As commander of the mission I was invited to the meeting. Advocates for each site presented their case, and then there was a fairly heated debate, at the end of which I was asked what I thought.

By far the more spectacular site, in my opinion, was Hadley Rille. It had more variety. There is a certain intangible quality which drives the spirit of exploration and I felt that Hadley had it. Besides it looked beautiful, and usually when things look good they are good.

It was a close call. But at the end of the meeting the decision was made: we would be landing at Hadley Rille. Exploring the Rille as well as the vast base of the Apennine Mountains was a spectacular opportunity.

CHAPTER 10

In the Footsteps of Captain Cook

1971

Colonel David Scott

Cape Kennedy, Florida.

Early each morning in the run-up to the launch of Apollo 15, I drove out to a deserted beach house near the launchpad at Cape Kennedy. It was a rough old place, to which only a few astronauts and the cook, Lou, had keys. Lou usually left a supply of orange juice in the icebox there. After changing and taking a drink of juice, I'd set off jogging along the sand with just a few seabirds circling overhead for company.

Often the only other sign of life I saw was the tracks of giant turtles which had hauled themselves ashore earlier in the day. Some time later I found out there were always one or two security guards posted at a nearby gatehouse keeping an eye on me, too. Still, it was nice out there, cool and fresh. It was a good time to get some peace and be on my own.

From the moment I sat down to breakfast, when I returned to crew quarters after my run, until late each evening the program was very intense: simulations, management meetings, geology briefings, mission rules and procedures, the flight plan, spacecraft checkout. Everything was progressing rapidly toward the launch deadline of 26 July 1971. There was no let-up. At every turn there was someone asking, "Hey, Dave, have you got a minute?"

But as I ran each morning I had a little time to think quietly about the mission, to reflect on the magnitude of the venture upon which we were about to embark. I sometimes drew inspiration from reading the experiences of great old explorers, in particular Captain James Cook. By coincidence, on 12 July 1771, almost exactly 200 years before Apollo 15 was due to soar into the Florida skies at the start of the first truly scientific expedition to the Moon, Captain Cook had dropped anchor in Deal harbor, England, at the end of the first truly scientific expedition by sea. For Cook it was the conclusion of a three-year voyage of discovery in the South Pacific with his seventy-strong crew aboard his ship the *Endeavour*. As a tribute to him we had named the Command Module of our Apollo spacecraft *Endeavour*, and we had arranged, courtesy of the Marine Museum at Newport, Rhode Island, to carry with us a small block of wood taken from the sternpost of Cook's ship.

Captain Cook was not the only explorer whose courage inspired me. Eighteen months earlier I had spent a week in Antarctica as part of a NASA delegation sent to observe scientific research in a hostile environment at what was then still a relatively new frontier for exploration. It was not a training exercise; I was there purely for observation. I had taken with me a gripping book chronicling the race to the South Pole by the British Antarctic explorer Captain Robert Falcon Scott and his Norwegian rival Roald Amundsen in the early 1900s. The small teams of men and dogs they relied on contrasted sharply with the enormous back-up and technical brilliance of the 400,000 gifted individuals working for the Apollo program.

Whereas Amundsen and Scott remained out of contact with civilization for many months, battling to overcome the unknown, we would be in contact with Houston throughout most of our mission. The isolation they must have felt would be echoed in some ways, however, by the physical isolation we would experience over a quarter of a million miles from home on the surface of the Moon.

Few places on Earth can prepare you mentally for living on the lunar surface, but that week spent in Antarctica in January 1970 was perhaps the closest I came to observing some of the challenges I would face. The reflection of sunlight on ice day and night during the southern summer would, in some ways, replicate the intense glare of the sun on the surface of the Moon. The difficulty of moving any distance in the Antarctic bore some resemblance, too, to the restrictions we would experience in lunar mobility.

That short trip to the Antarctic also stimulated in me an even deeper fascination with science. The excitement I witnessed among a team of geologists and paleontologists working on substantiating the theory of continental drift and plate tectonics at a temporary summer camp on the Beardmore Glacier really impressed me. Their enthusiasm when they found the clear impression, in metamorphosed shale, of a 200-million-year-old African leaf in support of the theory, was infectious. Though neither Al Worden nor Jim Irwin was able to make that trip, they shared my enthusiasm for our ambitious science program.

Jim, Al and I were to be the first all-Air-Force Apollo crew: hence the naming of the Lunar Module after the official mascot of the United States Air Force Academy, *Falcon*. I could not have wished for a better crew. As back-up crew to Apollo 12 we were already a close team. There is no one I would rather have spent time on the Moon with than Jim Irwin. He was a very bright guy, quiet, but constantly aware. He always had a pleasant demeanor, came up with good suggestions, and was very easy to work with. Al was a perfect match for the two of us. He was pretty independent. He had to be: he had to work through all the procedures and science he would conduct alone in the Command Module independently of Jim and me. All three of us worked well with our back-up crew, Dick Gordon, Jack Schmitt and Vance Brand.

Everything went pretty smoothly in the run-up to the mission, though the relentless grind of our training schedule did take its toll in some ways. During the final months Jim approached me one morning to discuss concerns over the stress his long absences from home were placing on his marriage. Fortunately he and his wife, Mary, managed to work things out. Al was not so lucky.

He had asked my advice about a problem concerning his marriage some months earlier.

"Things are just not working out," he said. "But I don't want to risk doing anything that might jeopardize my place on this mission."

The word at the time was that unless you were the all-American boy, with a perfect family, you weren't going to fly. Those experiencing marital difficulties were hesitant about divorce, although a few, who had already flown missions, had since divorced.

I sought Deke's advice. "As long as a guy is doing the job, that's all that counts" was his reaction. Al was certainly doing a great job, so, sadly, he did get divorced before we flew.

Since Lurton came from a military background, she had more of an understanding of the extreme demands our work placed not only on us but also on our families. She was fine about it. She even took an introductory class in geology at university in Houston so that she could join in conversations with the geologists who stopped by our house for supper. As for me, I was away less than if I had been a naval officer away at sea for long periods at a time. I did miss out on seeing some of the kids' after-school activities. Tracy was nine by then and very keen on ballet. Doug was seven and played Little League football. But he was a keen swimmer, too, and I swam with him at weekends. Besides, the kids were excited about their dad's "great adventure."

Several weeks before launch, Jim, Al and I were put into medical isolation to prevent us from being exposed to any germs or illnesses that would risk delaying the mission. "Flight crew health stabilization program" was the technical name for this. In effect it meant living behind huge panes of glass in the crew quarters and simulator buildings at the Cape. We still had direct contact with everyone in our support team and certain people in NASA management who went through daily medical screening and carried special badges to be allowed access to the restricted areas.

But the only way we could see our families from that point on was through the glass window of a conference room in the crew quarters when they came to wish us well before lift-off. They then went on to the big send-off party organized in our honor. We didn't, of course, but they told us it was a lot of fun and that we were missed. My brother came with his family. Al's and Jim's families and friends were there, too, together with a lot of NASA staff. All the kids went swimming at the beach.

As launch day drew close there were no last-minute hitches with health problems, as there had been on Apollo 9. Everything was running smoothly. It was a good feeling. I was totally focused. It was like training for a big sporting event; you could not afford to peak too early; but I knew our timing was right. We would never be more ready. Everything was coming together as it should. Again and again we had practiced every aspect of the mission to the highest possible fidelity. Nothing was left to chance: that was a lesson I had learned long ago from Neil Armstrong on Gemini 8.

All our geology field exercises were complete. All our brand-new equipment was ready and waiting for launch at the Cape. Gerry Griffin, a good friend of mine, was to be our lead Flight Director. That was good news. I had great confidence in him.

Everything was right. It was exhilarating. It felt as though we were on a giant wave rolling toward the shore that was launch day with a momentum of collective energy which was enormous, relentless.

At 4:19 a.m. Eastern Daylight Time on 26 July we were woken by Deke Slayton, judged to be well rested and then underwent brief physical exams. After being declared in excellent physical condition we sat down to a traditional hearty breakfast with members of our back-up crew and support team. From there we were taken to the suit laboratory, where our individual suit technicians began the lengthy process of fitting us into our spacesuits. There was a faint echo about the process of a medieval knight being helped into his armor, as there had been on Gemini 8 and Apollo 9, but this time our spacesuits were even more refined for the unique tasks that lay ahead of us. Those suits—known as "extravehicular mobility units" or EMUs—were amazing feats of engineering.

Once on the lunar surface Jim and I would add even more items to our suits. But at this stage all three of us donned essentially the same elaborate outfit, each individual piece of which had to be checked and verified before the next could be added. Our latter-day "armor" began with what we referred to jokingly as a "motorman's helper" after a device worn by tram drivers in San Francisco who could not leave their trains for long periods. Its technical name was a "urine collection device" or UCD, and it was a harness containing a yellow bag to which we attached ourselves with a condom-like device.

Once we were hooked up to these devices, a mesh of bio-sensors was attached to our bodies and connected to a bio-harness round our waists; the bio-harness also contained signal conditioners for the headset and microphones inside our helmets. On top of this we donned white, longjohn-type cotton underwear called a "constant wear garment," which—like every piece of our clothing and every other item aboard—was fireproof to prevent a repeat of the Apollo 1 tragedy. Then came the white outer spacesuit—the "pressure garment assembly," or PGA—which consisted of a neoprene-coated nylon inner-layer coverall which could be pressurized to maintain the body at the proper oxygen pressure. On top of the coverall came a multi-layered lightweight outer garment coated with beta cloth to protect against the impact of micrometeoroids.

The PGA was not an easy garment to get into. After stepping into the legs of the suit—they had moulded socks attached, over which yellow protective boots were placed until we reached the launchpad—the bottom portion of a zip, extending round the waist, up the chest and behind the neck, was closed. We were then helped to ease our arms into the sleeves, and our heads through the metal neck-ring to which our helmets would be attached. The bio-sensors were plugged in to outlets in the suit. We then donned brown and white cloth hats—bearing some resemblance to the leather helmets pilots wore in open cockpits—which we called "Snoopy hats." The cartoon dog had been chosen to represent superior workmanship in NASA's Manned Flight Awareness Program. Each hat contained two earpieces and two microphones near the chin-piece and was also connected to the bio-harness. Finally, our moulded gloves and inner polycarbonate plastic bubble-like helmet were connected to the suit and locked in place.

We were then hooked up to a "test-stand" where our suits were pressurized to make sure there were no leaks, that all the bio-sensors were working and that our medical read-outs were fine. We each had a back-up suit in case a fault was detected at any stage throughout this lengthy procedure, though were unable, for reasons of weight and space, to take them with us on the mission. During our mission the suits would be pressurized only while we were performing tasks outside the spacecraft, or if Apollo's cabin suddenly lost pressure. During take-off and landing our PGAs would be more pliable. But even when pressurized they were more comfortable than the suits Neil and I had worn on Gemini 8, since they had convolutes at the shoulders, elbows, wrists, hips, knees and ankles, which would stay in position when moved rather than springing back into a neutral position when flexed.

After this long "suit integrity check" we were at last able to sit down in a comfortable leather couch while we waited for the signal to go out to the launchpad. This was a brief moment of rest after the hard work of suiting up, and most guys, including me, took a little nap. It was just before dawn when the signal came and we boarded the bus bound for Pad A at Launch Complex 39.

Dawn broke to reveal a bright morning with a few scattered clouds and surface winds of about 10 mph. Conditions were perfect for launch, scheduled to take place at 9:34 a.m. Once again Guenter Wendt was ready and waiting in the White Room at the top of the launch tower. Al's back-up, Vance Brand, was there, too. He had been busy setting up all switches in the Command Module that needed to be set before launch; we would not be able to reach them once strapped into position. He also helped ease all three of us into our couches. Two hours and forty minutes before lift-off, first I, then Jim and Al maneuvered ourselves into the confines of the Apollo 15 Command Module perched atop the Saturn V booster rocket.

Over the course of the previous twenty-four hours approximately six million pounds of liquid oxygen and hydrogen propellant had been pumped into the launch rocket. For the next two hours, as we sat on top of this massive explosive device, we ran through a series of systems checks with Launch Control. If the rocket ignited prematurely or an accident caused the Saturn V to explode on the pad, it was widely recognized that it could take several days for the ensuing fires to burn out. Rescue teams would stand little chance of reaching those in the White Room, so an emergency contingency had been prepared for this apocalyptic scenario.

By the side of the elevator that had brought us to the top of the launch tower was a rapid-descent elevator, from the bottom of which a large chute fed directly into a deep underground bunker whose walls were lined with rubber to soften the blow of such a rapid descent. The bunker was stocked with enough food and supplies for the Apollo crew and White Room staff to survive for a month should an inferno be raging overhead and access be blocked by thousands of tons of debris.

But, as the countdown continued, we did not give such scenarios a second thought. Fifteen minutes before launch Apollo 15 was switched to full internal power. We felt a little nudge as the giant access arm across which we had walked before climbing into our gleaming spacecraft was retracted. Strong sunlight suddenly poured through the small porthole of the spacecraft's central window, which until then had been covered by the access arm. This was it, I thought. We were on our way.

Just three minutes before launch we received a wish of "God speed" from Launch Operations Manager Paul Donnelly. We thanked him and braced ourselves for lift-off.

8.9 seconds before lift-off and a deafening roar engulfed the launch tower as the five giant engines of the first stage of the Saturn V rocket—providing a total of 7.7 million lb of thrust—ignited.

"Three, two, one, all engines running ... Launch commit ... Lift-off." Right on schedule we started to lift away from the pad.

The sunlight pouring directly through the hatch window meant that, at one point, I had to put my hand up to shield my eyes so that I could read the instrument panels. But the launch didn't seem as noisy on Apollo 15 as it had been on Apollo 9. I could hear communications with Mission Control pretty well throughout. In the first second or so there was a little more lateral shaking and vibration than I had expected, but once we got away from the tower it was a smooth ride. The instant we cleared the tower of the launch gantry, flight control passed from the Cape to MCC in Houston.

"And the clock is running," I informed Houston, confirming that the mission event timer on the main display console in front of us—marking every second of our program in ground elapsed time, or GET—had begun counting.

All feelings were forgotten. All senses except sight subordinated. All my concentration was focused on hearing information from Jim, Al and Mission Control about the status of the spacecraft and Saturn V. I was the one who would have to take the decision to abort the mission if there was any indication of a major problem. From the very first moment my eyes were glued to one particular area of the instrument panel in front of us. Most critical were: the 8-ball, which indicated the attitude or orientation of the spacecraft; a small panel of eight orange lights and one red light indicating the status of the Saturn V engines; a "Q" meter showing the aerodynamic pressure on the spacecraft; another meter showing engine pressure in the launch vehicle; and the big, red Abort light.

Throughout these critical moments of our powered ascent there were various modes of aborting the launch in the event of a major failure of any aspect of either our Apollo spacecraft or the Saturn V gulping 15 tons of fuel per second beneath us. This was one of the riskiest times of the entire mission and one for which it had been particularly difficult to train. More crews "bought it" during launch simulations than in any other kind of sim. It was a matter of split-second decision making and very precise reflexes under conditions which were often extremely difficult to replicate in a simulator.

Few people realized it but if the launch vehicle guidance system were to malfunction the Saturn V could be "flown" using the spacecraft's guidance system. This could be effected by computer or else I could use the "joystick"—rotational hand controller—in my right hand to steer the Saturn V with its three separate stages. In the latter case, manual control was activated by turning the T-handle in my left hand in a clockwise direction 45 degrees. The Saturn would immediately respond to signals from the Command Module. I could then fly the vehicles using information from my instrument panel, especially the 8-ball, according to a specific flight profile based on pre-flight information as well as computer data. This technique was practiced many times during simulations, but never used on any Apollo mission.

If, however, a situation were considered serious enough, I could abort the mission. For this there had to be two independent cues of a problem. But the mechanics of initiating an abort took only a 45-

degree anti-clockwise twist of the T-handle. I could not afford to make a mistake because any error in moving this device could have catastrophic consequences for the mission. Yet the time-frame within which an abort could be activated measured only fractions of a second.

The first phase in which abort was possible occurred as the Saturn V climbed to a height of 10,000 feet in just 42 seconds. During this phase an abort would separate the spacecraft from the launch vehicle and ignite the launch escape system rocket on top of the spacecraft. The escape rocket would be jettisoned soon after the initial separation, and the spacecraft would then descend on parachutes to a normal splashdown. The second phase extended until just over a minute into the flight as we reached an altitude of just under 16 miles. Throughout these vital seconds Al and Jim were giving me constant status reports, Mission Control too. As we climbed to the 16-mile mark there were no signs of any malfunction. Three abort modes were defined to cover various contingencies until we reached orbit.

"OK. Looks good up here," I informed Houston. Less than two minutes later, as we approached an altitude of 38 miles—beyond the confines of the Earth's atmosphere—the first stage of the Saturn V rocket engines cut off and the second stage took over, followed by the third. Eleven and a half minutes into the flight (at 94 miles) the third stage of the launch vehicle shut down and we began to experience prolonged weightlessness. Once in orbit "heads down" we caught our first sight of the Earth. Yet we had far less time than on my two previous Earth orbit missions to absorb that breathtaking sight. We would perform only one full Earth orbit before lighting the engines of the third stage once more, to depart Earth orbit and head for the Moon. This increased our speed to 25,000 mph and we shot through the Van Allen radiation belts in less than fifteen minutes, minimizing our exposure to radiation. These belts, stretching from just under 2,000 miles above the equator to more than 8,000 miles, are composed of particles highly charged with energy from the solar wind, which remain trapped within the Earth's magnetic field. It would take three days of free flight—"translunar coast" or TLC—to reach a position near the

Moon from which we could light our big engine and enter lunar orbit.

After retrieving the Lunar Module from its garage three-and-a-half hours into the mission the combined Lunar and Command Service Modules separated from the spent launch vehicle, which continued on its own trajectory to a high-speed impact on the Moon between two major craters, Copernicus and Ptolemaeus.

Throughout the remainder of that first day and most of the second we had a number of scientific experiments to carry out. Perhaps the strangest involved Jim, Al and me sitting blindfolded, during each of three days of the mission, to monitor an unusual phenomenon of intermittent light flashes, seemingly caused by cosmic rays registering in the brain, which had been observed by previous Apollo crews during their journey to the Moon. Sitting there blindfolded giving a running commentary of when we saw these flashes was one of the most amusing experiences of the mission, though, inexplicably, all the data recordings from the first attempt at this experiment were subsequently lost.

During the first two days of the mission the only real problem was with the Delta-V thrust switch light unaccountably illuminating, wrongly indicating that the spacecraft's main engine— the service propulsion system or SPS—was firing. We worked closely with Mission Control to troubleshoot this fault, which it was thought could be due to a potentially serious short-circuit. If the short caused the engine to ignite, the spacecraft would be propelled into an unknown trajectory or the engine may have misfired when we needed it.

Trying to trace the exact location of the short was no easy task. It took careful coordination between all three crew members and the appropriate flight controllers in Houston. Mission Control obviously had very detailed diagrams of every aspect of the spacecraft, but on board we carried only simplified diagrams of each of its systems. This meant we had to follow precise step-by-step procedures worked out for us by people on the ground to set certain switches on or off and open and close certain circuit breakers while closely monitoring the response.

Most of the electrical circuit breakers were within easy reach of the right couch, so coordinating them fell to Jim. The Command Module cockpit was arranged in three sections. The left section closest to the commander consisted mainly of all the controls and displays that enabled him to fly, control and guide the spacecraft, including switching its engines on and off. Displays in front of the center section were devoted primarily to navigation and guidance; an additional area below the couch, called the lower equipment bay, included a sextant and telescope for navigation. The right section was surrounded by switches for the spacecraft's electrical and environmental support systems and also communications.

Very early in the Apollo program the spacecraft was designed for in-flight maintenance. The plan was for the crew to carry toolkits and certain replacement parts, which they would be trained to fit. But as Apollo evolved it became clear this was impractical, because of the extra weight it would entail. Instead, the emphasis was placed on making sure all systems were so reliable that such repairs were not necessary. Consequently, all circuits within the spacecraft were hermetically sealed and could not be adjusted, altered or modified by the crew.

After following the malfunction procedures very carefully, however, we were eventually able to trace the short to inside the Delta-V thrust switch mechanism itself. We rectified the problem by isolating the switch electrically and leaving it off.

Toward the end of the second day more serious problems emerged. Jim and I encountered the first when we entered the Lunar Module for the first time to power it up and check that all its systems were working. Almost as soon as we pulled ourselves into it we became aware of shards of glass floating around the cramped capsule that would serve as our home for the three days we were to spend on the surface of the Moon. It was soon clear that the outer glass face of one of the instruments had shattered. Though the instrument seemed unaffected and the inner pane of glass remained intact, this posed a potential threat to our own safety if we inhaled small fragments or if they got into our food supplies. But the problem was relatively easily solved. Most of the loose debris was sucked into the filters of the LM's environmental control system, which we then cleaned off using sticky tape.

A second problem, which emerged the following day, presented a far more serious threat, and could have signaled the end of the mission had we not been able to deal with it promptly.

Sixty-one hours and twelve minutes into the mission, as I was carrying out the small housekeeping chore of adding chlorine to our water supply, we noticed water droplets floating around the cabin.

"Hey, Houston, Fifteen," I radioed to Mission Control.

"Fifteen, go ahead," replied Karl Henize from the capcom station.

"You might take a look at this real quick and see if you can come up with any ideas on the thing," I said with a note of urgency, after explaining the problem. "It seems like we're accumulating a fair amount of water in the cabin right now." We had no idea where the water was coming from, despite Al clambering down into the lower equipment bay to try to locate the source. But Houston's reply was a striking example of how difficult it is for anyone with an Earthbound perspective to appreciate the effects of weightlessness.

"Can you give us an estimation of how many drips per second it is?" asked Karl.

Water does not drip in zero gravity; it just accumulates on a surface into big globs, until the surface tension of the growing mass is such that it floats free. This made the immediate source of the leak very hard to trace. It was a big problem, and would be a major safety hazard should water penetrate into the sealed system of electrical wiring. Most of our systems were cooled with water-glycol, and if our water supplies ran low this would seriously compromise the running of the spacecraft. Also, of course, water was used for drinking and food preparation. If the problem couldn't be fixed, we would not be going to the Moon. All we could do while Houston tried to figure out how to solve the problem was start trying to soak up the accumulating globs with towels.

Communications between us and Mission Control during these tense moments were transmitted via Honeysuckle Creek relay station in Australia (someone there told us later that Captain Cook's ship *Endeavour* had also sprung a leak on her voyage to Australia). Despite this slight delay in communications, it took Houston only six minutes to come back with a set of instructions they believed would solve the problem. It didn't. The water continued leaking at a steady rate. It was nearly a quarter of an hour after I first noticed the problem before Houston came back with another solution for tightening a seal in the spacecraft's chlorination system, which they believed was causing the problem.

"OK, Houston. It looks like that did it ... That was good thinking because we about had a small flood up here," I reported eventually, with huge relief.

Fortunately the flight controller for the electrical environmental and communications systems in the control center, otherwise known as EECOM, had written a procedure to correct the problem after a similar small pre-launch leak had been noticed. I also found out later that a technician at the Cape was driving home from work late that night when he heard on the radio that we were having this problem. He had pulled over, got to a telephone and called Mission Control to say he had detected a leak in one of the chlorination valves before launch, and had worked on a procedure for stopping this from happening. It was his procedure that was transmitted to us in space: an illustration, for me, of the total awareness and dedication of every individual who contributed to the Apollo program.

At last it looked like we were in good shape. All we had to do was clip a few towels out to dry. Over the next few hours our spacecraft took on the appearance of a laundry room as we hung them in the tunnel leading from the Lunar Module to the Command Module. Then it was time to get some sleep. The next day would bring us within lunar orbit. It was a big day. We needed to be well rested.

On the morning of the fourth day—seventy hours into our mission—came the wake-up call: "Good morning, Dave. It's time to rise and shine."

This was the day when we would pass beyond the western leading limb of the Moon, the leading edge, to its Far Side. Entering lunar orbit at a distance of 60 miles from the lunar surface later that day—at 078:22 GET—we lost all contact with the Earth for just over half an hour. As we were in the dark during the first part of the orbit, our only visual indication of the Moon was the lack of stars in the vast area of space it occupied.

Then came our reward. As we passed the terminator—the dividing line between the portion of the lunar surface in darkness and that in light—on the Far Side, we broke into sunlight and caught our first sight of the surface of that normally hidden portion of the Moon close up and fully lit. The Far Side is quite different from the Near Side. It has fewer giant dark craters and a more rounded topography. It looked spectacular.

At 078:56 GET we emerged from behind the eastern limb of the Moon to see our own planet ahead of us. This, too, was a most beautiful sight; looking back at Earth from lunar orbit for the first time was incredible—it really gave us a sense of how far we had traveled. We then resumed communication with Mission Control.

"Hello, Houston, the *Endeavour*'s on station . . . and what a fantastic sight!" I reported, unable even to begin to convey the wonder I felt at looking back at the Earth from this distance and also—more impressive still—at seeing so clearly and close the silvery globe we all spend our lives watching wax and wane in the night sky.

"Beautiful news," Henize radioed back. "Romantic, isn't it?"

"Oh, this is really profound," I said. "I'll tell you, it's fantastic."

This diversion from the normally clipped and controlled transmissions expected of us slightly exasperated some at Mission Control, I later learned. But the spacecraft was performing flawlessly and who could have failed to be awed by those incredible sights? As we orbited the change in lighting, depending on the viewing angle and the angle of the sun, lent the lunar surface a whole spectrum of color from gray through to a golden brown, throwing its surface craters and lunar mountains into stark relief. It was spectacular.

We had organized our schedule so that we would have time to sit and observe the landscape of the Moon passing beneath us. We could not pull our gaze away from the window. All three of us took many photographs.

The next twenty-four hours took us round the Moon a dozen more times as we prepared for the beginning of our separate missions the next day. After our final sleep period before descending toward the surface, we suited up and began the process of equalizing the pressure between *Falcon*, the Lunar Module, and *Endeavour*, the Command Module—a necessary prerequisite before Jim and I could transfer into *Falcon* and the two capsules could separate.

"Take care of home, Al . . . We'll be back in three days," I said, once this process was complete. Al closed *Endeavour*'s hatch behind us, and at 100:39 GET he threw the switch to separate the two vehicles. We were on our own. *Falcon* pulled away from *Endeavour*. After a final systems check and trajectory certification, Houston gave us a "Go for descent" call. We began preparations to ignite *Falcon*'s powerful engine which would take us on our descent toward the lunar surface.

This trajectory would take us looping behind the Moon before we came in to land at Hadley Rille on the Near Side, so again we would be out of radio communication with Mission Control. They could not monitor any problems we encountered during this final phase before our descent. Nor, in one particular case, did we want them to. An early indication that there might be a problem with *Falcon*'s environmental control system meant that, strictly according to mission rules, we might be forced to abort the landing.

It was not a serious enough problem to endanger our lives. It was my call. I was the commander. As it turned out Jim managed to fix the problem before our orbit brought us back to the Near Side. It appeared to have been a false indication that our cabin pressure was not secure, which might have meant we had to begin our descent and land in "hard" pressurized suits. That could be done, though it would make the landing very difficult and would make leaving the LM for surface activities even more difficult.

But, out of earshot of Houston, Jim and I had discussed it and I had made up my mind. Had the problem persisted we would have bent the rules and gone for landing anyway. We had not come this far to be stopped on a point of procedure written long before the launch. By the time we flew Apollo 15 there must have been 500-odd small-print pages of mission rules—an endless list of "what-ifs" and how to deal with them. You couldn't possibly learn them all and, as commander, you had to either discuss things with Mission Control, which we always did, or else, if time were too short, use your best judgment.

Our enthusiasm and curiosity about what lay ahead of us was too great to allow all but the most serious problem to deter us from reaching our goal at this stage. The view of the Moon as we began our final descent was stunning. As we caught our first close-up look at the lunar mountains below us, Jim lightened the atmosphere of growing tension and excitement a little with his usual dry humor.

"Make a great ski area if they'd just put some snow on it," he said.

"Looks like there is some in parts," I replied, adding, as I took a closer look at the horizon beyond, "The sky is just as black as the ace of spades."

"Don't think there's any atmosphere," said Jim, because, of course, on the Moon, there isn't. "I'm going to write me a joke," he went on, getting into his stride. "Astronauts come back from Moon; say it's great, but has no atmosphere."

"That's a good one," I chuckled. "You ought to save that one for the surface some time."

As we dropped through a gap in the lunar mountain range, toward the end of our descent, I had the surreal feeling that we were floating slowly over the mountains. No amount of simulation training had been able to replicate the view we saw out of our windows as we passed by the steep slopes of the majestic lunar Apennine Mountains. The plaster-of-Paris model of the Moon's surface we used during training was a relatively flat 15 feet by 15 feet. Mt. Hadley Delta to our left loomed 11,000 feet high. Jim and I had not expected to see peaks above us to left and right as our trajectory took us through a pass in the mountains. It made us feel almost as if we should pull our feet up to prevent scraping them along the top of the range.

Our descent was quite different from that of any previous Apollo mission. Not only was our spacecraft bigger and heavier, because of all the extra equipment we were carrying, but our landing site, beyond the ridge of the Apennine Mountains, meant we had to come in to land at a much steeper angle—twice that of previous missions. Rather than descend to a brief level-off, or step, during final approach as other lunar landing crews had done, I was determined to come down in as smooth a linear sweep as possible. This would conserve fuel and give us more time to hover and be more selective in our touchdown point.

But flying the Lunar Module during landing was rather like trying to run and turn corners on very slippery ice, which offers little friction to halt momentum: maneuvering in space meant starting a movement long before a change in direction or speed was required. The LM was supported vertically by the power of its descent engine, which would maintain it in a hover, unless tilted forward to allow some of the power to drive its forward motion. But since there was no air to slow the forward velocity by drag, slowing forward motion meant tilting the LM backward so that some of the power could be used as a brake. If power was used either for braking or to drive the LM forward, however, this had to be compensated for by increasing power—otherwise the vehicle would lose altitude. The same applied when moving left or right. This delicate ballet of forward–backward and left–right movements, coupled with handling the descent engine, was like no flying any of us had ever done; hence our insistence on training as much as possible with the LLTV or "Flying Bedstead."

During the early part of the descent the LM was tilted backward—windows facing up—so that the power of the descent engine would slow our velocity. But as our altitude dropped to around 6,000 feet I pitched *Falcon* forward by 30 degrees to take a look at where our trajectory was bringing us in to land. Although I could recognize the major features that lay below and before us, however, no photograph had been detailed enough to prepare us for exactly what we would find in this landing area. The highest-resolution photograph we had seen was at a resolution of 60 feet.

Craning my head toward the triangular viewing window I was momentarily baffled. Though I could see the giant meandering lunar valley of Hadley Rille out of the forward corner of my window, as I had been able to do during simulations, there were a number of craters and formations which had not shown up in photographs. I was pretty sure that, if we carried on the same flight path, we would land long and be to the south of our target area. As Jim calmly read off the numbers on our instruments indicating altitude and predicted point of landing, Mission Control confirmed what I suspected.

"*Falcon*, Houston," said capcom Ed Mitchell. "We expect you may be a little south of the site, maybe three thousand feet."

"OK," I responded neutrally. But I knew this might seriously compromise our planned program on the lunar surface. I had to adjust the flight path, and quickly. By carefully clicking the hand controller forward and to the side—eighteen adjustments in all—to alter our trajectory, I managed to nudge the LM on to the correct path. It was a pretty tight moment. Jim and I said little. We were totally focused. Then, fifty feet from the surface of the Moon, we lost all visibility.

The thrust from our descent engine had kicked up the very fine surface dust of the Moon's outer crust into a dense gray cloud totally obscuring the surface. It was like looking through a thick fog. From that point on I was flying on instruments alone. It was imperative that, as soon as the 10-foot-long probe sensors extending from the feet of the LM made contact with the ground, I shut the engines down, otherwise the engine bell of the craft risked being seriously damaged on impact. If the engine casing split and damaged the vehicle, we might become permanent residents of the Moon. So when a blue light signaled contact of one of the landing-leg probes less than a minute later, I immediately depressed the Engine Stop button.

The fog of dust dispersed almost instantly and our silvery, spider-like craft came in to rest with a firm thud. I knew exactly when to expect the impact, since I was at the controls and had anticipated a strong landing because of our extra ton and a half of equipment. But Jim was somewhat taken aback by the force. "Contact! Bam!" he exclaimed. Then everything went suddenly still.

Exactly 104 hours, 42 minutes and 29 seconds after lifting off from Cape Kennedy in Florida, we had touched down on the surface of the Moon.

"OK, Houston. The *Falcon* is on the Plain at Hadley," I reported. It was a small, but for me significant, tribute to my alma mater: the Plain is where we used to hold parades at West Point.

"Roger. Roger, *Falcon*," came Ed Mitchell's elated reply. We heard applause erupt in the background.

It was clear straight away that we had landed on uneven ground: the LM was tipped backward at a slight angle (it turned out later that one of the rear feet had landed in a shallow crater). But there was no time to reflect on this. We were spring-loaded to abort and leave again within seconds. We had to be ready to ignite the ascent engine and abort the mission immediately should any damage to *Falcon* or any major malfunction be detected. During those crucial moments, Jim and I, together with Mission Control, were totally absorbed with monitoring every aspect of our spacecraft and its systems to make sure that everything was in working order. We knew Al was anxiously awaiting news of the systems review, too. He had passed out of visual and communications range of our landing site by that time. He had a three-day mission full of scientific experiments to complete and had performed a pretty complicated maneuver just before our landing to get himself back into a higher orbit. If we had to lift off, he would need to prepare for rendezvous. The signal that all was well would be a "Stay" message from Houston.

"*Falcon*, Houston. You're stay," said Mission Control seventy-seven seconds after we had touched down.

It felt a little as if the long vacation we had been looking forward to so eagerly had at last arrived. Preparation for this moment had been long and extremely arduous, but now we were exactly where we wanted to be. It was time to take a look outside, and the prospect made me feel like a little kid waking up on Christmas morning, about to open his presents.

Every lunar crew before us had gone out on the surface very soon after touchdown because their time on the surface of the Moon was limited to just over a day. Our schedule, on the other hand, had been designed originally to include a sleep period before we started exploring. I had realized early on in our training the importance of maintaining our circadian rhythms in order to be able to perform at peak during our three days on the Moon. And, with so much adrenalin pumping, there was no way we could just pull down the blinds shortly after landing and simply go to sleep.

Though Deke Slayton and others in NASA management had

initially opposed it on the grounds that it would consume extra oxygen, the plan we had finally agreed on was for me to perform a half-hour stand-up EVA, or SEVA, shortly after touchdown. By climbing up on to the cover of the engine inside the craft, and opening the upper hatch—rather as if I was in the conning tower of a submarine or the turret of a tank—I would stand up and take a good look around. On geology trips I had learned the value of such an initial reconnaissance. But this would be a reconnaissance like no other. So, two hours after touchdown, after slowly depressurizing the cabin, I placed a protective outer helmet—a lunar excursion visor assembly, or LEVA—over my inner helmet, and clambered up to my vantage point in my bulky, but by now pressurized, spacesuit. Pulling my oxygen hoses and communication cables behind me, I slowly hoisted my elbows on to the rim of the hatch. I found I could support my own weight quite comfortably in the reduced gravity, and started to take in the view. What an awesome view it was.

Even with the protective filter across my visor, the sunlight reflecting off the crystal-clear features on the surface was intense, contrasting sharply with the deep, rich blackness of the sky beyond. The low sun angle of the early lunar morning laid long shadows over the spectacular scenery spread before me. It was like an exhibition of exquisite images by the great photographer Ansel Adams. There was no color, but great contrast between the brightly illuminated surface and the black shadow of the mountain slopes and craters where no sunlight fell.

"Oh, boy, what a view!" I exclaimed. "Beautiful."

Jim handed a bearing indicator and large orientation map up to me, and I began to give a number of precise location indicators and started to take a series of high-resolution connecting photographs of the full panorama that encircled us. Shortly after we had landed I had advised Houston, "Tell those geologists in the back room to get ready because we've really got something for them."

Now I was actually looking out at the magnificent moonscape we were about to explore, and I could hardly contain my excitement as I began a very detailed running commentary on what I could see. This was not only for the benefit of the man who had helped so much in preparing me for this mission, Professor Lee

Silver, and the other geologists in the back room at Mission Control. I also wanted to convey some of the beauty of this place to everyone listening to our transmissions on that hot July evening back home in Houston and elsewhere in the world.

"All of the features around here are smooth. The tops of the mountains are rounded off. There are no sharp, jagged peaks or large boulders apparent anywhere. The whole surface of the area appears to be smooth, with the largest fragments I can see in the walls of Pluton," I said; Pluton was one of the craters we planned to visit on our third and final day of lunar exploration. My description of the stunning virgin terrain continued for another ten minutes. "Trafficability looks pretty good. It's hummocky. But I think we can maneuver the Rover in a straight line . . . It looks like we'll be able to get around pretty good," I concluded.

"Sounds like we're in business, old friend," said Joe Allen, the astronaut assigned as primary capcom during our operations on the lunar surface.

"You're coming up on thirty minutes into the SEVA," Joe prompted after a few minutes, signaling it was time for me to climb back inside and close the hatch. I knew he was right. We had to keep on schedule. The clock was ticking.

"There's just so much out there," I replied. "I could talk to you for hours."

It was hard to tear myself away from such a spectacular view, but it was time to eat and then get some sleep. Our evening meal consisted of high-energy, low-residue reconstituted food. Tomato soup was big on the menu, as I recall. There was no hot-water supply in the LM as there was in the Command Module, so all our meals on the lunar surface were served cold, and we soon discovered that there was not really enough to eat, either. Exploring the Moon turned out to be hungry work, and we recommended subsequent missions be provided with additional supplies. But that was of little concern at the moment. We were just ready to sleep.

Stripping down to our longjohns, we were the first lunar crew to get out of our spacesuits while on the Moon. Previous missions had been too short, but, given how long we would be spending on the lunar surface, for us it was essential.

That first night we pulled down the blinds of the LM's two small triangular windows to block out the intense sunlight and strung up our hammocks across the cramped cabin—about the size of four telephone booths. We had had a lousy night trying to sleep in the hammocks in the simulator back at the Cape, but our reduced weight on the lunar surface gave them a much softer feel. With the background mechanical symphony of the LM's pumps and fans we popped in earplugs and were pretty comfortable as we settled down to sleep up there on the Moon that first night.

"We're all tucked in. We'll see you in the morning," I advised Houston.

"Roger, Dave. Good night, and don't fall out."

Besides excitement and pride at the prospect of becoming the seventh man to walk on the surface of the Moon, I felt huge relief that I would finally be escaping the confined quarters of the spacecraft when we woke the next morning and went through our last preparations before opening *Falcon*'s forward hatch and stepping outside. For the past five days we had been like exotic birds in an elaborate cage. Now we were about to fly. It was a great feeling.

After a good night's sleep, we awoke the next morning to a call from Houston as well as a small alarm clock I had carried, just to make sure we did not miss anything. It was still pitch black inside the LM when we turned the lights on and got out of our hammocks. I went to the window and raised the shade to let the morning sun in. And lo and behold, as I looked out, still in my morning fog (as many of us are, anywhere!), I saw this brilliant lunar surface—we were actually on the Moon! "Jim," I said, "raise your blinds, look out there—spectacular!" And it was time to get moving, but first we had to go through the lengthy process of suiting up: not only the elaborate procedure we had undergone prior to launch—and would now have to perform without help— but the addition of many extra protective layers. The first was a diaper-like pair of short pants worn with our "motorman's helper." After attaching our bio-sensors, instead of longjohns we donned the essential liquid cooling garment, or LCG, a nylon-Spandex knit outfit through which was threaded a mesh of plastic tubes carrying constantly cooled water to protect us from the intense heat of the Sun. We then helped each other into our PGAs, put outer silicone-soled moon boots on and strapped on our bulky backpacks— portable life-support systems, or PLSSs—containing systems to supply and recycle oxygen for breathing, another to control suit pressure and the water for the LGC.

On top of our clear helmets we each locked an outer plastic shell fitted with three eyeshades and both an inner and an outer protective visor to filter ultraviolet and infrared rays. It slightly restricted our lateral vision, but without it the inner helmet allowed us full visibility. Over our molded gloves we wore outer thermal gloves. After donning the suits we had to mount the backpacks and connect their oxygen, water and electrical hoses and cables to our suits. We then had to check and verify that everything was working. Donning all this equipment was the most arduous part of the mission on the lunar surface: it took over two hours.

Even then, getting out of the LM was no easy matter. Wearing all our extra equipment we could just squeeze out of the forward hatch and the only way we could do that was by bending down on both knees and crawling out backward on to *Falcon*'s front "porch." As I went through this awkward procedure, with Jim guiding me, I gave little thought to what I would say when I took my first step on to lunar soil. But after squeezing free of the front porch and almost hopping down *Falcon*'s open-rung ladder I was quite certain.

"As I stand out here in the wonders of the unknown at Hadley, I sort of realize there's a fundamental truth to our nature." I paused, my heart racing a little as I took in the enormity of this moment, for which I had trained so intensively for seven and a half years. "Man must explore." I paused again. "And this is exploration at its greatest."

After savoring the moment briefly, Jim backed out of the LM— with even more difficulty since there was no one to guide him out— and joined me. We began the process of unstowing equipment strapped to the outside of the LM. Most important was the spindly Lunar Rover, folded up with its wheels doubled under it, like an elaborate drawbridge. If the rover failed to work we would have to resort to the less adventurous walking traverses we had also trained for.

Once released, the rover—our "Moon car"—slowly descended to the surface; a brilliant piece of engineering with sealed electric motors in the hub of each wheel. After securing its joints with pins, I climbed aboard to test its operation. Almost immediately, I realized that there was a problem with the front steering. This was perfectly manageable with rear steering and eventually righted itself. Even so it was a pretty bumpy ride. After loading a rack at the rear with the geological equipment we would need for our first day of exploration we climbed aboard—something of a challenge since the suits did not bend easily into a seated position—and started the engines.

Driving the rover was actually more like flying an airplane, albeit with four wheels, than driving a car. Instead of a steering wheel, which would have been very difficult to grasp in our bulky suits, it was controlled with a joystick mounted on a control console between my seat and Jim's. Despite our maximum speed of only 7 or 8 mph, the reduced gravity and very irregular surface meant one or more of the independently suspended wheels lifted away from the surface every time we hit uneven terrain. No part of the lunar surface was totally flat or even. It was all rolling and irregular with a wide variety of small and large craters, some with debris inside.

Maneuvering around the larger craters and across this combination of rolling hummocks and fine- to coarse-grained lunar soil required intense concentration. Driving into the sun was the most difficult, since the glare caused a "wash-out" of the surface features. Though the rover could turn on a dime and had very good traction and power, the wire-mesh wheels kicked up impressive rooster-tails of dust, which were deflected by large fenders. It all made for a ride like a cross between a bucking bronco and a small boat in a heavy swell.

"This really is a rockin' and rollin' ride. Hang on," I told Jim as we headed southwest in the direction of the first of our destinations, Elbow Crater and the northeast flank of St. George Crater on the lower flank of Mt. Hadley Delta.

The spectacular moonscape reminded me a little of hilly terrain

blanketed with thick snow back on Earth. When we parked the
rover, our activities would be captured on a color television camera
mounted on its front end—the first time it had been done—and
transmitted on news broadcasts around the world, in addition, of
course, to allowing those in the science operations room to follow
our progress.

Each stopping point on our three days of lunar traverses had
been selected to allow us to examine contrasting aspects of the
Moon's complex geology, much of which was largely a mystery.
Our ultimate goal was to throw light on the origin of our nearest
galactic neighbor—which had long been the source of speculation
and serious scientific debate. Greater knowledge about the Moon's
genesis was expected to throw more light on the origin of our own
planet, of the Sun and of our entire solar system.

Over time three competing theories about the origin of the Moon
had evolved. The so-called "sister" theory held that the Earth and
the Moon were born at the same time and of the same cloud of gas
and dust. The "spouse" theory held that the Moon was born in
another part of the solar system and had been forced into wedlock
with our planet after moving into the Earth's gravitational field.
According to the "daughter" theory, the Moon was born of the
Earth, splitting off very early on.

During the following four hours, as we experienced the thrill of
maneuvering across the spectacular moonscape for the first time, we
traveled just under five miles. But most of our geology was done on
foot. The most effective way of walking turned out to be more of
a loping movement, which I can only describe as being something
similar to walking on a trampoline. Starting and stopping were
another matter. Because our bulky backpacks tilted us backward, I
found the most effective way of starting to move was to lean
forward as if walking into a stiff wind. Stopping required digging
both heels in sharply at the same time as leaning slightly backward.
It wasn't difficult, but it took some adjustment and kicked up quite
a bit of dust, which took a little while to settle.

In some places the surface was covered with dust as fine as
powdered snow, half an inch or so deep, resting on the harder crust
or regolith. The area around Hadley had a high density of randomly

scattered and rounded low-rim craters of all sizes, up to several feet in diameter. Occasionally, on less than 1 percent of the surface, there were fresh craters one or two feet wide, and we also found them on the slopes of the mountains.

To an "Earthling" one of the Moon's most striking features was its stillness. With no atmosphere and no wind, the only movements we could detect on the lunar surface, apart from our own, were the gradually shifting shadows cast to the side of rocks and the rims of craters by the Sun slowly rising higher in the sky. There were no other features: no trees, bushes, rivers, streams, flowers, grass, animals or birds—none of the signs of nature that human beings have evolved with and are used to. There was no sound, either, apart from the gentle humming of the equipment in our backpacks. There were no clouds, haze or mist, and there appeared to be no color. The sky was pitch black except for the deep blue and white of our own planet suspended high in the sky like a Christmas tree ornament.

The drive from the LM to our first geology station took about twenty-five minutes. After skirting the rim of the sinuous Hadley Rille, we arrived at a series of deep "fresh" craters where we took many photographs and started the exacting task of selecting the best rock samples we could find to illustrate the diversity of this site. We were soon able to identify basalt, breccia and some traces of olivine. Again and again I was reminded of the staggering fact that each rock we stored away carefully in collection bags attached to our backpacks had been untouched by atmospheric erosion—on the Moon there is no wind or rain to chisel and shape the topography— for the past 4.5 billion years. However, the Moon has had its own "gardening"—its surface has been molded and changed by a great flurry of bombardments from meteors, comets and other celestial debris. And we helped the process by kicking over a rock or two!

Our excitement at each new find was infectious, bringing regular comments of "Beautiful!" and "Atta boy!" and "Stupendous!" from the backroom guys at Mission Control.

As we prepared to wend our way back to the LM at the end of the first traverse, we rested momentarily to sip water from a tube attached to a small reservoir inside our suits. Fruit sticks stowed in pouches directly below our chins provided a little sustenance, too. But I could happily have eaten twice the amount. Once, when Houston prompted us to get moving and head home, I was so determined to pick up a very interesting black rock, which I could see not far away sitting all alone on the gray surface without a speck of dust, that I had to resort to subterfuge. I stopped the rover and pretended to readjust Jim's seatbelt so that I could stoop to pick it up. This beautiful rounded piece of scoriaceous basalt was later dubbed the "seatbelt basalt." We had another two hours' work ahead of us when we arrived back at the LM. We had to deploy a complex series of scientific experiments—known as the ALSEP or Apollo lunar surface experiment package—before we could rest. This would prove to be the toughest part of the day.

What should have been a relatively easy exercise turned into an exhausting and painful chore. The most difficult aspect involved using a battery-powered drill to bore two ten-foot holes into the lunar crust, extracting the soil and then inserting a pair of thermometers aimed at measuring sub-surface temperatures to determine the "heat flow" of the upper lunar surface. When I bore down on the jackhammer mechanism, it seemed something was preventing the drill bit from penetrating the crust to any great depth.

"I tell you one thing, the base at Hadley is firm. Boy, that's really tough rock," I muttered, almost under my breath, as the soft upper regolith gave way to much firmer material. After boring down to five and a half feet I simply had to stop. In addition to the firmness of the material, there seemed to be a serious problem with the drill. After consultation with Mission Control, it was decided that I would return to complete the task the next day.

By the time I climbed back inside *Falcon* that night and pulled off my gloves, the force of my fingertips pushing against the inside of the gloves as I applied force to the drill had made my hands swollen and painful, and blood vessels under several of my nails had burst. Part of the reason was that Jim and I had both laced the inner sleeves of our spacesuits to a length which made our gloves fit snugly when our arms were flexed. This gave us a greater degree of dexterity and comfort when we were collecting samples and working in close with the geological tools. But it was pretty uncomfortable when we had to exert any degree of pressure, as I had to with the drill. Jim was also suffering some slight ill effects. He had had a problem with his water pouch and had gone too long without quenching his thirst. He had a throbbing headache and seemed very tired.

There was no time to dwell on such minor discomforts, however. Before we could rest we had to struggle out of our spacesuits, which by that time were heavily soiled with fine-grained lunar soil. To prevent this dark grain and dust getting everywhere inside the LM, we stood in large cloth bags as we took our suits off, then stepped out of the bags and sealed them up until the next morning. The strong odor of this fine-soot-like dust surprised us both: the Moon turned out to have a slightly metallic smell, almost like gunpowder, which pervaded the LM for the remainder of our trip.

After tucking into our cold reconstituted supper, we prepared ourselves for some much-needed sleep. The next day we were due to drive to the eastern flank of Mt. Hadley Delta. What we would find at just that one site would secure the mission a unique place in the history books.

"First thing we've been concerned about, and I guess we'll start off with this, is that according to our data, you lost about twenty-five pounds of water during the post-EVA yesterday . . . We're wondering if you've looked about in the cabin and noticed any sign of that twenty-five pounds of water?"

At 138:04 GET it was not the sort of wake-up call we would have wished for on the morning of our second day on the Moon. We had been woken a little earlier than planned, because it appeared that water had leaked, apparently from a damaged filter mechanism, while we were sleeping. The slight slope in the spacecraft meant it had collected at the rear of the cabin in what I later referred to as a "little puddle." In fact, it was a great big puddle and given the maze of electrical wiring—albeit waterproofed—in the vicinity could have caused a very serious problem.

I was more than a little annoyed that Houston had not alerted us immediately when they became aware of the problem, even if it had meant waking me up. A commander always must be informed of the condition of his ship. The previous morning we had also been woken a little earlier than scheduled when a small oxygen leak was detected. The problem had been fixed quickly, but here we were with a similar annoyance. If we had fallen behind schedule in sorting out the problem either day, it would have prejudiced our busy program of exploration. As it was, Houston came up with a neat solution to our "little puddle."

"Use a used food bag as a scoop, if it's a deep enough puddle to scoop it up," they suggested. "And then . . . use utility towels to mop up the rest."

Fortunately, this "housekeeping" job did not take long, and we were able to get suited up for our second full day on the Moon pretty much on schedule.

Any irritation I had felt at the messy start to our day soon evaporated once we got under way on the rover at the beginning of a traverse which took us to the lower reaches of the magnificent Apennine Mountains. As we climbed the slope at the base of Mt. Hadley Delta, the spectacular panorama spread before us both took me by surprise and filled me with wonder. At sea level on Earth the horizon is roughly twelve miles away, but on the Moon it is only about a mile-and-a-half away because the curvature of that smaller globe is sharper. From our elevated position several hundred feet up, for the first time we could see much further than that.

In the distance was the gentle valley of Hadley Rille snaking across the landscape, surrounded by undulating crater-pocked terrain. In the foreground and to one side we could see our temporary home the silvery, spider-like *Falcon* squatting like a small insect on the vast Hadley Plain several miles away. Looming above us to the east was the majesty of the 15,000-foot Mt. Hadley Delta—something else entirely. The smooth flanks of the high mountains had taken on a golden hue as we moved slightly later into the lunar morning. Without the cycles of freezing and thawing to crack rock, the tops of the mountains were not rough as they would be on Earth, but were instead smooth and undulating, clearly framed by the dark dome of the sky.

"My, oh my! This is as big a mountain as I ever looked up," I said, as much to myself as to Jim or Mission Control.

How I wished I could explore higher up in the mountains. Though I had never been a mountain climber, I enjoyed skiing and knew the thrill of being able to reach great heights and look out over a vast panorama. Jim, also a keen skier, later likened the Apennines to the mountains back home in Sun Valley, where he loved to spend time with his family. But it was not possible to take the rover any great distance up the mountainside. The ease with which it had climbed the slope at the base of the mountain was deceptive. The gradient turned out to be steeper than it looked—about 15 degrees.

Just before we reached our main destination for that day, Spur Crater, we sighted a large boulder some distance away. It was the largest we had seen and one of the few on the side of the mountain. We were keen to take a closer look and I edged the rover carefully toward it while Jim carried on observing and commenting. But the further we drove, the softer and looser the soil became. It wasn't long before the rover began running into difficulty, its wheels sinking into the soft soil on the slope. This was getting pretty sporty! I tried to compensate by steering uphill to keep us moving along the contour line. We were determined to find out more and as we drew closer to the 10-foot-long boulder our curiosity was further aroused. We could see that a considerable amount of material had accumulated on its up-slope side, indicating that the boulder had been in place for a very long time. This meant it was almost certain to be significant geologically.

After bringing the rover to a halt above the boulder, however, I realized we might have real difficulty climbing back up the slope after examining the site because of the gradient involved. Our spacesuits did not allow us to bend our knees to any great extent, making it much more difficult to move up a slope than down it. I had no option but to start backing the rover up carefully, with Jim guiding my movement, in order to "park" it just below the boulder. This got the adrenalin pumping. Mission Control could see nothing of what was going on, since I had been unable to point the antenna for the TV camera. But it was clear they were growing concerned.

"Jim and Dave, proceed very carefully now, please," said Joe Allen in what records of the conversation note was "an exceptionally attentive" tone.

As I started to edge the vehicle down the slope, Jim made his way down on foot to start examining the site and take a set of photographs to record its exact location. Then I heard him catch his breath.

"Dave, I see a layer of green in that boulder," he said with a slightly puzzled air. Green? How could that be?

At first I challenged him. "It's a big breccia, that's all it is . . . I don't see any green, Jim," I said, looking toward the band of rock in the boulder he was pointing at. But then I began to make out its color more clearly—and it was definitely green.

This was something we had never been briefed or instructed to expect. We knew we had to get a sample and close-up photographs. Managing to park the "the old girl" about fifteen feet below the boulder, I got off the rover and moved toward the boulder. I was about to chip away some of the green material when I saw, to my alarm, that without our weight the rover was beginning to slide down the slope.

"Tell you what, Jim. We'd better abandon this one," was my first reaction.

"Afraid we might . . . lose the rover?" Jim replied. Mission Control still had no visuals, but alarm bells must have gone off at the mention of us losing our transport.

"Dave and Jim, use your best judgment here," was all Joe Allen said. But it was clear from his next comment that Houston were anxious for us to get going and leave behind what must have seemed a pretty hazardous site. "The block's not all that important . . . We'd like you to spend most of the remaining time at Spur Crater."

But we were so close to the boulder that we weren't going to give up so easily. Although it was difficult to work on the soft, steep slope, we were in no danger of a catastrophe occurring; we were playing it "real cool." After calling to Jim to position himself below the rover to keep it from slipping any further, I edged closer and eventually managed to chip off a fragment of the boulder and scrape away some of the mysterious green matter and bag it, before returning to join Jim so that we could get going again.

It was certainly a moment of high tension. But the risk was worth the gain. Years later some scientists concluded, after considerable analysis, that the green material was part of an original ocean of olivine which surrounded the Moon before a crust of anorthosite was formed. When the crust was pierced by the impact of the sort of projectile that formed the Mare Imbrium basin, the remnants of the olivine ocean were ejected to form great "fire fountains" of green olivine glass. This mixed with soil and formed the rim of the crater on which we stood that day on the lower slope of Mt. Hadley Delta.

Station 6a was the mission designation for this site. Not too exciting a name for what turned out to be the scene of a discovery that would be key to unravelling the mysterious formation of the Moon. Our next find, when we moved on to Spur Crater, was to lend further weight to the conclusions eventually reached about the origin and formation of our nearest neighbor in space. Both discoveries marked out that afternoon we spent on the lunar surface as one of the most significant chapters in the history of scientific exploration.

For Spur Crater also turned out to be a treasure trove of previously undiscovered lunar material. It did not present us with the same maneuvering difficulties as Station 6a, so left us freer to soak up the sheer pleasure of discovery. I can still taste the excitement. As Jim and I started exploring the site, casting around for the best rock samples to collect, we felt an almost child-like delight as we first set eyes on new riches. Once again we came across some fragments of green rock.

"Who'd ever believe it!" I said to Jim, scooping a little of it into a bag and moving on toward a larger boulder. "Can you imagine that these little rocks have been here since before creatures roamed the sea on our little Earth?" Just then some bright crystals caught the sunlight.

"Oh, man!" cried Jim, startled.

"Oh, boy!" I said, turning to look at a small white rock perched on top of a gray pinnacle, almost as if it had been placed on a pedestal to be admired. I picked it up with my tongs to look more closely and saw a sliver of white crystal to the side of it catch the sun again.

"Look at that glint," Jim exclaimed and in that moment I felt a bright flash of recognition.

"Aaaah ... Guess what we just found," I radioed Mission Control, barely able to contain my excitement. Unlike as regards the green rock, I felt pretty sure of what we had set eyes on.

"I think we found what we came for," I said, my heart pounding a little faster with a mixture of excitement and pride.

"Crystalline rock, huh?" Jim remarked in what was less of a question than a slow whistle of amazement.

"Yes, sir. You better believe it," I said, laughing out loud now and then for the benefit of the backroom boys. "Oh, boy, I think we might have ourselves something close to anorthosite ... What a beaut."

"Bag it up," radioed Joe Allen enthusiastically.

This gleaming sample was soon named the "genesis rock" by journalists hastily assembled for a press conference in Houston shortly after its discovery. Later analysis confirmed that it was indeed anorthosite, believed by many to be the principal constituent of the Moon's primordial crust. It was dated at roughly 4.1 billion years old. It was not until fifteen years after such lunar samples were returned to Earth, and after countless papers had been written about this lunar lode as well as the many other dicoveries from lunar exploration, that the science community finally reached a consensus on how the Moon was probably formed.

Its conclusion was that our Moon is the daughter of not one but two parents. During the early formation of the Earth, a collision with a Mars-sized object ejected part of the Earth's mantle. This joined with much of the object's residue in orbit round the Earth and the disc of mixed substances accreted to form the Moon. Such exciting discoveries lay in the future, of course, but I was certain even then that we had found what we were looking for.

"Joe, this crater is a goldmine," I radioed Houston.

"And there might be diamonds in the next one," he quipped back.

"Yeah, babe ... Hey, isn't that super!" I remained full of excitement as, over the course of the next ten minutes alone, we collected up a greater diversity of rocks than any previous mission.

"A jackpot," said Joe, summing up the back room's enthusiasm over our finds.

By this time, however, our oxygen supplies were running low. In addition to Houston very carefully monitoring our time away from the LM, we monitored the depletion of our oxygen by counting down on our wristwatches the period left in our tanks. Fortunately, I had brought two wristwatches along; at the end of this second day's EVA my first one broke. Its crystal popped off after overheating and the watch became filled with lunar dust. This was because our longer stay on the Moon meant we were being subjected to higher temperatures than previous missions; the Sun was rising higher in the lunar day. The higher "heat soak" to which our equipment was subjected was another new challenge. By the end of our third day, we would have to switch the temperature control in our liquid-cooled garments to a lower setting. But everything worked fine, except for the watch.

It was time to return to the rover, anyway, and make our way back to the LM. We stopped off a few more times to bag enticing samples.

But I knew we had to return to the ALSEP and the drill that had caused me such problems the day before.

Despite the painful swelling of my fingers, again I bore down on the drill to try and drive it into the lunar crust. It was laborious, backbreaking work, and the drill kept getting stuck—post-flight analysis revealed a design fault. Eventually, however, I succeeded in pushing it to a depth of ten feet. But when it came to pulling the core out it would not budge. There seemed to be a blockage. After much debate on the best way to tackle the dilemma, Houston advised us to leave it yet again and return to try and solve it once and for all the following day. After climbing back into *Falcon* at the end of our second day I was relieved once again to be able to pull my gloves off and massage my swollen hands. Stripping down to our longjohns, we ran through the long list of systems checks and were able to make contact with Al as he passed overhead in orbit, as we did each day during our stay.

While we were on the surface of the Moon, Al was busy collecting a wealth of scientific data from a series of instruments on board *Endeavour*. Just one of these experiments—observing the effect of solar X-rays striking mineral deposits on the lunar surface—later helped geologists develop a closer inventory of the distribution of mineral deposits on Earth. Al had also come up with the idea of marking his solo portion of our mission by regularly broadcasting in many different languages the message "Hello, Earth. Greetings from *Endeavour*." It was his way of emphasizing our conviction that what we were undertaking was being done in the name of all mankind, not just the United States of America.

"*Schön guten Tag. Wie gehts Euch?* [Very good day. How are you?]" Joe Allen woke us at 160:01 GET at the start of our third and final day on the surface of the Moon.

"*Guten Morgen, mein Herr. Ist gut,*" [Good morning, sir. Everything's fine]," I replied. This was Joe's and my way of paying a small tribute to Wernher von Braun and his Huntsville team, many of whom were also German, who had given birth to the giant Saturn V that had delivered us to the Moon.

Taunting us once more before we could set off on our third day of lunar exploration, however, was the drill. Jim and I were determined that it would not defeat us this time. Our first approach was to grip the base of the drill housing, but when it didn't move we knelt down and put our shoulders underneath the horizontal handle. As we fought our way upright, the drill at last yielded and the ten-foot core of crust grated out of the ground. We then had carefully to break the rocky column—in which it turned out were locked fifty-seven layers of material and millions of years of history—into segments for storage later in the LM. But here again a tool wrongly assembled before launch cost us extra time and effort.

The time we had to spend on extracting this unique treasure, however, had taken its toll on the plan for that day. We were working eighteen or so hours a day, with sleep periods of six hours or less. But still we could not extend the time we spent on the lunar surface beyond the time our oxygen, water and other supplies would last. And on that last day, approaching the depletion of all our "consumables," everyone wanted to be very conservative in case we had a problem with the remaining part of the EVA or preparing the LM for lift-off.

The original plan had been for us first to drive west to Hadley Rille and then to progress toward a group of hills and craters named the North Complex, which particularly fascinated me since it was believed they might be the remains of a cluster of ancient volcanoes. But our tribulations with the drill meant it had to be dropped from our program. It was a huge disappointment. In the years that followed I often wondered if the unique data revealed by the lunar core had been worth abandoning those that beckoned us in the North Complex.

Fortunately, the revelations that awaited us at Hadley Rille more than made up for our disappointment at the time. The traverse to the site took us over unexpectedly rough, undulating terrain, almost like lunar sand dunes; up and down we went. For the first time during our lunar explorations we lost sight of the LM before we reached the horizon. But as we drew closer to this vast gouge in the surface of the Moon we saw striking evidence of volcanic activity in the clear layers that lined the far upper wall of the valley. When we dismounted from the rover and began to take samples, we found that the lip of the Rille was a far gentler slope than we had expected and the footing was firm, though this was not how it seemed to those following our progress in Houston.

Ours was the first mission during which Mission Control could monitor our nearly every move via the television camera on the rover; the TV on previous missions was fixed near the Lunar Module, where the landscape of the Moon was inevitably much smoother. So those on the ground were seeing new vistas for the first time, and there was a distinct edge of nervousness in Joe Allen's voice when we left the rover to explore the area for rock samples. From Houston's perspective we seemed to be "standing on the edge of a precipice." In fact, the slope down which we descended was only about 5 to 10 degrees and the maximum slope of the Rille was only 25 degrees—not steep for such a canyon-like formation. And it was easier to negotiate than the slope at the base of Mt. Hadley Delta the previous day when we had been so determined to bag a sample of the mysterious green rock. The footing was much firmer.

The Rille's rim and upper slopes were covered in hard-packed regolith and small rocks, compared with the much softer and unconsolidated lunar soil we had encountered at the base of the mountains. The deep Rille held no fear as far as we were concerned. We did two hours' intensive work collecting rock samples at the site. But Houston was clearly relieved when we neared the end of our scheduled time there.

"Get ready to move out, Dave," came Joe Allen's slightly concerned-but-confident-sounding voice through our headsets.

As we were finishing up at the Rille we were told that the North Complex had been canceled, and so we reluctantly made our way back to the LM. We could not afford to fall behind schedule on that of all days. At 1:11 p.m. Eastern Daylight Time on 2 August 1971 we were due to bring our stay on the lunar surface to a close, to lift away from the Moon and rendezvous with *Endeavour*. As we rocked and rolled our way back to base camp, with the majestic mountains ahead of us, Jim revealed, for the first and only time on the lunar surface, his deeply held religious belief.

"Dave, I'm reminded of a favorite Biblical passage from Psalms. 'I look unto the hills, from whence cometh my help.' "

I was too focused on keeping the rover on track to reply. But then Jim did that for himself: "Of course, we get quite a bit of help from Houston, too."

After making sure all our equipment was safely stored ready for lift-off, I had a few items of more personal business to attend to. I wanted to conduct a simple scientific experiment solely for the enjoyment of all those back home tuning into our transmissions from the lunar surface. I wanted to prove the law, proposed more than three centuries before by the Italian astronomer and mathematician Galileo Galilei, that all objects fall with equal speed in a vacuum. Taking in one hand a falcon feather I had brought along for the purpose and in the other my trusty aluminium geology hammer, I positioned myself in front of the cameras, raised my arms and let both objects fall. Sure enough, they settled into the lunar dust at the same time. On the Moon there was no atmosphere to cause drag and slow the fall of the feather.

"How about that!" I said as applause erupted at Mission Control. No one there, except Joe Allen, had known I was going to do it. It was a little moment of levity, a nice visual image for those, especially the kids, watching back home. But it demonstrated an important scientific point, too.

"Proves Mr. Galileo was right," I said.

"Superb," Joe replied with satisfaction.

That falcon feather is sure to be still in the same place on the surface of the Moon in the spot that I dropped it over three decades ago. It remains as an example of Earthly fauna, together with a four-leaf clover we left to illustrate our planet's flora and a copy of the Bible we placed on the hand controller of the rover—evidence of one aspect of our culture on Planet Earth. With her battery recharged and no atmospheric conditions to have caused her to rust or deteriorate, I see no reason why the rover would not still run like new were anyone to revisit the Moon in future.

The next deviation from the checklist wasn't captured on camera, though I took photographs to mark the event. After informing Mission Control that I needed to do a little cleaning up back at the rover, I wanted to perform a more private ceremony. It had a much sadder note. Amid the euphoria surrounding the success of the Apollo program and our feeling of great personal accomplishment at having achieved what had once seemed such an elusive goal, we wanted time to reflect for a moment on the human cost of the race to the Moon.

In a small depression near the rover I placed a small statue and plaque dedicated to the fourteen American astronauts and Soviet cosmonauts who had lost their lives in pursuit of that goal. The fourteen names listed in alphabetical order on that plaque are Charlie Bassett, Pavel Belyayev, Roger Chaffee, Gyorgy Dobrovolsky, Ted Freeman, Yuri Gagarin, Edward Givens, Gus Grissom, Vladimir Komarov, Viktor Patsayev, Elliot See, Vladislav Volkov, Ed White and C.C. Williams. (Sadly, two names are missing, those of Valentin Bondarenko and Grigory Nelyubov. But at the time, because of the secrecy surrounding the Soviet space program, we were not aware of their deaths.)

Reflecting on their loss, I felt a strong sense of brotherhood with those men. Some had been close friends, some I had only seen in formal photographs alongside a brief announcement of their deaths in the Soviet press.

After positioning the rover and its camera far enough from the LM to be able to record our lift-off—the first time such an event would be televised—I made my way back to join Jim in the LM. As I pulled myself up the ladder and on to *Falcon*'s "front-porch" for the last time, I felt certain no other experience in my life would ever compare with these three days on the Moon.

My nature has always been to look for new challenges and I knew I would continue to do that. So I was sure there would be other great opportunities ahead. But I knew, even then, that I would never be coming back to the Moon. Only four more astronauts were due to make that journey. The Apollo program was winding down. I had no idea then just how definite an end to manned lunar exploration this would be—certainly in my lifetime, and who knows how far into the future? All I knew in those moments was that I had come to feel a great affection for this distant and strangely beautiful celestial body—in effect a small planet—constantly circling our own. It had provided me with a peaceful, if temporary, home. But it was time to return to my own home back on Earth.

Just before I climbed the ladder to reenter the LM and begin the final countdown to lift-off, Joe Allen transmitted to us a fitting summary of a poem written by one of the characters in Robert Heinlein's popular science fiction novels. It helped ease the pangs I felt as our spacecraft lifted away from the surface of the Moon.

As Rhysling, the blind poet of the spaceways wrote, "We pray for one last landing on the globe that gave us birth, To rest our eyes on fleecy skies and the cool green hills of Earth."

Falcon's ascent engines ignited at 171:37 GET, exactly on schedule, and we lifted abruptly away from the surface of the Moon in an impressive flurry of dust and debris, captured for the first time on camera and transmitted live to a worldwide audience. While Jim and I went through a series of checks to confirm that all systems were stable and that we were on course for our rendezvous with *Endeavour*, Al was unstowing the cassette player in the Command Module—it was carried as part of the official on-board equipment so that we could occasionally listen to music. He snapped in a tape recording we had brought along in anticipation of the success of this all-Air-Force crew.

The original plan had been for Al to play this just for the benefit of Houston, and he didn't realize that we had been switched into the same radio loop. So as we went through our checks we could hear the recording of the Air Force song. Unexpected as it was, it felt good to start our return journey to Earth to the strains of "Off we go into the wild blue yonder . . ." Although in this case the yonder was black!

Apart from this musical accompaniment, our lift-off from the Moon was quiet and soft—propelling us from one-sixth G to one-half G—compared to the launch from the Cape. Though fully suited up and helmeted so that all noise was muffled, as we took off I could hear a soft swishing sound and what seemed like the faint hiss of wind blowing through *Falcon*'s windows—though, of course, that was impossible. It was noise from the ascent engine.

Our rapid ascent took us right up over the Rille, and within three minutes we had reached an altitude of 30,000 feet. Two hours of delicate rendezvous and docking maneuvers eventually brought *Endeavour* and *Falcon* locked in unison as a single spaceship once more. After the cumbersome work of transferring all 170 lb of our carefully bagged lunar rocks and samples into the Command Module, we closed *Falcon*'s hatch and prepared to jettison her for our return journey to Earth in *Endeavour* alone. A small problem with the seal of our spacesuits led to a ten-minute delay in jettisoning, but at 179:30 GET Al confirmed he had separated *Endeavour* away from *Falcon*.

"Roger, copy," replied Mission Control. "Hope you let her go gently. She was a nice one."

"Oh, she was that," Al said wistfully.

The slight delay in jettisoning led Mission Control to feed us some confusing and incorrect data immediately after the separation, however. Had we performed the CSM-LM separation maneuver as they originally calculated, we would very likely have made physical contact again with the LM, which we could see out of the window right ahead of us. Mission Control recalculated the maneuver and we proceeded safely. But, as Jim and I had gone for nearly twenty hours without sleep by this stage, and had not eaten for eight hours, Mission Control seemed to think that we, not they, had got confused. They even suggested that Jim and I each take a Seconal tablet that night to help us sleep. Seconal—trade name for secobarbital, a barbiturate with sedative qualities—was routinely carried in case of a medical contingency aboard all Apollo spacecraft.

I was amazed at the suggestion and quickly dismissed it, assuming it had come from one of the NASA doctors monitoring our progress. They were always over-cautious as far as we astronauts were concerned. To my mind, though I did not say so at the time, taking a sleeping pill without good reason was ridiculous, given the extreme demands and level of alertness required of us in case of an emergency. Though physically tired, Jim and I were elated at how well our stay on the Moon had gone. We were relaxed and were certainly ready to sleep without pharmaceutical assistance.

When they suggested again that Jim at least take a sleeping pill the following night, he declined again. What we were not told was that both Jim and I had shown slight irregularities in our heart rhythms while on the lunar surface. The irregularities, technically known as premature ventricular contraction, were a result, it was discovered later, of a potassium deficiency caused by the rigors of our training regime prior to the flight. Subsequent crews stocked up on liquids fortified with potassium. I later became concerned that I had not been informed of this at the time and so could not take it into consideration when making my decisions as commander.

I was even more concerned when I learned after the mission that doctors had informed the Flight Director in Mission Control that EKGs registered Jim's heart as showing symptoms of a condition known as "bigeminy," an irregular rhythm whereby the heart skips a beat and then beats twice rapidly. This had happened several times during our EVAs and also just after we performed our rendezvous with *Endeavour* on completing our three-day stay. As soon as we were settled in the CM and Jim got some rest, his heart returned to normal. But, again, had I been informed of this at the time I might have been able to simplify his role during the EVAs to reduce the strain on him. Several months after our mission, Jim had a heart attack; and another followed a few years later. Sadly, he died of a third heart attack on 9 August 1991. I tried, unsuccessfully, for several years to get NASA to provide an adequate explanation of the effect of Jim's heart problems during the mission on the eventual decline in his health.

But that first night back on *Endeavour* and throughout our three-day return journey to Earth Jim and I slept just fine. We still had a fairly full program of scientific experiments to perform during our return journey. First, before leaving lunar orbit, we had to deploy a small hexagonal satellite, the first sub-satellite to be launched in space, which would continue to orbit the Moon and send back data on its magnetic field for over a year.

Then, still 197,000 miles from home, Al had to perform an EVA—the first ever performed in deep space—to recover film cassettes from the scientific instrument module, before it was discarded as we neared Earth. Another fascinating element of Apollo 15 was the range of sophisticated instruments inside a large bay of the SIM, which collected scientific data from lunar orbit as well as during the transit to and from the Moon. The SIM bay was primarily Al's responsibility and he operated an entire suite of experiments, mostly during the three days he was alone in lunar orbit.

One very interesting application of the SIM bay on the way back from the Moon was to study in detail the temporal behavior of pulsating X-ray sources, later to be known as "black holes." In coordination with an Earth-based observatory in the Soviet Union scanning the same part of the galaxy at the same time as we were, we took the first photographs in space of a series of suspected black holes, about which much less was understood at that time. It is nice to know we helped in this study.

The existence of one of them, Cygnus X-1, was only discovered shortly before our mission. I did not know then that black holes would hold a great fascination and professional interest for me three decades later. I now know, for instance, that a black hole was so named because nothing sucked into its center, including light, ever escapes. Cygnus X-1 is now believed to be a binary star system in which one star has collapsed and is sucking material from its neighbor on to a voracious rotating disc, where the frictional heating is radiated as X-rays. The black hole in the center of the disc is invisible.

On the twelfth day of our mission some time was taken up with a lengthy press conference conducted live from space, during which we fielded questions on everything from our problems with the drill to which moment during the mission we would most like to relive. My answer to the latter was clear. For me it was definitely that moment when Jim and I stood on the lower slope of Mount Hadley Delta, high above the Plain, and were able to absorb for the first time the wide panoramic view reflecting the diversity of our landing site on the Moon.

Perhaps the most touching communication we had with the ground on our return flight from the Moon was a brief exchange with Lee Silver, who had been following our mission intently from the geology back room at Mission Control. For a short while Lee was invited into the main control room to speak with us directly, the only time a geologist communicated directly with a crew during any Apollo flight.

Lee's words meant a great deal to me. "Hey, Dave, you've done a lovely job. You don't know how we're jumping up and down, down here," he said, full of enthusiasm and warmth after talking briefly about some of the sites we had visited.

"Well, that's because I happen to have had a great professor," I said.

"A whole bunch of them, Dave," was his typically modest reply, and then he said, "We think you defined the first site to be revisited on the Moon."

"I could just spend weeks and weeks looking and pick out any number of superb sites down there . . . To coin a phrase, it's mind-boggling," I concluded. "I hope one day we can get you all up here, too."

Twelve days after blasting away from the Cape we prepared our spacecraft for landing. This time, unlike the drama Neil and I had endured as Gemini 8 came down in unknown territory, *Endeavour* headed for landing on target north of Hawaii. My mind did flash back to that earlier mission for one heart-stopping moment during the descent, when one of *Endeavour*'s three main parachutes collapsed shortly after it was deployed. But the remaining two were sufficient to break our fall and bring us down safely—if a little harder than if the failed chute had remained open.

At 4:46 p.m. Eastern Daylight Time on 7 August—295:12 GET—we hit the calm waters of the Pacific, submerged briefly, then bobbed back to the surface.

"Recovery, Apollo 15. Everybody's in good shape," I confirmed a few minutes later by radio transmitter.

Within a very short time two swimmers were dropped into the water to attach a flotation collar and a sea anchor to *Endeavour* to secure our position. First I was winched into the helicopter hovering above, then Jim and Al were hauled up and we were flown to the nearby carrier USS *Okinawa*. As "Anchors Aweigh" struck up we saluted our recovery team and I made a short address both to them and to the general in command of the US Air Force in the Pacific there to greet us.

"It's been my experience that the Navy always makes a fine pick-up. Whether we land six miles from the carrier or six thousand," I said, smiling at the memory of the very different end to my mission aboard Gemini 8.

The sight of Jim, Al and me unshaven as we stood on the deck of the USS *Okinawa* might have surprised some. But before setting off for space I had made a small promise to my son and daughter that I would not shave while in space, as had most other astronauts on previous long-duration missions. Inspired by images of earlier explorers, like Scott, Amundsen and Ernest Shackleton, they had wanted me to return from my space travels looking equally rugged. So, as we stepped down from the recovery helicopter on to *Okinawa*'s deck and saluted, I smiled to myself that my kids would be pleased with this first sight of their dad back home, safe from his space travels. They would not have as long to wait to see us as had the families of previous Apollo crews, because we were the first crew not to be placed in quarantine on our return. From Hickham Air Force Base in Hawaii we were airlifted out almost straight away and flown back home to Texas. Doug, in particular, took great delight in rubbing my stubbly chin, before I finally shaved.

To my mind not placing us in quarantine was a mistake. Our twelve-day mission had taken its toll. Though elated by our success and exhilarated to be back, we were really tired. We should have been given some time to rest before fully engaging with the demands of life back on Earth. Our immune systems were low after living in such a carefully controlled environment for so long. We had lost weight—over 12 lb in my case—due to the exertions of the mission. But immediately on our return we went through endless debriefings about the science, engineering and operations of our mission, all of which was very interesting. Whereas in quarantine everyone comes to see you, we had to drive ourselves around from meeting to meeting. We had to undergo extensive physical exams, including monitoring on treadmills and electrocardiograms.

There was also the usual big press conference on our return. After a seemingly endless round of questions from journalists I concluded by paying tribute to all those astronauts and cosmonauts who had lost their lives in pursuit of our respective space programs. I then read out one of the quotes I made a habit of collecting, a favorite from the writings of the Greek historian and philosopher Plutarch, which I felt particularly fitting for the mission we had just completed.

"The mind is not a vessel to be filled," wrote Plutarch "but a fire to be lighted." Our journey to the Moon had certainly lit a fire of intense interest in the geological treasures we had brought back.

But unlike the careful post-field-trip analysis of the rocks we had collected, we were only given a brief opportunity to view our precious cargo. We had one dinner party with the geologists and one half-day viewing of the rocks, through a glass window of the "lunar receiving lab," before the samples were stored away for careful inspection by the scientists. It was as if we had given birth and were only allowed to view our babies through the screen of an incubator before they were whisked away for adoption.

That one viewing was really special, however. Dressed in white protective coats and caps we peered through the receiving lab window alongside geologist Jim Head, Lee Silver, Gordon Swann, Jack Schmitt and others. As Jim and I stood studying the green rock, in particular, we couldn't help feeling slightly awkward in front of Lee Silver when he asked what we thought had led to such an unusual rock formation. We had no idea and were wondering what to say when he broke the awkward silence.

"Well, you know, we've been looking at it for three days," he confessed, "and the best we can come up with is a friable green clod." It was to be years before its origin as an olivine ocean surrounding the Moon was identified.

Such moments of quiet reflection were rare in that period after the flight. Most nights we were thrown headlong into a round of parties, both private and public, held to celebrate our success. According to an old flying tradition whereby pilots are tossed in a pool following their first solo flight, Al and Jim received a public dunking. Some friends had proudly kept scrapbooks of all the media coverage of our mission and wanted to show us the results. Some wanted autographs. Kids wanted to show us the pictures they had drawn. Everybody was so congratulatory, so happy for us. Neighbors would drop by with cakes and want to sit and hear tales of our space travels.

Much as we wanted to meet friends and neighbors again, and appreciated their enthusiasm, warmth and hospitality, all we really wanted when we got back from the Moon was a little time to rest and be with our families. We were very tired, yet still had so much adrenalin pumping through our systems that it was sometimes difficult to sleep. When Al couldn't sleep he took to writing poetry, which was later published.

As for me, for more than a week after I returned I found myself still awake at two o'clock in the morning. The physical strain of working with the drill on the Moon had left me with a constant dull ache in my shoulder. Although NASA doctors pronounced me fit, I found a more sympathetic ear in NASA's long-standing flight nurse, Dee O'Hara, who arranged privately for me to undergo ultrasound treatment at a clinic in downtown Houston. This eventually solved the problem. But until it did I sought to ease the pain in my shoulder during the small hours of the morning with just one of the many comforts not available to us on the Moon—a long, hot shower.

CHAPTER 11

Cowboy from Siberia

1972–3

Colonel Alexei Leonov

Hall of St. George, the Kremlin, Moscow

There was a distinct thaw in relations between the Soviet Union and the United States in 1972. It was a rare period of calm in the Cold War. Both Leonid Brezhnev and the American president, Richard Nixon, were cultivating improved relations between our two countries. This atmosphere of détente became known in Russia as "*razryadka*."

People in the Soviet Union were in favor of such peaceful moves. We had suffered so greatly in the Second World War. Most were sick and tired of confrontation. People still remembered the time when American and Soviet troops had met at Torgau, on the River Elbe, in the spring of 1945. There was nostalgia for this time when our two countries had stood together against the common enemy of Nazi Germany.

In May 1972 Richard Nixon became the first American president to visit Moscow. Prolonged talks had prepared the way for our two countries to sign the first arms limitation treaty. This was in everyone's interests. Nixon understood very well that neither the United States nor the Soviet Union could afford to be brought to the brink of war. Both our countries had enough weaponry to destroy mankind. It was clear we could not carry on amassing arms at the rate we had done in the past.

Of great significance to the space programs of both the United States and the Soviet Union were talks, which had been going on between our two countries since the moon landing of Apollo 11, about future cooperation in space.

By spring 1972 the talks were advanced enough for Nixon and our premier, Alexei Kosygin, to sign a formal agreement committing both countries to a joint space project. This would be the first major international, cooperative space venture. It would involve an American Apollo and Soviet Soyuz spacecraft completing a rendezvous in Earth orbit and docking, to allow their crews to transfer between the two vehicles and "shake hands in space." It was to be a potent symbol of the potential that existed for cooperation between our two countries.

To mark the occasion a very grand reception was held for Nixon at the start of his visit in the Hall of St. George at the Kremlin. I was present, and at one point Brezhnev took me aside and asked me what I thought about the Apollo–Soyuz project.

"Now we have signed this agreement between our two countries, what do you think, Alexei?" he asked. "Do you believe in it? Do you think we can cooperate?"

I had first met Brezhnev in September 1964 when, together with Yuri Gagarin and Pavel Belyayev, I had taken Politburo members led by Brezhnev on a tour of Baikonur to show them the latest developments in space technology. This was while I was training for my Voskhod 2 mission, and I had demonstrated some of the maneuvers I was intending to perform during my EVA.

"Well done, guys," Brezhnev had said, slapping me on the shoulder. "Carry on the good work." From that point on he had always called me by my first name.

"Not only do I believe in it," I told Brezhnev, "but it is something I have always wanted to be involved in. We just need to concentrate on coordinating our efforts now."

Reassured and satisfied with this response Brezhnev turned his attention elsewhere and Kosygin introduced me to Nixon, who was about to make a formal toast.

"This morning the formal agreement was signed for the Apollo–Soyuz Test Project. There is no going back now," said Nixon, seeming very calm and sure of himself. "You know this is a very important undertaking for both our countries. The whole world will be watching, some with hope and others with skepticism. But I am optimistic. I am sure all will go well. Our two countries will do everything to support each other."

At that everyone raised their glasses and in the noisy atmosphere that followed I turned to Nixon and proposed my own toast, not just to the success of the project.

"To your health and to future cooperation between our countries," I said. "Not only in space."

I was to meet Nixon again when he returned to Moscow in 1974, a year before the Apollo–Soyuz mission was due to launch. He recognized me, smiled and asked how things were going. By that time I was the Soviet Union's key figure in the project and was able to assure him that everything was going according to plan. But, in 1972, at that very early stage in the venture, neither the American nor Soviet crews for the joint mission had been selected. I was still working intensively on the Salyut program.

Following the Soyuz 11 / Salyut tragedy the previous summer a state funeral for Patsayev, Dobrovolsky and Volkov had been held, after which their remains were buried in the Kremlin Wall. One of the pallbearers at their funeral was the American astronaut Tom Stafford, who I knew had not only flown two Gemini missions but also commanded Apollo 10. I later came to know Tom very well.

Colonel David Scott

The White House, Washington

With barely time to catch our breath after Apollo 15, NASA had us on the road on a goodwill tour. In Washington we were invited to address a joint session of Congress. After that, Lurton and I, together with Jim, Al and their families, were invited to the White House. This time we were invited to bring our children along.

While the concept of spaceflight had somehow come to seem routine to our kids, such post-flight trips were anything but.

While we adults dined downstairs with President Nixon, Vice President Spiro Agnew, and their wives and other guests, the kids ate upstairs in the president's private living quarters with the vice president's daughter. After supper Nixon gathered the eight children together and gave them a private tour of the White House. He was like the Pied Piper. He was very good with the kids. He delighted them by showing off a series of secret passageways and, at the end of the tour, handing out small gifts of pens and lapel pins. In contrast to the bitter recriminations and scandal that engulfed his presidency in later years, it was a very touching scene.

After Washington we were invited to New York, where we spoke to the general assembly of the United Nations and received the United Nations Peace Medal from Secretary-General U Thant. Then came a trip to Salt Lake City, where we spoke at the beautiful Mormon church, and Chicago, where fire-fighting barges on the Great Lakes pumped out streams of colored water in celebration. This was followed by a tour of several western European capitals, including Paris, Bonn, Brussels and London. Several weeks later we were invited on a ten-day goodwill tour to Poland and Yugoslavia. In the bowels of ancient castles, toast after toast was proposed to celebrate our success. It was quite something to be able to make such a visit behind the Iron Curtain.

The visit to Poland was especially interesting since, in Warsaw and Kraków, Lurton and I were provided with our first and only bodyguard, Major Sunday. He was a large, burly, good-natured chap, not only friendly but very protective. Over-enthusiastic crowds never got too close: his serious face and piercing eyes were enough to keep anyone at their distance. During our last night in Warsaw, the American ambassador hosted a dinner for us in one of the few local nightclubs. By then we had become very close to Major Sunday, despite his quiet and reserved manner. Still we were surprised when he asked Lurton to dance. On the way back to the hotel, Lurton and I were whispering to each other that we really should buy a gift for Major Sunday's wife before we left the next day. During a rather emotional departure the major presented

Lurton with a gift from his wife; it seemed he had heard every word we had said.

We continued these PR excursions until the end of the year, when we were the grand marshals at the Orange Bowl in Florida on New Year's Day, 1972. All the while we were thinking about the future and our next assignments. It was five months since Apollo 15 had returned, and the space program was changing course dramatically. Apollo 17, scheduled for December 1972, would be the last mission to the Moon.

The new Skylab space station program had only three missions planned, ending in late 1973, and all crews had been assigned. In the far distant future was the space shuttle, a combined airplane–spaceship designed to return from orbit to an airplane-type landing on a runway. But shuttle missions were years away—the president had not even approved the program. There was very little on the horizon, especially for highly experienced lunar astronauts.

Then one day in late January Al Shepard asked me to stop by his office. He was once again chief of the Astronaut Office and, recognizing that we had essentially finished our Apollo 15 post-flight activities, he offered me two options for my next assignment. Either I could back up Gene Cernan, commander of Apollo 17, or I could move into NASA management by joining the new Apollo–Soyuz Test Project as special assistant for mission operations working for Glenn Lunney, the project manager.

Apollo 17 was a dead end. I had no enthusiasm for going back through a training cycle with very little chance of being able to fly. But I liked Lunney. He had been a flight director on both Apollo 9 and 15, and Apollo–Soyuz sounded like a fascinating project. So I accepted and made the move, after spending two months supporting the first Skylab mission as technical assistant to the project manager.

Lunney had already assembled his team, and I had attended several meetings with the Soviets in Houston. My job would be to coordinate all the mission operations to ensure that America's Apollo Command and Service Module and the Soviet Soyuz and our two methods of spaceflight were compatible. The crews for the missions had not yet been assigned, so I would also represent the US side in working out how the US crew would operate and interface with their Soviet counterparts.

It was during this period, while on a trip to Washington, DC, that George Low, NASA deputy administrator, asked me to stop by his office. I had known George in Houston while he was Apollo spacecraft program manager and I had great respect for him. We spent a short while exchanging pleasantries, then George looked me in the eye and came to the point.

"Well, Dave," he said, "I have to tell you I was not very proud of my performance."

His comment was the culmination of what had become known as the Apollo 15 "cover incident." This involved some first-day postal covers which we had carried aboard Apollo 15, some of which were unfortunately sold by a German stamp dealer several months after the flight. As No. 2 in NASA, George had been heavily involved in NASA's handling of the incident and his comment referred to one of the poorest management performances the space agency had ever demonstrated—up to then.

The whole business had probably been building since Mercury, through Gemini, and into Apollo. People had a fascination with objects that had been carried in space, and they became more and more popular, and valuable, as the program progressed. Right from the start of the Mercury program, each astronaut had been allowed to carry on board a certain number of personal items in what were known as PPKs—Personal Preference Kits. Before each flight a list of the items was prepared for Deke Slayton, who had overall responsibility for approving them.

As the flights became more significant, the number and type of items increased. Aside from personal mementoes, each crew had carried a certain number of medallions, which they could hand out afterward as souvenirs. The number of medallions had grown steadily on each mission—eventually some crew members had been carrying several hundred each and their weight was becoming a concern. And, as always, commercialization began to creep in.

In the end, on Apollo 14 commanded by Al Shepard, it was alleged that the crew had carried some silver medallions on board, which were to be melted down after the flight and mixed with many other commemorative medallions by the Franklin Mint to be sold to the general public. The Franklin Mint had even advertised the proposed sale before the flight. After the flight, the deal was never consummated and all went quiet; nothing about it was printed in the media. But some members of Congress had heard about it and were unhappy with the situation.

Since we were busy training for Apollo 15 we knew nothing about all this. Had we known we would have been far more wary of what happened subsequently. All we knew was that Deke Slayton had halved the number of medallions we could carry. Shortly afterward Deke introduced Jim, Al and me to a long-standing friend of his at the Cape called Walter Eiermann. In retrospect this seems likely to have been more than a coincidence.

Deke invited us to join him at dinner at Eiermann's house one evening, several months before the Apollo 15 launch. Eiermann asked if we would like to make some money on the side by signing some stamps. "All the guys are doing it," he said. We agreed. We were also approached by several members of the Manned Spacecraft Center's stamp club, who asked us to sign many first-day covers before the flight for their members as well as for ourselves. At that time we could not buy life insurance. If we signed some covers, we reasoned, they could be held during the flight and act as limited life insurance for our families if anything happened to us.

But then Eiermann proposed we carry four hundred lightweight commemorative covers, a hundred of which would be passed on to a stamp dealer in Germany when we returned. All the covers would be franked on the day of launch and franked on the day of our return. Our understanding was that the dealer would hold his hundred covers to be sold at some future date, after the Apollo program was over. In return he would set up a $6,000 trust fund for each of us for the education of our children. In the months of intense activity before the flight we did not give this more than a moment of thought. But we agreed. In retrospect we made a mistake in even considering it.

In the months following our mission, we learned that the German dealer had begun to sell the covers. We let Eiermann know that we were opposed to the sale. Forget the trust funds, we said; we don't want any money. This is not what we had understood would happen.

But when senior managers at NASA learned about the covers being sold they were furious. Incomplete information was then provided to the press by NASA's Public Affairs Office and reports started to appear that we had smuggled the covers on to our flight. This was literally impossible. NASA personnel prepared us for every aspect of the flight—from our birthday suits out.

Everything we had carried had to be specially packaged by the flight support crew to make sure, apart from anything else, that it was fireproof. Everything in our PPKs was on a list, which was certified before launch. I was never aware of any rules about what could or could not be taken in these packs. Slayton was to approve the list, so as far as I was concerned it was up to management to keep us on the straight and narrow. I assumed if the list was approved that was fine.

Usually the list was certified by Deke. But before our flight, for some reason, he neither asked us personally for each of our lists, as was customary, nor signed off on the list personally. He said the flight-crew support team had already logged everything. Whereas we had purchased the covers ourselves, the Astronaut Office at the Cape had prepared the covers for the flight and had had them stamped and franked on the day of launch. Somehow, however, the support team had missed them when they prepared the PPK flight manifest.

It was also reported after the incident that we had been removed from the astronaut corps. Again, totally untrue. Unfortunately, NASA managers did nothing to dispel this false and misleading information. They just tucked their heads in and let the rumors run.

The reports infuriated Congress, not least because congressmen had to read such controversial information in the press before being informed by NASA, which is obligated to keep Congress fully informed of its activities. Recollections of the Apollo 14 medallion incident must have seeped into the minds of certain members of Congress, many of whom were not fans of NASA anyway.

When NASA started an internal investigation we were told by a senior official in the agency that we were on our own. We were advised to get our own legal representation. A Senate hearing into the matter had been called and a Justice Department investigation would follow.

It was turning into a witch-hunt. NASA, we were advised, expected us to keep quiet and take the Fifth Amendment at this hearing. We did not. We told it like it was. We had nothing to hide.

Eventually, the Justice Department concluded that we had broken some administrative rules, but had done nothing criminal. NASA confiscated the remaining covers, but they acted before discovering all the facts of the case. Following an investigation by NASA, the Justice Department and the Senate, the Justice Department concluded on 6 December 1978, in a "Memorandum Opinion for the Assistant Attorney General," that NASA had no claim to the remaining covers; that the covers were never intended for sale; that there was no attempt at concealment by the crew of the fact that the covers were to be flown on Apollo 15; and, finally, that the covers would have been approved to carry aboard Apollo 15 had a request been made.

We were reprimanded and we took our licks. But it was a very raw deal. NASA had hung us out to dry. Our bosses had abrogated their responsibilities, and we were left alone on a very wet day.

I spent many years trying to get full access to NASA and congressional records on the case. NASA refused to provide full disclosure. We subsequently found out there had been rumblings about profit-making on previous missions. A copy of a letter was released to me in which NASA admitted that ten other astronauts, who were not identified, had been involved in selling signed blocks of stamps and postcards for Eiermann. But the wave reached the shore on Apollo 15 and we were the ones who bore the brunt of the blame for such incidents.

Alexei Leonov

After the crew of Soyuz 11 was lost, the Soyuz spacecraft underwent a major redesign. A new spacesuit was also manufactured. This took time. Since Salyut 1 had been destroyed on reentry into the Earth's atmosphere, a second Salyut station would have to be launched. Again a Soyuz would rendezvous with the space station and transfer its crew to work aboard Salyut for three to four weeks. Again I was appointed commander of the two-man crew.

While we were preparing for this second Salyut mission, Brezhnev visited Star City for a second time. This time he brought Fidel Castro with him. The two men were scheduled to spend only two hours with us, but their visit extended to four.

Brezhnev was still strong and full of energy then. As we toured Star City, Brezhnev took off his jacket, unbuttoned his collar, rolled up his sleeves and kept making suggestions, telling us we should change this and that. The note he left in our visitors' book that day was remarkable and touching; "The hours spent here together with Fidel and you, dear friends, were the happiest." That autumn he had the first of several strokes, and his health diminished. His facial features became deformed, his speech became slurred, and the people around him started running the country. But my recollection of him on that day was of a man full of vigor.

By July 1972 we were ready for the second Salyut mission. Once again my belongings were loaded on to the space station. Once again we left for Baikonur. Once again it was not to be.

This time the mission was aborted within minutes, when the Proton rocket, which was to propel Salyut 2 into space, failed. The initial ignition system worked perfectly. But there was a fault with the control systems of the second stage of the booster rocket. Less than three minutes into the mission, the rocket and the Salyut it was carrying crashed back to Earth.

One member of the rescue party dispatched to the crash site to secure the wreckage later brought me back my initialled sleeping shorts, which he had found lying in the barren steppes.

"You may not have gone on this mission, but at least your sleepwear was briefly launched," he joked.

But there was no raising my spirits. After this failure I was extremely frustrated. I was really beginning to lose faith in the program.

Then in January 1973 the Americans announced their crew for the Apollo–Soyuz project. A few months later I was asked to command the Soyuz spacecraft for the joint mission. At first I was daunted by the prospect. I did not speak English. The rules of the mission dictated both crews needed to learn the other's language. I was told I would be put through an intensive language course. In truth, it was not really the language problem that worried me.

Despite all the frustrations and setbacks I had endured, I still really wanted to continue my work with the Salyut space station.

Eventually, I said I would agree to command the Soyuz craft for the joint mission—on one condition: that Valery Kubasov, who was due to have flown with me on Soyuz 11, would be the engineer on the crew. Unknown to me, he had also been approached independently to join the crew, and had agreed—on condition that I was his commander.

When I learned that Tom Stafford had been appointed commander of the Apollo crew, with Deke Slayton as pilot of the docking module and Vance Brand as the third member of the crew, I was delighted. I had already met Deke Slayton in Athens and Tom Stafford I came to respect greatly when he attended the funeral of the Soyuz 11 crew.

My first meeting with Stafford, after the two crews were selected, was neither in the United States nor the Soviet Union. It took place on neutral territory at the Paris Air Show in May 1973. A special pavilion had been erected to mark the mission. Photographs of the American crew had already been hung in the pavilion, but initially our photos were missing.

"How are you going to work with these Russian guys if they can't even manage to get their photos up on time?" one journalist asked at our first joint press conference.

There were always those who wanted to throw cold water on the project. Some argued that America should not cooperate with the Soviet Union because we were only out to steal America's technology secrets. But as far as we were concerned the Americans had as much to learn from us as we had to learn from them.

"I have full faith in my Soviet colleagues," Tom replied. "What do we need photographs for if the cosmonauts are here in person?" As he spoke our photographs were carried in and mounted behind the podium.

David Scott

Despite the tough time following the cover incident, I was still full of hope and optimism for the future of spaceflight. I was especially enthusiastic about the Apollo–Soyuz Test Project, on which I had begun to work. The prospect of cooperation in space had been raised as early as the summer of 1969, not long after the first lunar landing of Apollo 11.

In an address to the United Nations, President Nixon had spoken of the importance of countries cooperating in space. With the race to the Moon over, the United States would be taking steps toward "internationalizing man's epic adventure in space," he said. "An adventure that belongs not to one nation but to all mankind." It was this message of intent, which eventually evolved into the Apollo–Soyuz project. But at that very early stage there were no guarantees that such a joint mission would be possible.

Besides the major diplomatic hurdles it posed, NASA's budget was about to be cut. In March 1970, Nixon had issued a statement saying America's space program could no longer expect to command the resources it had when the race was at its height.

"We must recognize that many critical problems here on this planet make high-priority demands on our attention and resources," he stressed. Top of the list of priorities was extricating the United States from Vietnam. While publicly committed to withdrawing American troops, Nixon had ordered bombing raids on communist bases in Cambodia. The conflict had escalated further. Anti-war demonstrations at home had escalated, too, resulting in the death of four students when the National Guard opened fire on protestors in May 1970 at Kent State University in Ohio.

Yet, despite waning political support, NASA had forged ahead in its search for a new frontier in the wake of the Apollo program. Among the goals it identified were the launch of the relatively large, but temporary, orbital space station Skylab, and the start of a collaborative project with the Soviet Union, which evolved into ASTP.

From my point of view Apollo–Soyuz was a great opportunity.

On a personal level test pilots always like to fly the other guy's stuff, see how it works. From a wider perspective it would provide a rare chance for both sides in the bitter and dangerous Cold War to show they could cooperate if permitted to do so by their political masters. It would also give us a chance to learn more about how the Soviets did things: get more insight into how they trained, how they built their spacecraft, what their spacecraft looked like. This was not meant to be underhand spying on the Russians so that we could beat them in space—we were already in the lead, or so it seemed to us.

And yet the Soviet Union was still the other side, the other superpower. The more we could learn about them the better. Both sides still had countless inter-continental ballistic missiles pointed at each other. We both still had a clear order of battle to support a war against each other. This might be a period of détente in the Cold War, but who knew how long it would last? Not very long, as it turned out.

Working on Apollo–Soyuz, I realized, might also give me an opportunity to go visit the Soviet Union. That was a very exciting prospect. The USSR had been so isolated for so long that we knew very little about either the country or its people. The only images we saw of what life was like inside this closed and secretive society were those officially sanctioned for release, the majority of them relating to proud parades of military hardware.

I was driven by curiosity. My only contacts with the Russian cosmonauts up to that point had been the meetings with them in Paris in 1967 and 1969. They had left me hungry to find out more.

In the months that followed my assignment to ASTP there was a series of exchanges between our support staff and members of the Soviet team. Most involved reaching a compromise on the rules and planning for the mission, and also designing a docking system which would enable the two spacecraft to join together with a common, or androgynous, docking collar, to allow the American and Soviet crews to transfer between the two vehicles.

Less than a year later, in mid-June 1972, the opportunity came to visit Moscow. I was to lead a technical team of thirty-five negotiators who would define the configuration and operations for the Apollo–Soyuz mission. My main contact in the Soviet Union would be with Professor Konstantin Bushuyev, the Soviet project manager.

Although it was not explained exactly who he was at the time, I learned many years later that Bushuyev was a key engineer-manager in Sergei Korolev's Experimental Design Center, a highly respected engineer and designer who had been heavily involved in most aspects of the Soviet manned spaceflight program since the early 1960s. Bushuyev had been one of Korolev's key deputies during the early years of his OKB-1 at Podlipky.

Alexei Leonov

The program for the first visit of an American delegation to the Soviet Union, following the selection of both crews for the Apollo–Soyuz project, was very carefully prepared. The group of more than two dozen engineers and scientists from NASA were due to hold many meetings with our own designers and experts to reach important preliminary agreements in preparation for the mission, which was scheduled to take place in July 1975.

Kubasov and I had already begun studying English very intensively. The decision that the crews had to learn each other's language was, I believe, very important psychologically, because we had to understand how our partners in space were thinking. Kubasov and I would be visiting the Johnson Space Center in Houston and California later that summer, to see the plant where the Apollo spacecraft had been built.

I was looking forward to the trip a great deal. Although I had traveled widely outside the Soviet Union, I had never been to the United States. All I knew of it was what I had learned from textbooks at school and seen in Hollywood movies, and what I had read in books and the Western media about America's space program. Over the years I had met a number of American astronauts and liked them a lot, although our meetings had not always been easy.

Leading the American delegation to Moscow that summer was

an astronaut I had not met, David Scott. As he was the only astronaut in the group, I paid particular attention to him. I knew little about him apart from what I had read in *Life* magazine in the lead-up to his two Apollo missions. I had seen a photograph of his wife and knew he had two children.

As Soviet commander on the Apollo–Soyuz project, I was heavily involved in making many of the arrangements for each American delegation's visit to the Soviet Union prior to the mission. To mark this first visit we decided to take the Americans out of Moscow on their first free weekend, to Kaluga, birthplace of Konstantin Tsiolkovsky, the father of Russian space travel.

That morning I was seated on the bus beside David Scott. My first impression of him was that he seemed less open than some of the other astronauts I had met. Later, when I got to know him, I formed a different impression, but that morning he did not seem particularly comfortable.

As we drove west out of the city I pointed out the site where, during the Second World War, German forces came within sight of Moscow, before being turned back. I spoke of how the Nazis had plans to destroy our great city completely by flooding it, and building a citadel with a huge statue of Adolf Hitler at its center.

Shortly after the trip to Kaluga we invited David Scott and another member of the delegation to Star City to view some of our facilities and visit our homes. We hoped this might lead to a more relaxed atmosphere and better working relations.

As always, our wives took great trouble to prepare a good meal for our guests. We moved from apartment to apartment, introducing David and his companion to other members of the cosmonaut corps, and drank many toasts to the success of the Apollo–Soyuz mission and to further cooperation between the United States and the Soviet Union.

It was clear David thought it would be impolite not to accept any aspect of our hospitality, but as the number of glasses of vodka and whiskey we drank mounted he seemed to grow a little unsteady on his feet. The wife of one of our cosmonauts comforted him by saying that everyone feels a little unwell from time to time. But it was clear it was time to have our guests escorted back to their hotel.

Before the delegation left I decided to invite David to a more private meeting, at my painting studio on Ryleyev Street in the center of Moscow. In 1965 I had become a member of the Moscow Association of Artists, each member of which was provided with a studio. While most people left Moscow at the weekend and headed for their dachas in the countryside, I headed into town, whenever I could, to continue with my painting.

This third meeting with David would, I hoped, be an opportunity for us to talk more openly. As usual, I had a bottle of very good vodka laid out for the visit. As usual we began by drinking toasts to good relations between our two countries and to the success of the joint space program. We then spent several hours talking more seriously.

For me, at first, this was not a comfortable conversation. We talked a lot about how different life was in the United States and the Soviet Union. David felt America's system was better and I, of course, felt ours was the best. We had a tough exchange when we touched on international politics, particularly when the subject of America's involvement in Vietnam came up. I could not see why the United States had become involved in a conflict so far from home. David pointed out that the Russians had also sent troops, but I reminded him that this was only after the US had become involved.

When we moved on to discuss our space programs, from the way he spoke David gave the impression that he also thought Apollo was a much better spacecraft than Soyuz. I pointed out several aspects of the Apollo spacecraft I did not believe were very agreeable.

But as the evening wore on we both started to realize there was more that united than divided us. We were both professionals trying to solve problems which most people could not even begin to understand. We were both professional pilots first and foremost. We had undergone similar training, we had flown similar planes. Although I had been a cosmonaut longer, David had already achieved the goal I had treasured for so long: he had landed on the Moon.

Slowly, we started to talk about the relative merits of the Apollo and the Soyuz spacecraft. Their operating systems were very similar.

If the best aspects of Apollo and Soyuz were combined, we agreed, the end product would be a brilliant design. (In the event, it was many years before this was achieved with the International Space Station.)

As the evening drew to a close we both started to relax a little. Despite the harsh words we had exchanged, that night we felt the beginnings of a camaraderie which would grow. At the end of the evening I presented David with an emerald-green ceramic beer mug, on the bottom of which I signed my name. We would drink from the mug, I said, once the joint mission was accomplished.

"You know I just want this to be the best mission possible," David said as he was leaving. "I don't mean to hurt or offend you by anything I say."

"I understand," I replied. "We are very different and our systems are very different."

"Thank you for understanding," replied David. We parted warmly, slapping each other on the back.

David Scott

My first impression of Moscow was of how clean the city was. Very bleak and quiet, but clean. We had been picked up at the airport in a rickety old bus. As we drove in to the center of the city I was struck by how few people there were on the streets; not much traffic about, either. When our bus drew up on Red Square I looked across at the swallowtail outline of the Kremlin, with its golden domes and many towers. Above one tower a large red Soviet flag fluttered at full mast.

Our hotel was directly opposite the Kremlin. It was a monolithic building, just huge, called the Rossiya and it would be our home for the three-week duration of our stay. It was difficult not to be slightly unnerved by the constant presence of the elderly key ladies who sat hunched in the center of the corridor on each floor. They registered us leaving our rooms in the morning and monitored us returning at night.

There were stories that the rooms on either side of ours contained listening posts. At the very least we assumed our rooms were bugged. There was an atmosphere of secrecy, watching and listening. I didn't let it bother me too much. We were on a diplomatic mission. But there was a very real sense that we were in a mysterious and hidden world.

Each weekday morning we were picked up by the same rickety bus and driven out to a rather cold and sterile office building on the outskirts of Moscow. There a whole series of meetings was held to discuss how our systems could be synchronized for the project. Everything had to be negotiated and agreed, from what system of navigation we would use, to electrical systems and, particularly, the docking mechanism, a completely new concept, which was being designed for the mission.

My job was to coordinate all these meetings. At the end of each day I sat down with Professor Bushuyev to discuss what progress was, or was not, being made and then reported back to Glenn Lunney in Houston. Sometimes these were pretty tough talks. Our systems were very different. Bushuyev was a very nice guy, but very difficult too.

All our discussions were conducted with the help of interpreters. Mine was a wonderful elderly White Russian, Alex Tatistcheff, who became a good friend and "cultural counselor." I became very fond of him and heeded his advice carefully.

Despite the thaw in relations, times were still tense. We provided full, frank and open disclosures on our systems and I believe our Soviet counterparts did the same. But there was still a good deal of suspicion on both sides.

On the weekends, however, we were able to relax a little. We were taken on excursions out of Moscow. Our first trip was to Kaluga, southwest of the capital, the home of Konstantin Tsiolkovsky, the father of Russian rocketry. It was a wonderful outing, even though it did get off to a slightly rocky start.

We had been picked up from the hotel early that morning and told we would have breakfast on the way. I don't know what I expected, a small restaurant, I suppose. But after an hour or so the bus pulled over to the side of the road. We were led off the bus and into a small clearing in the forest.

"We'll be having breakfast now," one of the Russians announced. With that he started smoothing out a newspaper on the ground on which was laid raw sturgeon, a bottle of vodka and paper cups. It was a far cry from scrambled eggs and bacon, tough on the stomach. We'd been up quite late the night before, but we braced ourselves and started drinking vodka from the leaking cups. After that we were invited to play football, Russians against Americans.

Then we were herded back on to the bus and the journey continued. The scenery was beautiful. We were driven past white birch forests and fields where teams of farm workers were loading wagons with hay by hand.

That day I was seated on the bus beside Alexei Leonov. As the only astronauts in the delegation, I guess we were drawn together. Of course I knew who he was. I had met his commander, Pavel Belyayev, at the Paris Air Show in 1967, and I knew Leonov had performed the first space walk. More than that I did not know. The Russians did not release lengthy profiles of their cosmonauts to the world's press. But I was struck immediately by his good humor. I thought he was a really nice fellow, very amicable.

His English was pretty good and he started to explain the countryside through which we were passing. At one point, he asked whether I minded if he told a political joke.

"No," I said, a little surprised. "That's fine."

"Well, we're not too far from Borodino, where the Russians turned back Napoleon in the dead of winter in 1812," he began. "We're also not too far from the place where the Russians turned back Hitler, also in the dead of winter, in 1942," he continued. "You know the Moscow winters are very, very difficult."

"Yes," I said. "I imagine they are."

"You know we are advisers now to the Egyptians in the Middle East," he said.

"Yes, I know," I said. In the wake of the 1967 Six Day War there had been frequent clashes between Israel and Egypt. The United States supported the former and the USSR the latter.

"Well, you know what our latest advice to the Egyptians is, if the Israelis attack again?" Alexei asked with a broad, slightly crooked smile.

I waited.

"Fall back to Cairo and wait for winter."

That was pretty funny. We both rocked with laughter.

The next time I met Alexei was following a rather mysterious invitation to visit Zvyozdny Gorodok, Star City. That was a very big deal. All I knew of Star City was what I'd read in a *Time* article which mentioned the existence of this mysterious military facility. Who knew where it was? Everything was so shrouded in secrecy.

The visit, I was informed, would take place the following afternoon. I would be permitted to take two other people with me. I decided I'd take along the lead NASA engineer, and the lead engineer from Rockwell, the prime contractor on Apollo. The next afternoon we were picked up from the Rossiya and driven quite some way out of Moscow, along a country road.

After about an hour our car took a sharp turn off the main road and followed a smaller road which led into a wooded area. A guard post patrolled by military personnel then waved us through a barricade and we were driven through a spacious complex of low buildings and green open areas. It was summer. But it was already growing dark.

The dwindling light meant I didn't get a really good look at the place. I'm not sure how much the Russians wanted us to see. At first we were taken to a museum, where we were given a guided tour, shown mementoes which had once belonged to Yuri Gagarin and memorabilia from various spaceflights. We were then taken on a four-hour tour of the Soviet space program's training facilities conducted by Vladimir Shatalov and Gyorgy Beregovoy. We were shown the gigantic centrifuge machine used for cosmonaut training, simulators for the Soyuz spacecraft and a docking simulator, which I flew with Leonov, who then went on to show me a high-fidelity mock-up of the Salyut space station. It was one of the best tours I have ever had, but nothing very surprising. Then came cocktail hour.

We returned to the museum where hors d'oeuvres and vodka had been laid out. About eight very notable cosmonauts were present. I was impressed. I was presented with a beautiful shotgun, with my name engraved on it in Russian, a generous and touching gesture.

We drank several toasts to the success of our joint program. That was it, I thought; it must be time to leave. But then I was invited over to the cosmonauts' housing complex.

More vodka and hors d'oeuvres had been laid on in one of their apartments. The apartments were very well appointed—new, nicely furnished, and quite spacious. I was introduced to several of the cosmonauts' wives, and they were very gracious. I also met Valentina Tereshkova in one of the hallways. The toasts continued throughout the evening. I found it a little hard to keep up. Then it was time to eat and we moved to another apartment. It was an evening of great Russian hospitality. At the end of the evening, whenever that was, we were driven back to our hotel, quietly exhausted.

Toward the end of our stay I met Alexei a third time. One evening he invited me to visit another flat he had in the center of Moscow. It was a very comfortable place, in an old building. It was quite small, with a studio off to one side. But the main living area was covered in lovely old wood panels.

That evening it was just Alexei and I, together with his interpreter. I felt we knew each other a little better by then. First he showed me some of his paintings. I knew nothing about that side of his life, but it was clearly something of which he was very proud. Then we started to discuss our respective space programs. It was then that Alexei revealed that he had been designated commander of the first circumlunar mission and was expected to be the first Russian to land on the Moon.

It was amazing to learn the Russians had been that far along the path toward a lunar landing and to learn that Alexei was their key man. He also told me about how he and Kubasov had been replaced at the last minute on the mission to Salyut 1. It was all fascinating. We knew so little of what the Russians had been doing.

As the evening wore on Alexei talked a little about his great friendship with Yuri Gagarin. He also spoke of Sergei Korolev. I did not learn the full story of that extraordinary man's contribution to the Soviet space program until a dinner with Alexei many years later. But even to hear his name mentioned was, for me, significant. I had not been aware, until then, of the existence of this legendary Chief Designer.

Our mindset at the time was that the Russians did not tell anyone anything, so the openness with which Alexei and I talked that night was, to me, quite fascinating.

When it came time for us to leave Russia, several of the guys on our team were pretty worried about not being allowed out. Although I didn't give it a second thought, they feared our passports might be taken away at the last moment and didn't relax until we were on the airplane headed home. The Soviet Union was still a pretty spooky place.

At the end of the three-week visit I was very satisfied with the progress we had made. It was difficult to reach agreement on so many complex issues. There was no referee or judge; it was all a matter of trade-offs. But we had established a good foundation and the beginnings of some friendships, too. Soon after our trip to Moscow a Russian delegation was to visit the United States. A number of cosmonauts would be coming, too, among them Alexei Leonov.

Alexei Leonov

Shortly after David Scott came to Moscow, I made my first trip to the United States in July 1973. It was the first visit of a Soviet delegation to NASA headquarters in Houston since the crew of the Apollo–Soyuz mission had been selected. On landing in America I was immediately struck by the noise and activity. The noise of sirens wailing, car horns and constant traffic took me by surprise.

Our journey from Moscow to Houston included a short stopover in New York. I remember standing at the base of the Empire State Building and wondering how such a high building could have been constructed. Although I had seen them in the movies, the skyscrapers were so huge. Everything astounded me.

The number of luxury cars on the road was amazing. I had a Volga car, but it bore no comparison to the wide variety of large American cars I saw cruising up and down the streets. Everywhere we went there was noisy traffic, advertising billboards, bustling shops and skyscrapers.

In the Soviet Union it was difficult to get food, let alone a beautiful shirt, but in America you could buy whatever you wanted at any time of day. It really struck me, too, how clean and tidy everything was. The museums and art galleries were beautiful and sophisticated. It seemed to me that this civilization had been developing over a much longer period than three and a half centuries.

When we flew on to Houston we were given a tour of NASA facilities. We were shown the Apollo simulators and were invited to use some of the training facilities used by the American astronauts. I was struck by how little supervision the American astronauts seemed to receive from their training supervisors. In the Soviet Union we were treated as a privileged group of highly trained pilots. We had an intensive training program. Our diets were carefully supervised.

But the American astronauts seemed to do what they wanted, with nobody paying attention to what they ate or what physical shape they were in. I could not understand how, in such a rich country, so little attention was paid to the physical condition of their most elite citizens. Some of the astronauts did seem to work out regularly, among them David Scott. David was in great physical shape and it showed when we swam together in a swimming pool at NASA.

During this trip I paid my first visit to David's home. It was very elegant, nicely furnished. It really brought home to me then that we should not have opted to live communally in apartment blocks in Star City. But that was the system within which we were then living.

When I wandered around David's home and browsed along his bookshelves I was shocked to find Hitler's *Mein Kampf** among his books. In the Soviet Union it was strictly banned. As I took it down and opened it, I felt very uncomfortable. Our country had suffered so terribly from that evil regime that if you were caught reading such a book you would have been thrown into jail as a Nazi sympathizer. I came to understand later that an intelligent and educated reader should be able to read any book, but at the time I did not realize that.

*See p. 350.

This first visit to the United States was a real test of my English. Things did not always run smoothly. Our hosts were very patient with our early efforts, as we were when they first visited the Soviet Union. More often than not our mistakes led to great hilarity and served to break down the barriers of reserve between us.

On one occasion in Houston I stood up to propose a toast to our future cooperation. I meant to toast our future success, but it did not quite come out that way.

I raised my glass, cleared my throat and proposed a toast to "A sexful life in future." Everyone laughed and drank to that.

After our visit to Houston we were flown to Downey, California, to visit the Rockwell construction facility for the Apollo spacecraft. Following this formal part of the tour we were treated to a trip to Disneyland. It was pure fantasy. I was like a child, who wanted to see everything, try everything. Everywhere we went people seemed to know that we were Soviet cosmonauts. They broke into spontaneous applause, and many came and asked for our autographs.

It struck me how much more ready ordinary Americans seemed for cooperation than their sometimes-hesitant politicians. I had felt this very early on during our trip in a funny encounter when we were being driven from the airport in Houston to our hotel. It was late, very hot and humid. Some of the American astronauts who had come to meet us were thirsty and wanted to buy some beer for us all. They pulled the car over and went into a small restaurant by the roadside to buy some bottled beer. They didn't tell the guy there who they were.

He was sorry, he said, but it was after licensing hours and he could no longer sell liquor. Being American, they accepted his explanation: rules are rules. But, as far as I was concerned, there were no rules except the rules of hospitality.

"OK, let me try," I said, and I grabbed a bottle of Stolichnaya vodka we had brought with us from home. I walked into the restaurant and explained to the man that my friends and I were very thirsty and we'd really like some beer. When he asked who I was, I declared in clearest English, "I'm a cowboy from Siberia."

"Really?" he said. "And just how are you goin' to prove that?"

I held up the bottle of Stolichnaya and said if he'd sell me some beer I'd give him the vodka for free. This was top-quality vodka, difficult to get hold of outside the Soviet Union.

"You know, when your astronauts come to Russia we are very hospitable toward them. Our shops open their doors readily," I told him. "So what's the problem here? This country has to recognize its heroes."

When he realized who we were he slammed two crates of beer up on the counter and I paid for them.

Then he pulled out a third crate. "OK, cowboy. That's my present," he said. "Good luck to you."

David Scott

One evening, during the visit of the Russian delegation to Houston, we invited the group over for a cookout at my place. The cosmonauts brought lovely gifts for my wife and children. They were especially kind to the kids, amusing them by singing Russian folk songs. One cosmonaut had brought a traditional Russian instrument as a gift and played it for us. It was a very warm and friendly evening, although I was pretty busy out in the yard barbecuing a side of beef, so I didn't get too much time to talk to our guests.

Lurton and I wanted them to feel free, and left them to walk around the house and feel at home. I remember they were pretty taken with a snake my son, Doug, kept as a pet. After they had left, a couple of neighbors came round to ask how it had gone. Having a bunch of visitors from the Soviet Union in your house was pretty unusual. After all the competition of the Cold War, it was quite amazing to have "the Other Side" to dinner. But it was a really relaxed evening. We all felt very comfortable with each other. One of the cosmonauts, Vladimir Shatalov, had a daughter the same age as Tracy and the two girls became penfriends for a little while after that.

A few days later I traveled with the group to Los Angeles to visit the facility at Rockwell. Following the meetings there we went to

Disneyland. They seemed to really enjoy that. Besides the fun of the place, I think they were impressed with it as a spectacular feat of engineering.

But soon after this visit I started mentally shifting gears away from the Apollo–Soyuz project. In the summer of 1973 I had been offered the position of deputy director of the NASA Flight Research Center back at Edwards Air Force Base. It was a coveted job, overseeing a whole range of experimental flight tests on the most advanced new airplanes like the SR-71 Blackbirds.

The center was also, at this stage, involved in test flying lifting-body rocket-powered airplanes such as the X-24, a precursor to the shuttle, which was already under construction. It was a plum opportunity. It would mean a move back to the high desert—Chuck Yeager territory.

Edwards was a place I had always loved. I could continue to fly, too, at first the T-38s I was familiar with and then later the F-104 Starfighter. Besides, it was a chance to move ahead in management with NASA. When Apollo 17 had blasted away from the Cape at midnight in December 1972, it was "the end of the beginning" of our space program. Now that Apollo was finished, NASA was concentrating on new programs with, initially, limited potential for manned spaceflight.

The first Skylab was already in orbit, with three separate crews occupying it during 1973. But apart from this, and the Apollo–Soyuz project planned for 1975, no astronauts were expected to fly again until the first orbital flight of the space shuttle in the early 1980s. So in December 1973 we moved back to California.

This time we lived in a small community called Lancaster about forty miles from the Flight Research Center. Two years later I was appointed director of the center, later renamed the Dryden Flight Research Center after Hugh Dryden, one of NASA's founding fathers.

One of the highlights of my time at Dryden was the unique opportunity it gave me to spend time with Chuck Yeager once again. Yeager always considered Edwards Air Force Base—where he had first broken the sound barrier in a Bell X-1—his "home," even after his full-time posting there had come to an end. During his visits there he sometimes stopped by to talk with me and the seven really superb experimental test pilots working on my programs. We had acquired three additional F-104s from the German Air Force, which would fly chase missions including following the progress of our two Blackbirds as they flew past Mach 2 and Mach 3.

One morning as we sat chatting over coffee Chuck wondered how he could wangle a flight in an F-104.

"Well you know, Chuck," I said, easing myself back in my chair, "with all these new programs going on, we could sure use some expert advice on how to best keep things safe while still exploring the edge of the envelope." I paused for a moment, then leaned forward, studied the weathered face of this man, for whom I had such tremendous respect, and carried on.

"I wonder if you would consider consulting with us on an official basis, including chase flights with the F-104s?" We were both smiling broadly by this point. "With our limited budget I must explain," I added, trying and failing to adopt a grave tone, "that we would only be able to afford to pay you a dollar a year, but you would be welcome to come fly whenever you like." Chuck accepted.

After that he returned every couple of months to fly an F-104, though I never had the chance to fly with him again. Once I took over as director of the Dryden Center, however, I had far less time for flying. Though my flying performance didn't suffer, I just didn't have time to keep up with all the new rules and regulations, the revised emergency procedures and the constant tests you had to take to keep your flying status. That really hit me one sunny Sunday afternoon when I brought my plane in to land after a trip up to Fresno to play in a celebrity golf tournament.

"Boy, it feels good to be flying. But I haven't flown for three weeks," I said to myself. Come to think of it, I hadn't flown for three weeks before that, either. My landing that day was right on, very satisfying, but I knew it was time to quit. I clambered out of the cockpit, walked away from the runway and hung up my flying gear; my flying suit, helmet, gloves and parachute. That was it.

It was a sad feeling. But everyone has to quit some time. It was a very personal decision—a very subjective evaluation. Every pilot feels differently about it. Some, of course, are forced to quit for health reasons, or if they fail to keep up a certain flying standard or don't pass the periodic, but obligatory, written exams and tests. But I had to fly at 100 percent, with all the time and dedication I felt that required; for me 99 percent wasn't good enough. I'd had a great run. In twenty-three years I had flown over 5,600 hours in twenty-five types of aircraft—everything from a helicopter to rocket ships, from the Saber and the Starfighter to a lunar lander. I don't think I missed much. I was very fortunate to have had all those opportunities and there were just no more prizes—except, maybe, a First World War Spad plane. I was forty-four. It was time to move ahead with new challenges.

I had already retired from the Air Force, just before I took over as director at Dryden. The Flight Research Center was a civilian facility and could not have a military man as its head, so I had put in my papers. Some day, I thought, I might go back into the military, but I did not really believe I would. Although I had drawn an Air Force salary throughout my time as an astronaut, I had not really been with the service for a long time.

When I handed in my papers I was asked if I wanted a retirement parade. "No thanks," I said. "I don't need a parade."

It was time to move on, move forward.

*David Scott points out that he has never owned the book *Mein Kampf*. Alexei is probably remembering seeing the book *The Rise and Fall of the Third Reich* by William Shirer. The original jacket has a swastika on both the front cover and the spine, which is probably what caught Alexei's eye. The swastika symbol would have been startling for a Soviet to see at that time and the Shirer book would not have been permitted or published in the USSR.

CHAPTER 12

Smooth as a Peeled Egg

1973–5

Colonel Alexei Leonov

Zvyozdny Gorodok, Moscow

Near the homes where we once lived in Star City there is a long avenue of birch trees, each one marking the achievement of a Soviet cosmonaut in space; the tradition of planting commemorative trees began very early in our space program. A little further away, along the shores of one of Star City's artificial lakes, stand three trees planted to commemorate the work of the American astronauts on our joint Apollo–Soyuz mission.

Those trees stand alongside a special hotel in Star City I helped design to make the American crew and their support team as comfortable as possible when they visited Moscow for joint training sessions. The three-story Hotel Cosmonaut was not quite ready when the American crew paid their first visit to Moscow in autumn 1973. So, during this first trip they stayed at the giant Intourist hotel in the center of Moscow.

In the beginning there was some aggravation between our two teams. I learned later that the Americans often complained that they were monitored the whole time they were in the Soviet Union. A lot of this paranoia was cultivated, I think, by the American intelligence services. The mistrust was mutual. The first time I visited the United

States I used to clap my hands loudly every time I entered my hotel room at night.

"Attention, please," I'd say, for the benefit of those I believed to be bugging my room. "Let's go."

It seemed the Americans did not at first understand the trouble we had taken in attending to every detail of their stay. Sometimes they did not turn up for breakfast or supper, for instance, even though exact numbers had been catered for and so food went to waste. As head of all matters concerning training at Star City, this meant I had personally to reimburse our canteen for the meals missed. It was obvious there was a great difference between American and Soviet concepts of hospitality.

Throughout the time we spent training together it never ceased to amaze me that the American team had no idea what, where or when their astronauts ate. Nor, apparently, did they monitor their crew's physical training. We continued to have a strict physical regime. We never dreamed of complaining about it. As far as we were concerned we were aiming at perfection.

But, slowly, each side came to understand the other better. Trust developed. Parties we organized in our homes for the astronauts and the hospitality their families showed us, when we visited the United States, went a long way toward cultivating better relations.

From the very inception of the Apollo–Soyuz project, it had been agreed that, in the interests of mutual understanding and respect, we would speak English throughout the mission and the American astronauts would, in turn, speak Russian. It was a symbolic gesture, but it was also important psychologically. We had to understand how our partners in space were thinking.

In addition to the many hours spent learning each other's language, the crews had to learn exact details about the operation of each other's spacecraft. Over the next two years we visited the United States four or five more times, and the Americans traveled to Moscow several more times. Each of us spent a month or so at a time training with our respective counterparts.

As commander of the Apollo crew, Tom Stafford and I developed a close working relationship and became good friends. Among the astronaut corps Tom was often referred to as "Granddad." This was not because he was old, and he did not then have grandchildren. Rather, it was because of his considerable experience of spaceflight. Before Apollo–Soyuz he had already completed three space missions, Gemini 6, Gemini 9 and Apollo 10. It was also because of his geniality. He was strict and sober-minded, but invariably good-humored.

I also felt a particular affinity with Deke Slayton, whom I had met in Athens. I knew he had had to endure a great deal of frustration during his time as chief of the astronaut corps. After doctors declared him unfit to fly because of a heart condition, he spent years training hard physically in order to prove them wrong. Whenever he was asked how he managed to combat his illness his reply was the same: "I never was ill. It simply took ten years to prove to the doctors that I'm fine."

As to Vance Brand, whom we nicknamed Vanya, he endeared himself to all of us with the amount of effort he put into mastering the Russian language. As it was for Deke, the Apollo–Soyuz mission was to be Vanya's first space mission, though he had served as back-up crew on several Apollo missions before that.

For all of us the chance to fly in space with this joint mission would not come a moment too soon. According to the program, our Soyuz would be launched into space from Baikonur seven hours before Apollo lifted away from the Florida Cape. Once both vehicles were orbiting in tandem, the plan was for the two spacecraft to rendezvous and dock, allowing crew members to transfer back and forth. The two craft would then separate and conduct a series of complex maneuvers in close proximity, culminating in a further rendezvous and docking, before the two crews returned to Earth in their separate landing modules.

Our launch, which had been set two and a half years in advance, was scheduled for 15 July 1975.

Midway through our two years of training, however, there was a significant change in the management of the Soviet space program. On 15 May 1974, Vasily Mishin was removed from his position as head of OKB-1. Mishin's health had been failing for some time, and he had made enemies in high places as a result of the continual delays and failures in our space program.

Mishin was replaced by Valentin Glushko. The decision was taken to merge OKB-1 with Glushko's design bureau, which until then had been primarily responsible for the design of rocket engines. The giant enterprise would eventually be known as the Energiya Scientific Production Association.

Despite suspending the Soviet Union's manned lunar program after Apollo 8's circumlunar flight in 1968, Mishin had continued to pursue an unmanned lunar program with the successful landing of the second Lunokhod lunar rover on the surface of the Moon in January 1973. He had also continued trying to rectify the problems of the flawed N-1 rocket, with the first test launch of a redesigned rocket planned for August 1974. But when Glushko took over all those plans were dropped.

Feelings in the cosmonaut corps about these changes were divided. The civilian cosmonauts, most of whom had been recruited from Mishin's design bureau, were not pleased at all. But those of us with military backgrounds had never had particularly good relations with Mishin, so for us the change was welcome. We had long since abandoned hope of Soviet cosmonauts being sent to the Moon. Our goals had changed, our focus shifted.

At that time I was totally concentrated on what we hoped to achieve with the Americans in July the following year. After all the frustrations I had endured throughout the previous years, I felt strangely little emotion the closer we came to this date. So much was riding on this joint mission. I was proud I had been chosen as first commander. But I was ready. I had been ready for a long time.

The cosmodrome at Baikonur was a very different place in 1975 from what it had been ten years before. A new hotel had been built there with a swimming pool and tennis courts. There was a theater in the town, too, many new shops and a fleet of air-conditioned buses available for crews in training. At the end of April that year, less than three months before the Apollo–Soyuz launch, we invited the American astronauts to visit Baikonur.

Before showing the Americans the launch facility we took them on a brief tour of the Soviet Uzbek cities of Tashkent, Bukhara and Samarkand. When we arrived at Baikonur a big party had been arranged to welcome the Americans. Some local people of this remote region of Kazakhstan arrived by camel, weighed down with regional dishes and specialities. Sitting cross-legged in a traditional Kazakh tent we drank toast after toast to the success of our joint mission.

This was the last opportunity our two crews would have to spend any time together before the start of the mission. The next time we would shake hands and propose a toast to each other's good health would be while orbiting the Earth at 30,000 kph.

During the two months that followed I sometimes spoke to Tom Stafford on the phone. The evening before the launch he called me to ask how everything was.

"Everything is going as smoothly as a peeled egg," I told him. "How about you?"

"As smoothly as three peeled eggs," Tom replied in his broad Oklahoma twang. "See you in space."

Some of our team had already been dispatched to Mission Control in Houston for the duration of the mission, while members of the American team had arrived at a new Mission Control center we had by then developed at Kaliningrad on the northern outskirts of Moscow. All had undertaken foreign language instruction, as we had. Everything was set.

The night before our Soyuz spacecraft was due to lift off from Baikonur, in addition to all the usual traditions prior to a launch, which had evolved over the years, a more recent custom was observed. We all sat down to watch *White Sun of the Desert*, a classic Soviet film about the country's civil war. It put us in the perfect mood. It both inspired us with its patriotism and relaxed us with its gentle humor. After the film we opened a bottle of champagne, drank a little and put it to one side to finish when we returned.

The next morning, as we were leaving our hotel, we signed our names on the doors of our rooms along with the date. On our way to the launchpad we all, once again, gathered in a semicircle and pissed on the wheel of the bus.

Our launch was scheduled for mid-afternoon on 15 July. Apollo would also launch mid-afternoon local time in Houston. Orbital dynamics and the rotation of the Earth determined this would bring us into parallel trajectories

It was hot, with clear skies and light winds, when we blasted away from the launchpad at Baikonur that Tuesday afternoon. It was the first launch of any space mission broadcast live on television in the Soviet Union. Fortunately, it was flawless: no hiccups at all. The only hitch in the first hours was a technical problem with the system of television cameras aboard Soyuz. No signal from our module was being received back on Earth.

For a mission whose significance was to demonstrate to a watching world that cooperation in space was possible, this was a problem that had to be solved quickly.

We had no choice but to dismantle a major part of our orbital section in order to gain access to the wiring for the system of five cameras connected to the switchboard, and fix the problem by disconnecting the switchboard from the circuit. It was a long and painstaking task. It took us many hours, during which we had been scheduled to sleep.

During our joint training sessions with the American crew, my crewmate Valery Kubasov had earned a reputation as an expert "handy-man." "If anything breaks down, Kubasov can weld it together," they used to joke. And it was true. During the Soyuz 6 mission with Gyorgy Shonin in 1969, Valery, as flight engineer, had done the first ever welding in space.

Our gradual progress in solving the problem was followed in live transmissions broadcast on Soviet radio. On our return to Earth this prompted a hilarious mailbag of requests from fellow Soviet citizens wanting Kubasov and me to come and fix their television sets.

As we were finishing off this complicated task we picked up our first radio transmission from the Apollo spacecraft after it launched. It was Tom speaking in Russian. "*Vyo normalno* [Everything is OK]."

And then we heard Vance Brand's voice. "*Miy nakhoditsya na orbite* [We are in orbit]."

They were on their way. Everything was going according to plan. It was an exhilarating feeling.

We were not due to rendezvous until our second day in space; at the moment our two vehicles were still orbiting on opposite sides of the Earth. During this time the American crew had their own technical hitch to deal with. Listening in to their transmissions with Houston we understood that, at the end of their first day in space, they were having some difficulty in opening the hatch leading from the orbital section of the Apollo spacecraft to its docking module.

After a few hours' rest we followed events aboard Apollo carefully and realized that, like us, they had ironed out their initial difficulty. Vance Brand, "Vanya," had managed to disassemble the docking probe, and Deke had been able to move into the docking module to check that everything was working as it should.

By the morning of 17 July it was time to move toward each other. Until that point Apollo had been circling the Earth in a higher orbit. We could hear the voices of its crew in our headsets, but could not see it. By lowering their apogee, and so increasing their speed, Apollo moved closer to us.

As our orbit took both vehicles high above the European continent, I suddenly caught sight of the American spacecraft's beacon out of my viewing porthole. There it was, right in front of us. At first, from a distance of about 25 km, it looked like a bright star. Then, as it came closer, I could see the clear outline of the silver spacecraft.

"Apollo, Soyuz. How do you read me?" I transmitted when I heard Deke Slayton wishing us "*Dobroye utro* [Good morning]."

"Alexei," said Deke, "I hear you excellently. How do you read me?"

"I read you loud and clear," I replied.

The maneuvers that followed, bringing the vehicles closer and closer, though conducted at speeds of over 30,000 kph, seemed like choreography from a graceful celestial ballet. Eventually, as the two spacecraft drew to within a few meters of each other I could make out a face in one of Apollo's windows. It was Tom. He was smiling.

Fifty-two hours after we had lifted away from the launchpad at Baikonur, our spacecraft were given the go-ahead by Houston and Moscow to move together for final contact. The new androgynous docking mechanism that had been specially designed to allow

Apollo and Soyuz to join and lock together glided smoothly into place.

"We have capture," Tom reported.

"Soyuz and Apollo are shaking hands now," I replied.

It would be several hours before we could open the hatches of the docking mechanism and see each other face to face. The difference in pressure between the two craft had to be equalized first, in order to prepare the vehicles for the transfer of crews. We had been slowly lowering the pressure inside Soyuz for some time. Now the American crew had to increase the pressure inside their docking module by adding nitrogen to its almost pure oxygen atmosphere.

During this time we received a message of congratulations from the Politburo. It was the second time Leonid Brezhnev had addressed me while I was orbiting the Earth. This time I was more prepared to conduct a conversation with the general secretary from space. This time I was not walking in open space at the time but sitting more comfortably inside the spacecraft.

"The whole world is watching with rapt attention and admiration your joint activities," said Brezhnev. "Détente and positive changes in Soviet–American relations have made possible the first international spaceflight."

Then he spoke of his hopes that such cooperation between our two countries would continue once we had returned to Earth. It was something I believed in and wished for very profoundly.

Once the pressure between our two craft was equalized, we were ready to open the hatches separating Soyuz from Apollo. First I opened the hatch of Soyuz and eased myself into the joint docking module, surrounded by a tangle of life-support cables. Then, watched by millions around the world, the Apollo hatch opened and, for the first time in history, a Soviet cosmonaut and an American astronaut came face to face in space. Tom gave me a big smile.

"Very, very happy to see you," I told him as I stretched out my hand and started pulling him across the dividing line between our two craft to give him a big bear hug.

"*Tovarich!* [Friend!]" Tom replied, grabbing me by the arms.

At that moment I felt that everything I had been through in my career as a cosmonaut—all the disappointments and very difficult years—had been worth it. This was the highlight of the mission. Few experiences before or after have been able to touch the elation I felt then.

Tom moved across to Soyuz and Deke Slayton and, briefly, Vance Brand followed. Each move and exchange between the two vehicles had been carefully worked out in advance. All, that is, except for a few surprises I had planned.

As Tom and Deke maneuvered themselves around a metal table in our cabin, they listened intently to a congratulatory message coming across their headsets from the American president, Gerald Ford. What was supposed to have been a short message turned into a whole series of questions from the president to each of us in turn.

How was the food in space? he asked Kubasov. With neither beer nor seafood on the menu, food was better on Earth, my crewmate replied.

"As the world's oldest space rookie, do you have any advice for young people who hope to fly on future space missions?" Ford asked Deke.

"Decide what you want to do," Deke replied, "then never give up until you've done it."

As the American crew spoke with their president, I pulled out the first of my surprises and watched carefully to see how they would react.

Before we left Baikonur I had peeled the labels from several tubes of borscht and blackcurrant juice and replaced them with labels from famous brands of Russian vodka.

"Before we eat we must drink to our mission," I told Tom and Deke, handing them a tube each.

"There are many people watching everything we do," Tom said, looking a little concerned and referring to the television cameras we had spent so long repairing.

"Look, Tom, don't worry," I said. "I'll show you how it's done."

With that I took one of the tubes, squeezed its contents into my mouth and swallowed quickly. Tom gave me a broad smile, a wink and followed suit.

"Why it's borscht!" he said, eyes wide and slightly disappointed.

It was time for my second surprise. During our joint training sessions I had made sketches of Tom, Deke and Vanya, which I had intended to present to them when we met in space. "O Brave New World that has such people in it," I had written on each sketch and then at the bottom, "Welcome to Soyuz—come again."

The men seemed genuinely touched by this gesture. Tom in turn presented us with a packet of spruce seeds, as a present from the American people, in line with our tradition of planting trees to commemorate a mission. We gave the Americans a selection of pine, fir and larch seeds, which were later planted near Mission Control in Houston.

We then signed a number of different certificates, as pre-arranged, exchanged flags and teamed up our respective parts of a special commemorative medal. As we then at last ate supper we discussed some details of the remaining part of our joint program, which would last two days.

The following day we were invited to share a meal with the American crew aboard Apollo. There followed joint scientific experiments—one involving welding—and televised tours of our respective vehicles for the benefit of the millions of viewers in the Soviet Union, the United States and elsewhere around the world.

When the two craft had remained linked for forty-four hours it was time to undock and perform a further series of rendezvous maneuvers, followed by a second docking. Deke Slayton would be at the controls of Apollo this time.

The second docking did not go as smoothly as the first.

After seizure and during contact, Deke inadvertently fired one of Apollo's side roll thrusters, which had the effect of pushing both vehicles off center, folding them toward one another. There was a real threat of damaging the joint docking mechanism and the possibility of a catastrophic depressurization of our orbital module. Fortunately, the mistake was quickly corrected and a major accident avoided. A series of checks by Mission Control in Moscow reassured us that no serious damage had been done.

We never spoke about the incident afterward. It would not have been very diplomatic for us to reveal how close Apollo had come to crippling Soyuz. We treated it as an internal affair. But we did receive an apology from Mission Control in Houston for the mistake.

The next three hours were taken up with a series of complicated flying maneuvers with the two vehicles both docked and undocked. Then, after five days in space, two of them spent joined together, it was time for Soyuz and Apollo to go their separate ways. After this Kubasov and I would spend one more day in space before returning home. Apollo would remain in orbit for three days longer. For the Americans these would be the last days of manned spaceflight for six years.

Our own manned space program would continue to progress steadily over the following years and I would continue to play an integral part in that program. But, for the Americans, this Apollo mission would be the last manned spaceflight until the maiden voyage in 1981 of its first space shuttle.

We had been well aware since the early 1970s that the Americans were developing a reusable orbiter, which became the space shuttle. Several years before, I had been involved with a group of students from the Zhukovsky Academy in drawing up designs for our own multi-use space vehicle.

Under the tutelage of Professor Sergei Belotserkovsky, in our final months before graduating from the academy in 1968 we had been set the task of developing a "winged spacecraft of a shuttle type." We had presented a twelve-volume dissertation on the subject and sent it with a letter to the Central Committee of the Communist Party, suggesting that development of the vehicle should begin immediately.

But the Central Committee forwarded our proposal to the Ministry of Defense, which dismissed the project and returned our dissertation to us with the words "fantasy" and "mind your own business!" scrawled across it. Our plans were put in a drawer and forgotten. More than ten years later, when the Americans were about to launch their space shuttle *Columbia*, we were summoned to the Central Committee and asked why no such project was being developed by the Soviet space program.

"As early as 1968 we put forward our proposals for such a vehicle," we informed the committee. "We were not taken seriously. Now we see the consequences."

Only then was the decision taken to start work on our own shuttle, which later became known as Buran (Blizzard). There was one major difference between the American space shuttle, however, and our own Buran. While the Americans brought the shuttle in to land under pilot control, Buran was brought back to Earth by an automatic guidance system, which was a great deal more costly and time-consuming.

Glushko had insisted on the automatic guidance system, arguing that it would allow the vehicle to land under any conditions. We tried to argue that the automatic system should be used only in emergencies, not as a matter of course—this was an extension of the argument we had had with engineers for years about cosmonauts having more control over their space vehicles. I had always believed that the manual control system should be primary, not secondary; after our emergency on Voskhod 2, the importance of manual control became very clear. But the engineers won. Buran was designed with an automatic system. We lost time because of this, and eventually the program was canceled through lack of funding.

But at the time of the Apollo–Soyuz mission such frustrations lay in the future.

Our last day in space was spent conducting a series of experiments and preparing for reentry into the Earth's atmosphere. I was able to spend a little time furthering my passion for painting as part of a scientific experiment. Once again I had taken my crayons and paper with me. This time I had also taken a special device which allowed me to measure and record, very precisely, the colors of different parts of the surface of the Earth.

My intention was that cartographers could use this information in future to color their maps more accurately. I started by making a record of the very different colors of our planet's oceans and smaller seas. The Black Sea, for instance, really is the darkest sea in the world and I was able to record its true color. I wanted to make the same assessment for the Earth's main mountain ranges, but unfortunately I ran out of time.

After six days in space, on the afternoon of 21 July, it was time for us to return to Earth. As with our launch, this was the first time a landing would be broadcast live on television throughout the

Soviet Union. Millions of our countrymen and -women gathered round their television sets to watch our landing and recovery.

At 1:09 p.m. Moscow time the retro-engines on Soyuz fired in accordance with the landing schedule. Nine minutes later, as our descent module separated from the orbital section, helicopters and rescue personnel moved toward the location where we were expected to land. At 1:40 p.m. cameras aboard rescue helicopters began to pick up pictures of our spacecraft descending gracefully under a billowing parachute.

Within meters of the ground our landing rockets fired to break the impact of our touchdown in a flat wheat field in Kazakhstan. Three minutes later Kubasov was opening the hatch and clambering on to firm ground. I followed, smiling broadly and waving, before boarding the helicopter that would fly us back to Baikonur.

A message of congratulations on our safe landing was transmitted to us mid-flight from Tom Stafford and the rest of the Apollo crew. Their own landing would be more problematic. We did not learn the full details of what had happened until afterward. All we knew when they splashed down northwest of Hawaii on 24 July was that they had had a serious problem during their descent.

It turned out that the Apollo landing module had filled with toxic nitrogen tetroxide gas during its descent. This was due to a failure to shut down a series of Apollo's thruster engines at the correct height. As a result, when the module's parachute system deployed, the toxic fumes seeped through the ventilation valves into the cabin and nearly choked the crew to death.

Later it was explained to us what had happened. But we did not reveal publicly what we knew of the problem.

After a brief stopover in Baikonur for medical checks, we were flown to Moscow where I was to present my report on the Apollo–Soyuz mission to the State Committee. What happened next was quite extraordinary.

When our plane touched down in Moscow the pilot was ordered to stop at the end of the runway and wait for the delivery of a parcel. As we looked out of the window of the plane, we saw a military jeep speeding along the runway. One of its passengers leaped out and hurried up the steps of the plane. Through the open door I could hear him shouting cheerfully.

"Give this to General Leonov," he said, handing over a large packet.

"There is no General Leonov on this plane," a member of the crew replied.

My military rank was colonel. Although I had been doing the work of a general for some time, there were so many generals in our country that it was almost a matter of waiting for one to die before another officer could be promoted to this highest rank.

"I repeat, give this to General Leonov," the officer snapped back. "That's an order."

When the packet was handed to me, I peeled back the paper and saw a general's uniform inside. I can hardly describe my feelings of pride. It was clear a great deal of thought had been given to this in advance. Yet I was touched by the informal manner in which the promotion came. This same day I was also awarded a second star of Hero of the Soviet Union and Lenin Award, the highest honors the Soviet Union could bestow.

As the plane taxied down the runway I quickly changed into the new uniform. When the plane came to a stop and its door opened, I saw the astonishment on the faces of many of those waiting as I descended to the tarmac. Among the waiting crowd was the chairman of the State Committee to whom I had to report immediately. Everyone was expecting that I would slip up when stating my rank as I addressed him. I did not.

"Crew Commander General Leonov reporting, Comrade Chairman," I said with a salute. "Mission accomplished."

Everyone applauded. It was a thrilling moment. Svetlana was there with our two daughters, who ran toward me when I had finished the formal side of my duties and clung to my side throughout the celebrations that followed.

From the aerodrome we were driven to Star City along a motorway lined with crowds tossing flowers at our car. Following the assassination attempt on Brezhnev, in which I almost lost my life, such motorcades had been banned for fear there would be another terrorist attack. But this reception was spontaneous, incredible.

In the hours and days following the Apollo–Soyuz mission I began to question what I had done to deserve so much attention.

As a boy I had once written in my diary, "My fate is me." The extreme trials and rigors of my childhood had forged in me a feeling of utter determination. This has always been my greatest strength. I achieved what I set out to do.

I had known fame at an early age. After my Voskhod 2 mission I received a great deal of acclaim when I traveled the world. But I was still young then, and everything always seems good when you view it through the prism of youth; the apples seem tastier, the nights shorter, the sun bigger. I had not been expecting such attention and it had seemed almost like a dream.

By the time of Apollo–Soyuz I was more mature. I think I came to appreciate much more what I had achieved. Looking back, I view this mission as the real highlight of my life.

In the months that followed, protocol called for the American crew to undertake an extensive tour of the Soviet Union and for us to pay a return visit to the United States. The tour of the Apollo crew, together with their families, took us from Moscow to, among other places, Leningrad, Kiev and Novosibirsk. From the freezing temperatures of Siberia we flew to the Crimea, and from there traveled to Sochi for a four-day rest on the shores of the Black Sea.

Everywhere we went we were followed by American and Soviet television crews. Despite the goodwill our mission had engendered, one small incident during the tour illustrated how little trust there really was between our two countries, and was blown out of all proportion by the American media.

As we were walking toward a waiting plane at Kiev airport, a police motorcade rushed toward us, sirens wailing, and we were signaled to stop. A young Ukrainian policeman marched up to us, stood to attention in front of the Americans and began reading from a document.

When Americans saw live television footage of their astronauts being approached by the police and stopped, they immediately assumed they were being arrested.

The statement the policeman read to the astronauts was as

follows: "During the three days you have spent in Kiev you have proved to be good Ukrainian citizens. It has, therefore, been decided that you will be given Ukrainian driving licenses and the right to wear police helmets and use police whistles."

The hue and cry over the incident in the United States was just one example of the way in which good intent can be misinterpreted.

I did not personally feel any mistrust among the American people, however, when we returned to the United States after the Apollo crew's tour of the Soviet Union. We were made honorary citizens of nearly every city we visited.

In Washington, together with our families, we were invited to a meeting in the Oval Office with Henry Kissinger and President Gerald Ford. This was not just protocol. We were very warmly received. We spent most of the afternoon together. It was a great honor.

After that we were welcomed very warmly in New York, Houston, Chicago, Los Angeles, San Francisco, Lake Tahoe, Nashville, Huntsville, Salt Lake City, Miami, Las Vegas and Birmingham, Alabama.

In Houston, we were introduced to Frank Sinatra, who sang for us and made a very nice speech. He was clearly a big fan of the space program. In Chicago our reception was quite extraordinary. Over a dozen marching bands led our motorcade through the streets of the city, and helicopters circled overhead dropping photographs of the American and Soviet crews into a cheering crowd. As we snaked along the shores of the Great Lakes, tugs moored offshore shot colored streams from water cannon.

While in Los Angeles, I was touched that David Scott took the trouble to travel to congratulate us on the success of our mission. I had not seen him since our first visit to the United States in the lead-up to Apollo–Soyuz. He had clearly driven some way in the heat of the day to find us at the hotel where we were attending a big party.

When I received a message that David was in the lobby I left the party briefly and tried to persuade him to come and join us. He said he had to get back to his base. But, in addition to congratulating us, he said he also wanted to thank me for keeping quiet about our problems during the mission.

"Both crews did a brilliant job," he said. "But you also turned out to be our great buddies. I am not so sure, had the Soviet spacecraft experienced the same problems that Apollo did during the mission, the matter would have been kept so quiet. So thank you for being so generous and discreet."

I knew David was very ambitious and self-confident, and realized that to say such things had taken great effort. I did not forget his honesty.

After we returned to Moscow I continued as deputy director of the Cosmonaut Training Center and commander of the cosmonaut corps. I worked on the Salyut program and was also appointed director of the Inter-Cosmos. This was a program responsible for selecting cosmonauts for training from Eastern-bloc countries such as Czechoslovakia, Hungary, Poland and Romania, and also from countries further afield, such as Vietnam and Syria, which had close relations with the Soviet Union

As it turned out, the brief thaw in relations between the Soviet Union and the United States did not last long. Our mission had paved the way for future cooperation in space, but it was to be many years before our two countries again undertook a joint space venture.

In the winter of 1975 the Cold War once again deepened. Strategic arms limitation talks faltered and, at the end of the following year, Gerald Ford was defeated in elections by Jimmy Carter. Carter's election campaign had heavily criticized Ford's policy of détente, and it was not long before he canceled all future cooperation in space. This was a huge disappointment. All the work we had done in developing common systems was lost. All talk of détente or *razryadka* disappeared from the political vocabulary of both our countries.

EPILOGUE

1975–2003

Major General Alexei Leonov

Zvyozdny Gorodok, Moscow, 2003

The Moon has been a source of fascination for every civilization since the dawn of man. Every culture has its myths about our nearest neighbor in the universe. The ancient Greeks believed the Moon provided a resting-place for souls destined for paradise among the stars. Nordic hunters saw there a girl held captive after going in search of water. Pacific islanders looked up and saw an immortal woman weaving a white sail with which to move across the sky.

As a young boy I used to look at the Moon and believe I saw a horseman riding across its silvery face. Years later, when I began to study the Moon in detail, I learned to distinguish its many craters from its so-called lunar seas. I came to know many of the beautiful and mysterious-sounding names: the Lake of Dreams, the Marsh of Decay, the Bay of Rainbows.

Throughout the years I spent training for lunar missions my focus had been the Moon, as well as the astral path that would take me there. Yet the first time I flew in space it was the sight of our own planet from afar that held me in both joy and awe. Looking back at our blue globe from such a distance profoundly changed my vision of space and time.

For most of us it is an act of faith that the Earth is round. But I have seen that with my own eyes. I have seen how thin the atmospheric layer that surrounds our planet really is. From space it

seems like a thin blanket in which the Earth is wrapped. It is all that
protects us from harmful ultra-violet rays and the impact of falling
meteorites. This experience has given me a different perspective on
life.

While it was conflict between two competing superpowers, two
ideologies, that propelled me into space, in the moment when I
looked back at the Earth it struck me very forcefully that our planet
is home to just one human race.

Following our Soyuz–Apollo mission in the mid-1970s, global
alliances once again sank back on either side of the bitter East–West
divide. When Ronald Reagan came to power in the United States in
1980, talking of the Soviet Union as the "evil empire," the sense of
comradeship between our two countries engendered by the joint
project seemed to disappear.

Brezhnev's successor, Yuri Andropov, had a very good grasp of
how the world was changing and how our country needed to move
with the times. He called for a freeze in nuclear arsenals and
initiated the very beginning of private enterprise in the Soviet
Union. Had he been elected sooner, our country would have started
moving in that direction years before. But he was old and died in
1984 after less than two years in power.

It was not until Mikhail Gorbachev was elected general secretary
of the Communist Party in March 1985 that a lasting thaw in
relations between the United States and the Soviet Union set in.
Undoubtedly Gorbachev's greatest achievement was to improve
relations between the Soviet Union and the rest of the world. It is
this for which he will best be remembered. In overseeing the end of
the communist regime, however, Gorbachev destroyed our country's
economy, which paved the way for the eventual break-up of the
Soviet Union. Those aspects of the system that had benefited our
society—such as high-quality free education and health services and
low-cost utilities—were destroyed, too.

Gorbachev did no favors for the Soviet space program. Under his
leadership the government's attitude to the program began to
deteriorate. This was a sharp contrast to his predecessor, Andropov;
I do not count Konstantin Chernenko, who during his brief tenure
(1984–5) did nothing for our country. But Andropov was an

enthusiastic supporter of the program. He always encouraged the Association of Space Explorers, which I had founded, to step up contacts with those involved in the space programs of other countries. "Let the governments catch up later," he used to say. But under Gorbachev the space program took a sharp down turn.

This was not, I believe, because of lack of interest or animosity on Gorbachev's part. Whenever Gorbachev met a member of the cosmonaut corps he always behaved in a civil manner. But there is no question he was overwhelmed by what needed to be done in our country. I think his wife, Raisa, was partially responsible for this change in attitude. She had great influence over her husband. Though popular abroad, in her own country she was disliked by the majority, principally because she tactlessly tried to flag up her participation in every aspect of government.

Raisa Gorbachev was quite cold toward the cosmonaut corps. She seemed jealous of the attention we attracted. She wanted that attention for herself. She never established any contact with Valentina Tereshkova, for instance, whereas the wives of Khrushchev and Brezhnev had treated Tereshkova as a daughter.

Under Gorbachev we had to keep insisting on the importance of the program, which had started to receive less and less money. But it was of little use. In the early 1980s work had begun on our space shuttle program, Buran, which I had been involved in designing. But only one unmanned Buran mission was ever flown, in 1988. After that Gorbachev canceled the program.

All that remained of our program when Boris Yeltsin was elected president of the Russian Federation in May 1990 was the Mir space station, which had been in orbit since 1986, together with a severely curtailed Salyut and Soyuz program and a paper commitment to the planned international space station.

Following the Apollo–Soyuz mission, I had continued in my position as deputy director of the Gagarin Cosmonaut Training Center, and director of the Inter-Cosmos program. As head of cosmonaut training I had slowly dropped some of the more severe physical tests for cosmonauts to which we had been subjected in the early days and which I considered unnecessary. I no longer, for instance, required them to sit in the isolation chamber at such

extremely high temperatures and for so long. Nor did I subject them to such extreme forces in the centrifuge machine, which had led some early recruits to lose consciousness.

As time went on, the criteria for selecting cosmonauts were gradually relaxed. Even so they were harsh and it was not easy to find the right people. None of those recruited during these later years had suffered the severe deprivations and hardships we had had to endure during the Second World War. They were of a different caliber from those of us recruited in the early days—whom I refer to as "the children of war."

By 1991 I was in discussion with the minister of defense, Dmitry Yazov, and the deputy chairman of the Defense Council, Oleg Baklanov, to take over as director of the Gagarin Cosmonaut Training Center. But both men went on to play a key role in the coup against Gorbachev, who by that time had become the first president of the USSR. When the coup failed both men were imprisoned. Shortly afterward I received the order that my military service, and with it all duties with the cosmonaut corps, was being terminated.

I was given no warning. This was totally out of keeping with normal military protocol. When I demanded an explanation from the new head of the Air Force I was simply told these were new times and new people were needed. I reminded him of an official note he had sent me on my birthday a short while before, in which he had congratulated me and praised my work as the founder of the Soviet school of cosmonaut training. I thanked him for appreciating my work and stressed that this training had been my life's work since the death of Yuri Gagarin.

"If you thought I was not competent to do my job I would understand this decision," I told him. "But you can't say that, can you?"

I knew it was simply a matter of political intrigue and that there was no point in further discussion.

"Goodbye," I said. "I'm leaving."

It was a stab in the back. I felt betrayed. Everything I had dedicated my life to seemed to count for nothing. The cosmonaut corps got up a petition to appeal against the decision, but it was of no use. In the months that followed Gorbachev resigned as Soviet president and successive Soviet republics declared their independence. Overwhelmed by what was happening in our country, President Boris Yeltsin clearly had no time to consider my position.

Until that point, under the old system of the USSR, power had been so concentrated that it did not allow the vast majority of its people to lead decent lives. The state apparatus was structured to impose limitations on the lives of ordinary citizens. Throughout the history of the Soviet Union my fellow countrymen had sacrificed themselves for the state. But the state did little for its citizens. A country can never be rich if its people are poor.

As a result of my travels in other communist countries, I had long been convinced of the limitations of a one-party state. Yet our country has always tended toward extremes. So, when new power structures were created, instead of two or three rivals to the old Communist Party being formed, dozens of new political parties were created. This number of parties was both confusing and counter-productive.

In many ways, the transition to democracy was simply too quick. Most were not ready for it. The process of transition was, and still is, very painful. We could not achieve overnight what it had taken democratic countries centuries to build.

Almost immediately, however, many new financial and economic organizations were created, and I soon found myself being offered interesting positions within a number of them. First I was invited to work as director of the space program of a private technological company. Then, in 1992, I was offered a post as president of an investment fund, Alfa Capital Foundation, created by the large, private Alfa Bank, of which I later became a vice-president. Both organizations were full of young people who had no connection with the old regime. They were exciting places to work.

The move into the private sector was a new challenge. My whole life I had been used to solving problems on my own, taking the initiative, doing what had to be done. I applied the same skills when it came to private enterprise.

The investment fund, for instance, took over ownership of some two hundred large enterprises, including major glass, cement and food manufacturers. The directors of most of these companies were men of my own age, who found it very hard to trust someone younger to help them through the difficult restructuring necessary for transition to the private sector. Throughout my time with the space program I had been used to dealing with the heads of major industries.

Unfortunately, I had to sack some of these directors. That was one of the most unpleasant experiences of my life, but I enjoyed being able to help companies turn themselves round, especially those which, before we became involved, had been on the verge of declaring bankruptcy, with directors carving up the spoils among themselves. Many people became involved in private enterprise purely for personal profit. Corruption was rife. Many were killed in Mafia shootings. If ever we felt that a business in which we were investing had any such connections, we pulled out immediately.

Although I was no longer part of the space program, from the early 1990s I continued my work with the Association of Space Explorers and followed closely our country's more limited involvement with space. While the decline of the program had already started under Gorbachev, the collapse of communism meant central funding was even more limited. Those running the program were ill equipped to seek independent financing. This was particularly detrimental to the Mir program.

The original expectation was that Mir would stay in orbit for just five years. In fact it lasted for fifteen and became the world's longest-manned orbiting space station. During that time it served as a laboratory for over 20,000 scientific experiments conducted by over a hundred cosmonauts and astronauts from more than fifteen countries. It was a unique project. It provided invaluable knowledge about long-duration spaceflight after Sergei Avdeyev broke all records by living and working on the station for a total of 748 days. It set the stage for future international cooperation and long-duration missions aboard the International Space Station currently under construction.

But the final years of Mir were marred by catastrophes: collisions, fires, toxic leaks and computer breakdowns. As interest in, and commitment to, the International Space Station grew the resources available for Mir declined. Yet it still seemed there was some hope of saving the crippled space station. As founder and co-chairman of the International Association of Cosmonauts and Astronauts I, with a number of friends, wrote to President Yeltsin in the late 1990s, making many suggestions of what we thought should be done to secure the future of Mir. What we had to say still carried some weight, but eventually the cost of maintaining Mir became too much. The decision was taken to decommission the space station. The most important consideration, then, was returning it to Earth safely, without loss of life.

As I watched the remains of the space station carve an arc across the evening sky and plunge into the Pacific Ocean in March 2001, it was like watching a beautiful ship sink beneath the waves. It was a terrible spectacle.

America's *Challenger* disaster before then had evoked similar, but even worse, feelings because of the human tragedy involved. Like most people, I watched with horror the television images of the shuttle breaking apart and plummeting back to Earth. The sight of the families of the crew screaming when they realized what was happening broke my heart.

Since then there has been much debate about the direction in which space exploration is heading. Ever since the Californian millionaire Denis Tito paid $20 million to be flown into space aboard one of our Soyuz spacecraft in late spring 2001 and spend six days aboard the International Space Station, there has been much criticism of the exploits of space tourists. But, as far as I am concerned, what Tito did was a positive thing.

Tito had worked hard for his money. Why should he not be allowed to spend it as he wished? He was well prepared physically and mentally. Although the program he carried out in space was limited, it did not interfere with the main purpose of the mission on which he flew. He brought much-needed funding to our space projects. Three space tourists like Denis Tito each year would have been enough to secure Mir's financial future.

This space voyage of an amateur, I believe, also went some way toward rekindling the romance of the cosmos in the mind of the general public. Sadly, the public passion that existed in the early days of space exploration has largely evaporated. Maybe future plans for a mission to Mars will once again ignite enthusiasm for this most extraordinary sphere of human endeavor.

Sooner or later, I believe mankind will begin to look more seriously for alternative planets on which to live. This might be through a wish to explore or out of necessity. The latter is less likely, since our planet still has enough resources to support us for many generations to come.

Venus is now known to be too inhospitable to support human life. The cool surface temperature on Mars and the discovery of water, suggesting the presence of oxygen, mean the planet could present future opportunities for human habitation. But the Moon still appears to be the most suitable other planet in the galaxy on which human beings could survive.

The greatly reduced surface gravity of the Moon makes fewer demands on the internal organs of the human body, the heart, for instance. This could make it an attractive proposition for future habitation on medical grounds. As long ago as 1967 I was involved in designing a lunar living module, which could be buried beneath the surface of the Moon to avoid the impact of meteorites.

In reality, however, rockets with sufficient power to carry such a lunar station to the Moon no longer exist. America's giant Saturn V rockets have long since been relegated to museums, and our own Energiya booster, last used for the one and only flight of our Buran space shuttle, has also been mothballed.

It would take investment of a much greater magnitude than any country is now prepared to make to realize a lunar venture. Many believe such investment would only be made in future in response to a threat which brought into question the survival of the human race; a threat such as a giant meteorite with a trajectory bringing it toward the Earth. This has been the stuff of Hollywood movies for many years. I prefer to concentrate on the more immediate future. Here, I believe, there is cause for optimism. The degree of international cooperation involved in the International Space Station, for instance, is very positive.

I have long dreamed of the possibility of an orbital space station with a sophisticated central dwelling module around which a necklace of other modules can dock and between which cosmonauts and astronauts can travel on a sophisticated form of space motorbike. Arthur C. Clarke foresaw such a space station long ago in his acclaimed novels of science fiction.

Clarke has always had a big following in the Soviet Union and became a friend of mine when he visited Moscow in 1978 and I was able to show him around Star City. Several years later he published his sequel to the novel *2001: A Space Odyssey*. Entitled *2010: Odyssey Two*, the book centers on the voyage of a spaceship, which Clarke named after me. He also dedicated the book both to me and to the dissident physicist Andrei Sakharov.

Shortly after its publication the book was serialized in a Soviet youth magazine. But halfway through the story the serialization was canceled and I was hauled before a meeting of the Communist Party's Central Committee.

"How could you allow this to happen?" they demanded to know. I had no idea what they were talking about.

"The crew of the spaceship *Alexei Leonov* consists of Soviet dissidents," they informed me.

"You're not worth the nail on Arthur C. Clarke's little finger," I told them, and I left in disgust.

The small-mindedness of such people had always amazed me. Pettiness and lack of appreciation of true talent and creativity by certain Party members was one of the factors that crippled our system.

As I look back on my life now I have few regrets. I regret that I am getting old. And I regret, of course, that I was not able to go to the Moon. Whenever I look through my telescope at night it is that silvery disc which draws my gaze the longest. Throughout my life I have drawn great pleasure from painting. Sometimes I paint scenes remembered from my time in space; more often I paint seascapes or landscapes; exhibitions of my work have been held in many countries.

Although I have achieved a lot in life, I often wonder if there is more that I could have done. But in those moments I remember some words written by Yuri Gagarin after he became the first human being to look back on Earth from space. "After circling our planet in a spacecraft and realizing how beautiful it is, I implore mankind to preserve it and enhance its future, not destroy it."

Yuri orbited the Earth only once, but it was enough to give him a true perspective on what it is to be a human being. Every manned spaceflight since then, I believe, has served to reaffirm this feeling of a unified human race.

For my own part, the unique experience of spaceflight has given me a different perspective on the passage of time. Through the ages man's sense of time has been determined by the distance that could be covered in a day; first by foot, then by horse, then train and airplane. But, to have circled the Earth in just an hour and a half has given me a different appreciation of how quickly time passes.

I still remember with affection my father casually repeating the expression he picked up from a pilot as he cut the grass all those years ago at the military aerodrome in Kaliningrad.

"Don't wait until the end of the runway to apply the brakes, and don't leave love affairs too late in life to enjoy." Put simply, it translates in Russian as the expression "*Lovi moment!* [Seize the day!]" It is the motto by which I believe life should be lived.

Colonel David Scott

London, 2003

As we watch the Moon tracking its path across the night sky, no matter where we are on Earth, or whether the Moon is waxing or waning, we only ever see the same features of our satellite's Near Side. Since it takes the Moon the same time to rotate fully on its axis as it does for it to complete one full revolution of the Earth— a phenomenon known as "synchronous rotation"—the same dark *mare*, such as Tranquillity, Serenity or Crisis, constantly face our planet, contrasting with the lighter lunar highlands to form the familiar Man-in-the-Moon features we learn to recognize in childhood.

Looking back toward the Earth from the Moon, however, our own planet's aspect is constantly changing. While the position of the

Earth in the dark star-studded sky—as observed from the surface of the Moon—does not differ, the face of our revolving planet alters hour-by-hour. As the Earth turns on its axis it is possible to see how interrelated are the physical and climatic features of our luminous blue globe, the great beauty and true aesthetics of which can never be fully conveyed in two-dimensional photographs.

As Norman Cousins, former editor of the literary magazine *Saturday Review*, declared in a congressional hearing on the future of space exploration, "What was most significant about the lunar voyage was not that men set foot on the Moon, but that they set eye on the Earth." Certainly from the perspective of the lunar surface it is startlingly clear how dynamic our own planet is, how alive and constantly evolving. But it is also striking how limited and fragile is our only home in the universe.

At one point during my three-day stay on the Moon, in the summer of 1971, I remember standing close to our Lunar Module, *Falcon*, and raising my hand toward the point where the Earth hung in the black sky above. By slowly raising my arm until my stiff, gloved thumb stood upright, I found my thumb could entirely blot our planet from view. One small gesture and the Earth was gone.

Many things come to mind when I think of that: everything from how we got there to what it means and what will happen next. Foremost is the conviction that, unless we humans protect and nurture our planet, everything with meaning—science, history, music, poetry, art and literature—will disappear in a relatively short period, just as our globe disappeared with one small twist of my thumb. All of us who have walked on the Moon, or explored it from orbit, share this important vision, no matter what our chosen, and very varied, paths were afterward.

Whenever I step up to a podium to speak about my time as an astronaut, I am invariably asked how I have been "transformed" by the time I spent on the Moon. How did it feel? And why did some astronauts turn to religion on their return? As one of only twelve men to have set foot on the Moon, I understand why people want to ask me. My answers depend on who is inquiring.

If it is a young boy asking how it felt to walk on the Moon, I can say "Fantastic" or "Just marvelous" and know he will go away with his eyes bright, and perhaps dream that night that he, too, will set foot on our celestial neighbor. For those expecting a more elaborate answer I describe the majesty of the lunar mountains, the layers of volcanic lava in the sides of Hadley Rille or the beauty of the sparkling crystals in the rocks. They may sometimes feel disappointed that I don't eulogize the experience at greater length. But my initial instinct, as I soared away from Earth's gravity for the first time aboard Gemini 8, still holds true: only an artist or poet could convey to others the true beauty of space. Alexei's paintings do, I feel, help capture the awe such an experience inspires.

As to transformation, maybe I was transformed and I didn't notice. Reflecting on this theme I am sometimes struck by the difference between the subsequent lives of the six of us who commanded lunar landing missions and our six Lunar Module pilots. Call it my "left-seat, right-seat" hypothesis if you will; commanders occupied the left side of the Lunar Module during descent to and take-off from the surface of the Moon, while Lunar Module pilots occupied the right. CDR.'s flew; LMP's monitored the instruments.

The post-Apollo lives of the six men in the right-hand seat on lunar missions tended to follow more unusual, and sometimes difficult, paths than one might expect of pilots and engineers. Buzz Aldrin's personal struggles following Apollo 11 have been well documented. After Apollo 12 and a mission with Skylab, Al Bean achieved his ambition to become a professional artist. After Apollo 14 Ed Mitchell founded the Institute of Noetic Sciences, devoted to researching the mind and the nature of consciousness. Jim Irwin founded the evangelistic High Flight Foundation after Apollo 15, and Charlie Duke became a born-again Christian and motivational speaker after Apollo 16. Only Jack Schmitt, in the right seat on Apollo 17, stayed within his professional field of geology after returning from the Moon, although he did also serve as a senator in the US Congress for six years.

This may be coincidental. It may perhaps reflect the frustration some of these men might have felt about not being in command of their own lunar missions. Or maybe it had something to do with not being burdened with the same level of responsibility as their commanders and so could spend more time thinking about and absorbing the extraordinary experiences we all went through. Perhaps now they have their own "flight."

Whatever the reason, the fact is those of us in the left seat went on to pursue more straightforward careers after returning from the Moon. Neil Armstrong went into business and academia after Apollo 11. Pete Conrad went on to fly Skylab 1 and then went into senior management at the McDonnell-Douglas Corporation after Apollo 12. Al Shepard returned to banking after Apollo 14. John Young continued working for NASA after Apollo 16, and after Apollo 17 Gene Cernan went into business.

There never has been a reunion of all twelve men who walked on the Moon, so we have not had a chance to sit around and discuss such matters as how our experiences on the lunar surface affected us differently. It may surprise some people that the twelve "moonwalkers" never got together as a group; three members of this unique band have subsequently died. But while strong individual friendships have endured, as a group we were always more of a loose amalgam of over-achievers than a military "band of brothers" sticking together through thick and thin. When the Apollo window closed and the shades were drawn on America's lunar program we all went our separate ways.

We no longer live near each other, for instance. We are spread out across the United States—from California, to Ohio and Florida—and even in Europe. I am struck by how this contrasts with the post-space program lives of the Russian cosmonauts, several of whom still live close to one another: I once watched Valentina Tereshkova playing with her grandson in the garden of her home neighboring Alexei's beautiful new house on the outskirts of Star City.

As for me, after six years in management with NASA, the last two as director of the Dryden Flight Research Center at Edwards, I began to realize toward the end of 1977 that my future lay beyond the space agency. By this time I had overseen a dozen exciting aircraft flight research projects at Dryden and as far as I was concerned it had been one of the best jobs at NASA. Among these were the space shuttle approach and landing tests, during which the shuttle was released from the back of a converted Boeing 747 and came in to land on the airfield's long, dry lake bed, demonstrating that the shuttle could indeed land like an airplane. But those tests were scheduled to end that November. Budgets for aeronautical research were difficult, and the frequent trips I was required to make to NASA headquarters in Washington were becoming tiring and boring. Up to that point it had been fun and exciting and great management experience. But I felt it was time to move on.

Yet the skills I had developed as an astronaut were unique to the goals I had been set. It was not like being a professional sportsman reaching the end of his career; there was no place to go and coach, no advisory position on future lunar exploration. Looking for new challenges I spent some time discussing options with an old friend and colleague from MIT, Al Preyss, who had been running a comparable flight research center at Wright-Patterson Air Force Base. Al and I decided it was time to get out of government service and set up our own consulting business in the private sector.

Among the interesting projects we developed was a very early single-string digital recording system for a group of four partners at the Cherokee Studios in Hollywood, one of whom was Ringo Starr. We then designed and attempted to develop a system of storing data optically using holograms, and began technical and management consulting for several companies in England.

Most significantly, we also began research and development of a unique optoelectric sensor—known as a structural information detector, or SID—which could monitor and measure minute vibrations. I was eventually asked to continue developing this device by looking at the way it could be applied to monitoring the structural integrity of aircraft and their possible impending failure, and also the possible structural failure of bridges, buildings and offshore platforms. By this time, however, I had taken over sole responsibility for the SID project with the help of a number of former colleagues and technical experts. Al had decided he wanted to become a full-time consultant and we had flipped a coin to see who would take over our joint company.

At the same time as developing these projects I was approached by the assistant secretary of the Air Force to set up a top-secret program training Air Force officers as astronauts to fly "blue shuttles," shuttle spacecraft the Air Force was planning to build for itself. After recruiting six of the best former Apollo NASA engineers for the Manned Spaceflight Engineer, or MSE, project, I began assisting in the selection and training of the first two groups of "MSEs." Their training took place in Los Angeles and several later flew on regular, not "blue," shuttle missions. But once again the Air Force reversed its decision regarding participation in manned space-flight, bringing back memories of its canceled Dynasoar, X-15 and Manned Orbiting Laboratory projects. The MSE program was canceled, as was the "blue shuttle" program.

As this window closed, however, another opportunity arose. In early 1984, NASA announced that it wanted the private sector to develop an "upper stage" for the shuttle. This would be a rocket vehicle purchased commercially by NASA to deliver satellites from the low Earth orbit reached by the shuttle to higher orbits required by specific satellites. It was an ideal commercial opportunity to combine our knowledge of the shuttle with experience in raising funds for research and development from the private sector. Two years later, under contract and with our funding, British Aerospace completed the development of a full-scale, nonflight prototype of our Satellite Transfer Vehicle, or STV.

When the *Challenger* shuttle was lost in 1986, however, there was a presidential decree that the shuttle must no longer carry commercial payloads and the STV needed to find a new launcher. All commercial satellites would have to be launched on "expendable launch vehicles," so the following year I spent two trips to China negotiating a memorandum of agreement for the purchase and support of six Long March launch vehicles to launch the STV into space with its satellite payload. But investors had no interest in a product from China and our STV project was dissolved.

At roughly the same time the primary bank and major lender to our company over the previous eight years began to fail. Almost instantly all loans were called and we found ourselves in a difficult financial situation through no fault of our own; our major project had dissolved and our major source of cash flow disappeared. As a result the company was consolidated to a consultancy on aerospace projects. Among these we serviced several companies in England, as introduced by my good friend Peter Bloomfield, with whom I continue to work on new and challenging opportunities. At this point I was approached by the Air Force once again.

This time they wanted my assistance with a project to demonstrate the scientific—peaceful—application of the so-called "Kinetic Kill Vehicle," or KKV, developed as part of the Star Wars program. This small self-propelled spacecraft had been designed to destroy an object—expected to contain nuclear warheads—on impact in space, hence "killing" it with kinetic energy. Proving such a vehicle could be used for scientific purposes was one way of trying to improve the public image of this highly controversial program.

The aim was to adapt a standard KKV—a self-contained spacecraft just 4 feet long and about 2 feet in diameter including miniature rocket engines similar to those used on Apollo and Gemini—into a planetary landing vehicle which could carry one or more micro-rovers for surface exploration. After enlisting the services of MIT Professor Rodney Brooks, a leading expert in the field of artificial intelligence, we designed a small six-legged micro-rover—just 8 inches long and weighing 6 lb—which would carry multiple miniature sensors and behave very much like an ant using "behavior control" software. This micro-rover could be placed in a small cocoon which would replace the target sensor package on the front of a conventional KKV.

A year later we successfully demonstrated our prototype in an "advanced concept architecture test" conducted in a large hangar at Edwards. To the amazement of the assembled audience our small lander, fitted with four legs and the rover cocoon, was launched inside the hanger and maneuvered to touchdown on a simulated lunar surface. The cocoon was then ejected; the micro-rover automatically opened its legs, exited the cocoon and conducted several minutes of surface exploration. This proved that such micro-rovers could be delivered to the Moon, Mars or other planets by modified KKVs and used to explore the surface almost as effectively as a human being—minus, of course, the spirit of human adventure.

While working on this project I regularly used to take a two-hour drive from our home, then at Manhattan Beach, Los Angeles, out to Edwards. This gave me plenty of time to think, and one afternoon I got to pondering the dilemma of a valuable communications satellite, which the Navy had lost in early March 1993 after its launch vehicle placed it in an incorrect orbit. The satellite was undamaged but useless, and an insurance claim of $180 million for it was paid out. I realized that if a "tug" spacecraft could dock with the lost satellite and, using its own propulsion system, move the satellite into its correct orbit, the satellite could become fully operational. Not only that, but if the vehicle docking with the stricken satellite carried extra fuel, the length of time the satellite could orbit would be extended and its commercial potential increased.

During my days at Dryden we had developed a concept of flying high-risk experimental aircraft from a remote cockpit in the basement of an administration building. This meant that a test pilot could sit in an exact replica of the aircraft's cockpit, with full controls and displays, but would not be exposed to flying an experimental aircraft—often beyond the "edge of the envelope." If this concept could be applied to flying such a small "tug" spacecraft as I envisaged, it would eliminate the need to include all the requirements of a manned vehicle and would reduce both its cost and the risks attached. Applying what I had learned from the remotely piloted research aircraft at Dryden and my years of experience in rendezvous and docking techniques as an astronaut, I began a long period of research and development into the concept.

The commercial potential of such a "tug and remote cockpit" rescue vehicle—the cost of which would be far less in both time and money than building and launching a replacement satellite—could, I realized, be enormous. It would be a win–win–win situation, with the satellite owner having an operational satellite during the planned market window and insurance companies cutting the cost of claims in half by paying the new company I formed to implement this concept, Vanguard, to conduct the "recovery." I immediately filed patent applications on several forms of this "invention" and, during the years that followed, with the assistance of several former NASA colleagues and friends in the space insurance community, I

refined the concept and began to seek financing for its
development and operation. My patents for the scheme were
eventually bought by a German company, which continues with its
development.

While my work on the scheme was in its early stages I also
received a call one day from Jim Lovell, commander of Apollo 13
and a good friend. Jim wanted to know if I would like to assist in
the making of a movie based on his book about the harrowing
experiences of his ill-fated Apollo 13 mission, which narrowly
escaped disaster when an oxygen tank exploded on the way to the
Moon in 1970. The monumental efforts of Mission Control in
saving the crew, played out in real time on television, had held a
worldwide audience fascinated. Jim lived in Texas and was busy
promoting his book; I lived about thirty minutes from Universal
Studios in Hollywood. Within a few minutes of telling Jim I was
interested, I received another phone call. This time it was Ron
Howard, the film's director, asking if I would be his technical
assistant, working directly both with him and with the four actors
who were to play lead parts, Tom Hanks, Kevin Bacon, Bill Paxton
and Gary Sinese.

It was the start of one of the most pleasant, rewarding and
exciting periods of my post-NASA career. It was my first experience
of the movies and took a little adjustment. On the first day of
filming, for instance, I turned up smartly dressed in a suit, only to
find that no one wears suits on a Hollywood set—except "the
suits," the men who keep the accounts. Over the course of filming,
not only did I get involved in all aspects of the movie-making
operations, but I also met, and found outstanding, some of the
leading personalities in Hollywood.

Soon after the "wrap" of *Apollo 13* Tom Hanks invited me to
lunch. He was about to start filming a ten-part television mini-
series, to be called *From the Earth to the Moon*, which would
tell the story of spaceflight from its earliest days right through to
the final Apollo mission. Tom offered me the role once again of
technical adviser. But the breadth of the series meant I soon found
myself becoming involved in all aspects of its production; from
reviewing screenplays to coaching actors on how astronauts would

speak and behave in given situations, to advising directors and even post-production editing. One full hour in the series was devoted to Apollo 15 and involved me coaching the actor Brett Cullen to play my part, which was pretty interesting. The series turned out to be highly successful. It was the most accurate depiction I have ever seen of the highs and lows of the American space program, and went on to win the television Emmy Award for best mini-series of the year. Far removed from my conventional, government culture background, it was a fascinating experience and led to lasting friendships with many in the film industry.

In recent years I have turned my attention to environmental protection and, together with my good friend Arthur Miller, who is a professor of physics, have developed a concept for the permanent disposal of "high-grade" nuclear waste in space. Nuclear energy has been identified as one of the best alternatives to counter the problem of global warming posed by the continued use of fossil fuels if a safer method of disposing of nuclear waste than underground burial can be found. Major expansion in the commercial launch vehicle industry offers new options for launching nuclear waste into space. Once in space, it can be sent on trajectories departing the solar system for destinations such as black holes.

Aside from such interests, however, both throughout my time as an astronaut and in the years since leaving NASA, I continued to be intrigued by how our old Cold War rivals in the space race were faring. As the old world order that had kindled that race began to change with the collapse of communism, I found myself wondering more and more about the man I had grown to like so much during my first visit to Moscow in the summer of 1973, Alexei Leonov. With old rivalries gone and none of the restrictions Alexei and I had felt about what we could say as we sat in his wood-paneled studio so many years before, the idea of chronicling and comparing our personal recollections of such extraordinary times took hold. These interwoven memoirs are the result.

Our vividly contrasting experiences are a stark illustration of the fact that the race to the Moon was politically motivated. It was the result of a bitter clash between two superpowers, each attempting to demonstrate to the world that its technology and political ideology was superior, each dedicating vast resources and its most brilliant minds to becoming the winner. In the words of President John F. Kennedy, it was a battle between "freedom and tyranny" and freedom won. Neil Armstrong's first steps on the Moon were a clear demonstration of the technological strength, scientific expertise, peaceful intentions and openness of a free society.

Not everyone will agree that it was worth the cost, or that the net results were beneficial. Some will argue that the money should have been spent on social welfare programs, others that the development of space technology will lead to war in space. Regardless of the results, none doubt that the race to the Moon has powerfully shaped the world in which we live.

I firmly believe that apart from its immense political implications this valiant race opened up two important new frontiers in the mind of mankind. It marked the first steps of the human race to explore new worlds beyond our own, and it made us reflect more earnestly on the ways in which our planet can be best protected and sustained for future generations.

Looking to the future of space exploration, I do believe, however, that the next major advance in spaceflight will be by the military; science will continue to explore the planets with robots. In the not too distant future—the next ten years or so, perhaps—I believe the American military, probably the Air Force, will establish some form of "space patrol" to inspect, monitor and track satellites from other countries. These small inspector spacecraft are not likely to be manned by astronauts, but rather will be "flown" by highly qualified space pilots from a remote cockpit located in underground control centers. In essence they will be unmanned fighter aircraft, operating in a similar way to the emerging concept of unmanned aerial vehicles, taking aggressive action against hostile or threatening satellites or other space objects, should this be necessary. In the meantime it seems increasingly likely that some form of Star Wars defensive shield against nuclear attack by intercontinental ballistic missiles will be deployed by the United States.

New types of propulsion for planetary transfers will finally be developed and put to use. But chemical rocket propulsion will be the preferred method of launch from Earth for years to come. As

Walter Dornberger, one of the early German rocket pioneers, observed, "Three really great inventions have or will influence the history of man . . . the wheel with which man conquered the Earth, the screw with which he conquered the seas and the air and . . . rocket propulsion which will help men conquer space and push forward to the stars."

The future of manned space exploration is far less certain. The question of when humans will again walk on another planet is very difficult to answer. The tragic loss of space shuttles *Challenger* and *Columbia* and their multinational crews have served to highlight the risks of any space venture, albeit in the case of the shuttle in low Earth orbit. As the soul searching continues at NASA in the wake of the disasters, one could argue that even the current status of robotic technology precludes the need to risk humans exploring the surface of an extra-terrestrial body in situ.

The remarkable advances in computer science and robotics, including software which produces human-like capabilities, does indicate that artificially intelligent robots could soon replace artificially robotic humans (astronauts in spacesuits) when it comes to space exploration. One need only picture those somewhat intelligent Apollo beings of the mid-twentieth century moving around on the Moon in their stiff and bulky pressure suits to be reminded of the latter! Whether robots will ever be able to convey the high adventure of exploring new frontiers in space is less certain.

Some argue that explorations of the magnitude of the manned lunar landings occur because technology has advanced to a point where it can readily be used to expand new horizons. Others believe it results from a confluence of events at a unique time; certain individuals with strong reasons, the necessary resolve and sufficient resources drive events forward, even though the requisite technology does not yet exist. I believe our experience proves the latter.

By today's standards the technology available to us as we embarked on the Apollo program nearly four decades ago was crude; the total memory capacity of computers on board our Apollo spacecraft, for instance, was just 36K. But this technology was used to optimum capacity by 400,000 gifted individuals inspired by the vision, intellect and leadership of two unique individuals, President John F. Kennedy and Wernher von Braun, who were prepared to press forward in the face of significant hardships, failures and losses. As far as Alexei and the rest of the cosmonaut corps were concerned, as he has movingly described here, it was the brilliance and drive of their Chief Designer, Sergei Korolev, and his determination to succeed against all odds that drove them to excel.

Whether such individuals exist today is open to question. Questionable too is whether any nation or group of nations have sufficient reason, resources or resolve to embark on renewed human exploration of the Moon, Mars or any other planet. In contemplating events of the past and prognosis for the future, however, we humans tend to overestimate the near term and underestimate the longer term. Our Earth–Moon system has been in existence for about 4.5 billion years. If such an eon were compressed to a 45-minute span, each minute representing 100 million years, man can be seen to have roamed the Earth for just the blink of the eye.

One special quality of Apollo should be considered, however, as we contemplate the future of human exploration of the planets and moons of our solar system. Science fiction was a major factor in the emergence of human spaceflight. And of all the science fiction writers who wrote about the exploration of the moon, none, as far as we can tell, predicted that the whole world would watch the first steps of man on television! This was a remarkable image, and it left an indelible impression on those who watched, some of whom were only small children. Almost everybody I meet, who was old enough to be placed in front of a TV, tells me that they watched Neil Armstrong take the first step on the Moon. And for those 400,000 people who actually worked on Apollo, it was a major milestone in their careers, regardless of their involvement. Thus, the Apollo exploration of the Moon actually became a "shared experience" by the remainder of the humans on the Earth. It was, one might say, an experience of "virtual reality," they were there virtually and it essentially became real in their thoughts and emotions, it became a "real" adventure. With the ongoing advance of technology, this "shared experience" will be ever more prominent, and it might even be a driver in committing humans to actual planetary exploration. Pretty exciting thought.

When I was on the Moon I performed a simple, but meaningful experiment—simultaneously dropping a falcon feather and a geology hammer to prove Galileo's classical law about falling objects in gravity fields. I doubt if Galileo Galilei, with all his magnificent vision of the future, would ever have predicted his hypothesis would finally be proven by a man on the Moon.

Taking the long view, and drawing comparisons with terrestrial equivalents, it seems to me that inter-planetary exploration is barely post-Columbian in its development. From this historical perspective we are at the stage with spaceflight where Christopher Columbus has just returned to Spain after "discovering" the "New World" and furious debate continues over whether the Earth is or is not flat. But the fleet of Fernando Magellan has yet to set sail from Seville to make the first circumnavigation of the world, and Captain James Cook of the *Endeavour* has not even been born.

Before setting off for new planetary destinations space explorers of the future, together with those inspired by the wonders of worlds beyond our own, can reflect here on the high drama and adventure marking the lives of those who pushed forward the early frontiers of space exploration. Ambitious dreams of manned space exploration will return. But, though we must learn from the past, we must not live in the past. We must dream of the future, but not sleep through our dreams.

Sometimes, while walking along the seashore or a riverbank at night, more than three decades after the three most memorable days of my life, I look up at the Moon riding bright and clear in a crisp dark sky and my eye picks out the largest circular marking on its surface, the Mare Imbrium or Sea of Rains. As I move my gaze to its eastern rim where I once landed in a spaceship, I often feel a pang of nostalgia. For when I look at the Moon I do not see a hostile, empty world. I see part of the Earth, a radiant body where man has taken his first steps into an endless frontier.

ACKNOWLEDGMENTS

David Scott

One summer evening in Moscow in 1973, during a brief thaw in the Cold War, I sat in the well-appointed studio / flat of Alexei Leonov discussing the exploration of the Moon. It was twelve years after the Cuban missile crisis and almost twenty years before the demise of the Soviet Union. We had been brought together by the Apollo–Soyuz Test Project (ASTP), for which I was leading a US delegation to Moscow to negotiate all technical aspects of the forthcoming mission between our two countries. As we sat alone, except for Alexei's interpreter, I heard for the first time the fascinating story of the Soviet thrust to land a man on the Moon before the United States. Alexei would have been that man. The Iron Curtain between us had precluded me from learning even the general aspects of this major political and technical Soviet program.

As his story unfolded, we compared the Soviet approach with my recent Apollo 15 mission to explore the Hadley Apennine region of the Moon. There were so many similarities and yet so many differences, that we both thought it was a story that should be told some day. But the visit was too brief and we both had to press on with our respective and independent lives. Alexei went on to command the very successful ASTP mission in 1975, and I went on to NASA management as the director of the Dryden Flight Research Center at Edwards AFB, California.

Twenty years later we had dinner together in New York City. Alexei's stories had continued to evolve. By then the time seemed right to go back and recount the amazing space race between the two political superpowers of the time. This book, then, is the culmination of ten years of discussions (with some lengthy gaps!)

between Alexei and me. But the stories go back much further than that, to our very beginnings. So many people contributed to the eventual result that another entire book would be required to thank and acknowledge them all. We hope we will be forgiven for missing those we haven't mentioned, you are in our hearts and minds nevertheless and we are deeply grateful for your friendship, advice and contributions to our lives and the stories told here.

First, I would like to express my sincere appreciation to Alexei for joining me in this adventure—truly a partnership in space and time. I am saddened that he did not have the opportunity to walk on the Moon. He would have been brilliant. But he was the first man to walk in space, and he was brilliant—for that he will always be remembered. Our friendship has spanned many years, and we are all appreciative of the exceptional insight he has provided into his fascinating life and that of the Soviet space program. I also appreciate the long hours, patience and perseverance he dedicated in recounting so many interesting and exciting events during those heady days of racing to the Moon. It's a pity we could not include them all. But thanks, Alexei; your contributions to space exploration as well as to this book are exceptional and timeless.

And an enduring thanks to Neil Armstrong for not only writing the thoughtful and meaningful foreword to this book, but also for your leadership, skills and rapid response to the trials of Gemini 8. Your overall contributions to the space program are unlimited and appreciated by all. Special thanks also to Tom Hanks for bringing the book to life with his perceptive and stimulating introduction. I am also indebted to Tom for his warm friendship as well as his consideration in inviting me to be part of his exciting world of filmmaking.

Our writer, Christine Toomey, with her superb journalistic and writing skills really pulled it all together. Through countless interviews—a most difficult task with Alexei speaking in Russian and I in English—she wove our two lives and cultures together into a unique narrative which highlights both the great contrasts as well as the many connections between Alexei's experiences and mine. Thanks also to her daughter, Ines, for her patience during Christine's absence while listening to our recollections.

And our sincerest appreciation to Ed Victor, our agent, who really made it happen. He enthusiastically accepted our approach to combining two autobiographies and suggested and found our writer, Christine Toomey. He then selected our publishers and established essential and effective relationships with our editors. He has encouraged and inspired us all the way, many thanks for all you have done, Ed.

We are also grateful to several individuals who reviewed draft manuscripts and offered valuable and useful comments. My daughter Tracy and her husband Tim were a constant source of comments, insight and suggestions, and followed the development of the book week by week; so, many thanks Tracy and Tim. My former wife Lurton filled in many gaps and clarified my memories of many events and activities. We are especially grateful to David Harland, a noted space historian, who provided a "technical" review of events, technology, dates and other historical facts. His colleague David Woods, editor of the Apollo Flight Journal, also provided valuable comments. And special thanks to Eric Jones, editor of the Apollo Lunar Surface Journal who ensured that our exploration of Hadley was accurate and thorough. Content, comprehension and composition were reviewed by my good friends Peter Bloomfield, Professor Arthur Miller and Dr. Ian Perry, all in London. And I am forever grateful to Annabelle Brotherton who reviewed and commented in detail, provided valuable insight into how to tell our stories, created several chapter titles and was instrumental in the design of the jacket. Bobbie Ferguson from NASA headquarters was enthusiastically helpful and almost overloaded us with photos from NASA archives. Many thanks to all; I deeply appreciate your intellectual and emotional support—you kept us focused, consistent and comprehensive.

Among other key individuals, we are all deeply grateful to our translator (written word), Marianna Taymanova and her husband Pavel, as well as our several interpreters (spoken word), Anatole Forestenko, Vladimir Danilin, Nikolai Douplenski, Alexander Stepanenko and Anna Chistova. Through many interviews our interpreters would pass Christine's English questions to Alexei in Russian, pass his Russian reply back to Christine in English, from which she would draft the narrative in English. Marianna then translated this into Russian for Alexei's review and his Russian comments came back through Marianna for translation so Christine could incorporate them in the final English manuscript. But it all worked! And many thanks to Natasha Filimonva, Alexei's personal assistant, who was a vital link in all these transmissions. And without Alex Robson, Alexei and I would probably never have had the opportunity to really discuss the making of the book. Alex ("Sasha" to Alexei) accompanied me to Moscow in the summer of 2001 to meet with Alexei and discuss the book. Her superb knowledge and practice of the Russian language and culture paved the way for meaningful discussions during a wonderful evening of Russian hospitality from Alexei and his lovely wife, Svetlana.

Alexei and I would both like to express our sincere gratitude to our families who supported and encouraged us during the high-tension and complicated days of preparing and flying our missions. Lurton, Tracy and Doug were there for me every moment of the way; especially providing a comfortable and fun family life during our infrequent holidays and when I could get home on the weekend. Alexei's wife, Svetlana, and his daughters, Vika and Oksana were likewise the foundation of a stable and loving home.

And so many thanks to our parents and siblings as we grew up and began to direct ourselves toward our ultimate goals. My dad pointed the way to flying school and continuously encouraged me as challenges arose. My mother was always there for me and provided the loving warmth that only mothers can. And my brother Tom has been my strongest supporter all along the way. Thanks so much to our loving families.

I would also like to express my gratitude to several other unique individuals who guided me and inspired me along the way. Without them I would not have taken the path to the Moon. Three swimming coaches taught me the elements of competition and fair play, and stimulated my physical and mental growth: Charles Halleck in high school, Matt Mann at the University of Michigan and Gordon P. (Slim) Chalmers at West Point. Interestingly enough, Coach Chalmers had been a student of my dad's in flying school. To land on the Moon, one must know how to fly, and with great skill and precision. I first learned how to fly from one Chauncy P. Logan, a true "barnstormer" before the Second World War, and a classic stick-and-rudder pilot who taught me to "feel" the airplane. "Boots" Blesse taught me how to be a fighter pilot during the wonderful days of the 32nd Fighter Squadron in Holland. He wrote the book of aerial tactics, *No Guts, No Glory*, and he passed it on well. And finally, the world of a test pilot was another step above, and I was fortunate to have benefited from the master himself, Chuck Yeager, with whom I flew and assimilated the touch with great admiration.

At MIT I learned how to learn, especially about space guidance and navigation, under the tutelage of the masters, to whom I owe special thanks, particularly Professors Richard H. Battin, Walter Wrigley and Holt Ashley. And as my career began to require some knowledge of science, especially geology, I was tutored by perhaps the best professor on the planet, Prof. Leon T. Silver. I will always be indebted to his enthusiasm and inspiration, not to mention his exceptional teaching skills. From him and his colleagues, especially Gordon Swann, Jim Head, Bill Muehlberger, Farouk El-Baz, Hal Masursky and many others, I learned to appreciate and truly enjoy the geological exploration of the Earth and the Moon. Apollo 15 would have been bland without each and all of these leaders and teachers who were so influential and so patient.

I was fortunate to fly the three best space missions on record, at least in my opinion. Each was an entirely different challenge and separately rewarding. And on each I had crewmates who carried me all the way. Neil Armstrong was my commander on Gemini 8, his leadership and skills were invaluable to both our mission and the overall Gemini and Apollo programs. On Apollo 9, Jim McDivitt and Rusty Schweickart were not only teammates, but also close companions for over two years as we prepared and flew the intricate and demanding "connoisseur's test flight." Al Worden and Jim Irwin joined me for Apollo 15, called by NASA "the most complex and carefully planned expedition in the history of exploration." Their performance and skills were nothing less than exceptional. We were saddened to lose Jim in 1994, he always pushed himself to the limit in everything he did, to the benefit of all.

Many thanks, crewmates, without you there would be no stories to tell. Although we do not see each other much these days, we are lifelong friends and share some of the most memorable events in the history of space exploration.

I am also grateful indeed to our back-up crews not only for their support and enthusiasm but also for being an essential check and balance as we pressed forward to achieve our goals: Pete Conrad and Dick Gordon on Gemini 8; Pete, Dick and Al Bean on Apollo 9, and Dick, Vance Brand and Jack Schmitt on Apollo 15. And special recognition must go to the troops of Mission Control and all of its supporting and planning elements. We are especially grateful to our flight directors who kept us on track and made sure we had everything on Earth that we needed: John Hodge and Gene Kranz on Gemini 8; Gene Kranz, Gerry Griffin and Pete Frank on Apollo 9; and especially Gerry Griffin, Milt Windler, Glynn Lunney and Gene Kranz on Apollo 15. We relied on you explicitly minute by minute and you were always there for us—so many thanks. During the missions our capcoms were essential, we appreciated their constant vigilance and cool communications, especially Jim Lovell, Jim Fucci and Keith Kundel on Gemini 8, the team on Apollo 9 and the great group on Apollo 15: Joe Allen, Ed Mitchell, Karl Henize, Bob Parker and Gordon Fullerton. Thanks also to Jim Lovell for managing the "science room" during Apollo 15, your coordination between science and operations was exceptional. And my dear friend, Charlie Bassett, who died just before his Gemini 9 mission, must be remembered for his intellect, inspiration and many contributions to the success of the program.

So many others at NASA and its contractors deserve special recognition and appreciation. The success of the Apollo program was due to the tireless and dedicated efforts of over 400,000 people. The Soviet lunar program required an equal number. We would like to thank them all individually and collectively. Alexei and I were merely at the peak of the pyramid and carried their banners.

Many individuals in NASA were instrumental in the success of my missions. I shall always appreciate the guidance and decisions of them all, especially the superb NASA management team, with particular remembrance to those with whom I had frequent personal contact: George Low, Rocco Petrone, Bob Gilruth, Chris Kraft, Sig Sjoberg, Jim McDivitt, Joe Shea, Deke Slayton, Bill Tindall, Cliff Duncan, Paul Donnelly, Jim Harrington, Chet Lee, Lee Scherer, Bill Bergen, Dale Meyers, Joe Cuzzopoli, Joe Gavin, Tom Kelly, Davy Hoag, Jim Nevins and our Apollo 15 flight crew support team leader, Jim Smotherman. You were the leaders and decision-makers; I offer my thanks with respect and admiration. Others who were so helpful during those years include Dee O'Hara, our NASA nurse, and Phil Smith, of the National Science Foundation, our guide during a memorable 10-day visit to the Antarctic, many thanks. The efforts of all individually and collectively are sincerely appreciated for achieving our mutual and multiple successes.

On the other side of the race to the Moon, special recognition and appreciation goes to Pavel Balyeyev, Alexei's commander on Voskhod 2, and Valery Kubasov, his flight engineer on ASTP. I met several other cosmonauts along the way who sparked my interest and curiosity in the Soviet program, especially Pavel Belyayev and Konstantin Feoktistov as well as Vladimir Shatalov and Alexei Yeliseyev at the Paris Air Shows in 1967 and 1969, respectively. And I have great admiration for Professor Konstantin Bushuyev who was my ASTP counterpart during my Moscow visit in 1973, his patience and persistence are sincerely appreciated.

After Apollo was a challenging period, and many people influenced and guided my career and opportunities. Among these, special thanks to Glynn Lunney for including me in the management of Apollo–Soyuz, Lee Scherer for giving me a top management position at the Dryden Flight Research Center, and Al Preyss, my first partner in business who helped me learn from the bottom up. And then when I shifted into an entirely new world of movie-making, I am very grateful to those who gave me the opportunity for this wonderful experience, in particular Tom Hanks, Ron Howard, Brian Grazier, Tony To, Graham Yost, David Carson, Brett Cullen, Terry Odem and especially Amy McKenzie who kept it all together. And Jim Lovell who introduced me to Ron Howard for the making of Apollo 13, the movie. Thank you all for making my post-Apollo life so interesting and meaningful.

This book reached its final form through the tireless efforts of our editors, Amanda Harris, Cassandra Campbell, Edwina Barstow and Helen Simpson; without them we would still be drafting, many thanks to you all.

Our final thanks go to John F. Kennedy, Wernher von Braun and Sergei Korolev. President Kennedy committed the US to "landing a man on the Moon and returning him safely to Earth" and thereby set the stage for this magnificent adventure. Wernher von Braun and his team designed and developed the most powerful launch vehicles in history, truly an amazing feat, especially given the full success of the venture. And Sergei Korolev, the Chief Designer, was the mastermind behind the Soviet program, a true visionary and always a step ahead in the race to the Moon. Many others contributed to these ventures, but gentlemen, you were the movers and shakers, and for that we who participated are all very, very grateful. May you be recognized and applauded throughout history.

DAVID SCOTT
London, February 2004

GLOSSARY

ADIZ	Air Defense Identification Zone
AFB	Air Force Base
Agena	Upper stage rocket used as a target vehicle for Gemini rendezvous
ALSEP	Apollo Lunar Surface Exploration Package
ARPS	Aerospace Research Pilot School
ASTP	Apollo Soyuz Test Project
astronautics	the discipline involving movement through space
anorthosite	a granular igneous rock composed almost exclusively of plagioclase
attitude	orientation relative to a given reference, usually the local vertical or horizon
basalt	volcanic rock
breccia	mixture of different types of basalt within a matrix
capcom	capsule communicator
CDR	Commander, occupies left seat in the CM and left station in the LM
CEO	Chief Executive Officer
CM	Command Module, houses 3-man Apollo crew, contains spacecraft operating systems
CMP	Command Module Pilot, navigator, occupies center couch in CM
CSM	Command and Service Modules, Apollo lunar orbit spacecraft
CSQ	*Coastal Sentry Quebec*, NASA tracking ship
Delta-v	Engineering term used for a change in velocity

de-orbit	a reduction in velocity by a rocket engine so as to bring the spacecraft out of orbit into a descent
Dynasoar	early spaceplane program developed jointly by the Air Force and NASA, canceled
downlinked	communications and telemetry sent down to a ground station from a spacecraft
EECOM	Electrical, Environmental, and Communications flight control position in MCC/MOCR
EMUs	Extravehicular Mobility Units, comprised of the spacesuit and the PLSS backpack
EVA	Extravehicular, outside a spacecraft
GET	Ground Elapsed Time, the time elapsed since lift-off
IAU	International Astronomical Union
JFK	John F. Kennedy, President of the United States
JSC	Johnson Space Center, Houston, named the MSC until 1973
KKV	Kinetic Kill Vehicle, a space projectile designed to destroy a target upon impact
KSC	Kennedy Space Center
LCG	Liquid Cooled Garment
LEVA	Lunar Extravehicular Assembly, outside helmet
LLTV	Lunar Landing Training Vehicle
LM	Lunar Module, two stage spacecraft designed for lunar landing, lift-off, and rendezvous
LMP	Lunar Module Pilot, systems engineer, occupies left couch in CM and right station in LM
LOI	Lunar Orbit Insertion
LRV	Lunar Roving Vehicle, or "rover," car designed to be driven on the Moon
MCC	Mission Control Center, Houston, Texas
MiG-15	Soviet fighter aircraft
MIT	Massachusetts Institute of Technology
MOCR	Mission Operations Control Room, central control room within the MCC

MSC	Manned Spacecraft Center, Houston, name replaced by JSC in 1973
MSE	Manned Spaceflight Engineer, Air Force astronaut program, canceled
N-1	Soviet lunar launch vehicle, comparable to the US Saturn V
NASA	National Aeronautics and Space Administration
NKVD	Soviet intelligence agency
Orbiter	spacecraft designed for orbital operations, e.g., the Space Shuttle Orbiter
PGA	Pressure Garment Assembly, spacesuit
plagioclase	a feldspar having calcium or sodium in its composition
PLSS	Portable Life Support System, backpack providing oxygen, cooling, and communications
PPK	Personal Preference Kit
RCS	Reaction Control System, small rocket engines used to control attitude and small manoeuvers
Rille	canyon on the Moon
reenter	return to the Earth and its atmosphere from orbit or transearth trajectory
regolith	loose surface material on the Moon derived from a series distribution of rock fragments
Salyut	Soviet space station
Saturn	Apollo launch vehicle
SEVA	Standup EVA
Seconal	medication to assist sleep
SM	Service Module, support and supply spacecraft attached to the CM
SPS	Service Propulsion System, main engine on the Service Module
Starfighter	F-104 fighter aircraft
Super Saber	F-100 fighter aircraft
scoriaceous	basalt with high proportion of vesicles, or gas bubbles
sim	simulation

Soyuz	Third generation Soviet spacecraft, designed for crew of 2 or 3 in Earth orbit
Space walk	EVA, conducting operations outside a spacecraft
station-keeping	maintaining a steady position relative to another space vehicle, formation flying
thruster	small rocket engine, normally used in groups to maintain spacecraft attitude or motion
TEI	Maneuver used to inject a spacecraft into a transearth trajectory from lunar orbit
TLI	Maneuver used to inject a spacecraft into a translunar trajectory from Earth orbit
transearth	travel from the Moon to the Earth
translunar	travel from the Earth to the Moon
UCD	urine collection device
Voskhod	second generation Soviet spacecraft, designed for 2 or 3 men in Earth orbit
Vostok	first generation Soviet spacecraft, designed for one man in Earth orbit
Zond	Soyuz spacecraft modified for lunar operations

BIBLIOGRAPHY

Aldrin, Edwin E., *Return to Earth*, New York: Random House, 1973

Allen, Joseph P., *Entering Space*, New York: Stewart, Tabori & Chang, 1973

Apollo 15 Preliminary Science Report, NASA SP-289, Washington, DC: Government Printing Office, 1971

Bainbridge, William Sims, *The Spaceflight / Revolution*, Malabar, Fla: Krieger Publishing Company, 1983

Baker, David, *The History of Manned Spaceflight*, London: New Cavendish Books, 1981

Bean, Alan, *Apollo*, Toppan Printing Co., Ltd., Japan: The Greenwich Workshop Press, 1998

Borman, Frank, *Countdown*, New York: Silver Arrow Books, 1988

Brooks, Courtney G., *Chariots for Apollo*, NASA SP-4205, Washington DC: NASA, 1979

Burrows, William E., *This New Ocean*, New York: Random House, 1998

Carpenter, M. Scott, *et al.*, *We Seven*, New York: Simon & Schuster, 1962

Cernan, Eugene, *The Last Man on the Moon*, New York: St. Martin's Press, 1999

Chaikin, Andrew, *A Man on the Moon*, New York: Viking Penguin, 1994

Clarke, Arthur C., *The Exploration of Space*, New York: Harper & Row, 1951

Clarke, Arthur C., *2010 Odyssey Two*, London: Granada Publishing, 1982

Collins, Michael, *Carrying the Fire*, New York: Farrar, Straus and Giroux, 1974

Collins, Michael, *Liftoff*, New York: Grove Press, 1988

Compton, David, *Where No Man Has Gone Before*, NASA SP-4214, Washington DC: NASA, 1989

Cunningham, Walter, *The All-American Boys*, New York: Macmillan, 1977

Doran, Jamie, and Bizony, Piers, *Starman*, London: Bloomsbury, 1998

Ertel, Ivan D., *et al.*, *The Apollo Spacecraft: A Chronology*, NASA SP-4009, Washington, DC: NASA, 1978

Farber, David, and Bailey, Beth, *The Columbia Guide to America in the 1960s*, New York: Columbia University Press, 2001

Glenn, John, *John Glenn*, New York: Bantam Books, 1999

Godwin, Robert, *Apollo 9: The NASA Mission Reports*, Toronto: Apogee Books, 2001

Godwin, Robert, *Apollo 9: The NASA Mission Reports*, Toronto: Webcom Ltd, 1999

Godwin, Robert, *Apollo 15: The NASA Mission Reports*, Toronto: Apogee Books, 2001

Godwin, Robert, *Apollo 15: The NASA Mission Reports*, Toronto: Webcom Ltd, 2001

Hacker, Barton C., and Grimwood, James M., *On the Shoulders of Titans*, NASA SP-4203, Washington, DC: Government Printing Office, 1977

Hall, Rex, and Shayler, David J., *The Rocket Men*, Chichester: Praxis Publishing, 2001

Harford, James, *Korolev*, New York: Wiley, 1997

Harland, David M., *Exploring the Moon: The Apollo Expeditions*, Chichester: Praxis Publishing, 1999

Heiken, Grant H., *et al.*, *Lunar Sourcebook*, Cambridge: Cambridge University Press, 1991

Irwin, James B., *Destination Moon*, Portland, Oregon: Multnomah, 1989

Irwin, James B., and Emerson, William, *To Rule the Night*, New York: Hodder & Stoughton, 1973

Jones, Eric, *Apollo Lunar Surface Journal* (Internet only: www.hq.nasa.gov/alsj/), July 2003

Kelly, Thomas J., *Moon Lander*, Washington, DC: Smithsonian Institution Press, 2001

Kraft, Chris, *Flight*, New York: Dutton, 2001

Launius, Roger D., *Frontiers of Space Exploration*, Westport, Conn: Greenwood Press, 1998

Launius, Roger D., NASA: *A History of the U.S. Civil Space Program*, Malabar, Fla: Krieger Publishing Company, 1994

Lee, Wayne, *To Rise From Earth*, New York: Facts on File, Inc., 1995

Leonov, Alexei, *Spaziergänger im All*, Stuttgart: Deutsche Verlags-Anstalt, 1971

Light, Michael, *Full Moon*, London: Jonathan Cape, 1999

Lovell, James, *Lost Moon*, Boston, Mass: Houghton Mifflin, 1994

Lunine, Jonathan I., *Earth*, Cambridge: Cambridge University Press, 1999

McDougall, Walter A., *The Heavens and the Earth*, Baltimore: Johns Hopkins University Press, 1985

Mackenzie, Dana, *The Big Splat or How Our Moon Came to Be*, Hoboken, NJ: John Wiley & Sons, 2003

MacKinnon, Douglas, and Baldanza, Joseph, *Footprints*, Washington, DC: Acropolis Books, 1989

Mitchell, Edgar, and Williams, Dwight, *The Way of the Explorer*, New York: G. P. Putnam's Sons, 1996

Murray, Charles, and Cox, Catherine Bly, *Apollo: The Race to the Moon*, New York: Simon & Schuster, 1989

Orloff, Richard W., *Apollo by the Numbers*, NASA SP-2000-4029, Washington, DC: NASA, 2000

Schefter, James, *The Race*, New York: Doubleday, 1999

Schirra, Wally, *Schirra's Space*, Annapolis: Naval Institute Press, 1988

Seamans, Robert C., Jr, *Aiming at Targets*, Beverly, Mass: Memoirs Unlimited, 1994

Service, Robert, *A History of Twentieth-Century Russia*, London: Penguin, 1997

Shayler, David J., *Disasters and Accidents in Manned Spaceflight*, Chichester: Praxis Publishing, 2000

Shayler, David J., *Gemini*, Chichester: Praxis Publishing, 2001

Shelton, William, *Soviet Space Exploration*, New York: Washington Square Press, 1968

Siddiqi, Asif A., *Challenge to Apollo: The Soviet Union and the Space Race 1945–1974*, NASA SP-2000-4408, Washington, DC: Government Printing Office, 2000

Slayton, Donald K., *Deke!*, New York: Tom Doherty Associates, 1994

Stafford, Thomas P., *We Have Capture*, Washington, DC: Smithsonian Institution Press, 2002

Stuhlinger, Ernst, and Ordway, Fredrick I. III, *Wernher von Braun: Crusader for Space*, Malabar, Fla: Krieger Publishing Company, 1994

Swenson, Loyd S., *et al.*, *This New Ocean*, NASA SP-4201, Washington DC: NASA, 1966

Tindallgrams, NASA Kennedy Space Center Microimaging Facility, 1967–1970

Wendt, Guenter, and Still, Russell, *The Unbroken Chain*, Ontario: Apogee Books, 2001

White, Frank, *The Overview Effect*, Boston, Mass: Houghton Mifflin, 1987

Whitehouse, David, *The Moon: A Biography*, London: Headline Publishing, 2001

Wilhelms, Don E., *To a Rocky Moon*, Tucson: University of Arizona Press, 1993

Wolfe, Tom, *The Right Stuff*, New York: Farrar, Straus and Giroux, 1979

Woods, David; Apollo Flight Journal (Internet only: www.hq.nasa.gov/office/pao/History/ap15fj/), July 2003

Worden, Alfred M., *Hello Earth*, Los Angeles: Nash Publishing, 1974

Yeager, Chuck, *Yeager*, New York: Bantam Books, 1985

Young, John W., *America, Russia and the Cold War*, New York: Longman, 1993

INDEX

Scott, David *(continued)*
  awards, 181
  back-up to Commander in Apollo 12
    mission, 243, 250–52
  on backup crew of Apollo 1, 183
  capcom for Gemini 4 mission, 126–28
  childhood and upbringing, 11–15
  close calls while flying, 43–46, 74–75
  consulting business, 381, 382–83
  as director of Dryden Flight Research
    Center, 348, 380–81
  ending of flying career and retires from
    Air Force, 349–50
  enters Air Force, 26–29
  family background, 12–13
  feelings about being on the Moon, 378–79
  first flight in airplane, 19
  flight hours, 95, 349
  and Gemini 8 mission. *See* Gemini 8
    mission
  goodwill tour after Apollo 15 mission,
    325–27
  home in Nassau Bay, 92
  interest in black holes, 317–18
  interest in environmental protection, 386
  interest in geology, 95, 271
  and Kinetic Kill Vehicle project, 383
  learning to fly, 26–29
  Life contract, 92–93
  marriage, 46–47, 278
  meetings with Leonov, 337–39, 341–44,
    345, 366–67
  meets Soviet cosmonauts in France, 202–5
  observing of scientific research in
    Antarctica, 276–77
  public speaking tours, 131–32
  reaction to Russian space walk, 123–
    24
  reaction to Sputnik launching, 36–37
  research into a "tug and remote cockpit"
    rescue vehicle, 384–85
  and Satellite Transfer Vehicle project, 382
  schooling, 15, 19–20
  selected for NASA space program and
    training at Houston, 82–85, 89–95
  setting up of training program for air
    force officers as astronauts to fly "blue
    shuttles," 381–82
  stationed as fighter-pilot in Europe,
    32–34, 41–47
  studies for graduate degree in
    aeronautics/astronautics at MIT,
    51–53, 61, 63
  studies mechanical engineering at
    University of Michigan, 20
  swimming career, 19–20
  technical adviser for movies and TV,
    385–86

  at test-pilot school at Edwards, 65,
    69–76, 84–85
  training as "space pilot" at Aerospace
    Research Pilot School (ARPS), 71–75
  visit to Moscow, 335–44
  at West Point, 20–24, 26, 51
Scott, Douglas (son), 91, 278, 320
Scott, Lurton (wife), 47, 75, 91–92, 148,
  179, 223, 243–44, 278, 325–27
Scott, Marian (mother), 13–15
Scott, Captain Robert Falcon, 276
Scott, Tom (brother), 14
Scott, Tom William (father), 11–16, 18–19,
  20–21, 23, 28
Scott, Tracy (daughter), 51, 237–38, 278
Sea of Tranquillity (Moon), 246
"seatbelt basalt," 302
See, Elliot, 148, 180, 230, 313
  death of, 158–59
September 11, 2001, events, 248
Seregin, Vladimir, 219
Severny, Andrei, 260
Shatalov, Vladimir, 229, 342, 347
Shattuck, David, 15
Shea, Joe, 187, 194
Shepard, Al, 62, 135, 207, 258, 327, 380
Shonin, Gyorgy, 356
Shuttle. *See* space shuttle
Silver, Professor Lee, 272–73, 295–96, 318,
  321
simulators, 186–89, 268–69
Sinatra, Frank, 184, 366
Sinese, Gary, 385
Sirius, 53
Skylab space station program, 327, 334, 348
Slayton, Deke, 146–47, 182, 207–8, 213,
  254–55, 258, 271–72, 278, 279, 294,
  329, 330
  and Apollo-Soyuz project, 333, 353, 357,
    359
  as director of Flight Operations Crew, 84,
    129–30, 135
"slick chicks" (spy planes), 41–42
Smirnov, Leonid, 196
sound barrier, 46, 50
Soviet space program
  decision not to continue with
    circumlunar and manned lunar-landing
    program, 253
  disasters experienced by, 55–56, 206–7
  discussions with U.S. on cooperation in
    space, 125–26, 324–25, 334–36
  downturn of, under Gorbachev, 369–
    70
  Korolev's influence on, 53–57, 145–46
  mission controls, 59
  official support for, 99
  recruitment of civilians, 146